Biosemiotics

Volume 28

Combining research approaches from biology, semiotics, philosophy and linguistics, the field of biosemiotics studies semiotic processes as they occur in and among living systems. This has important implications and applications for issues ranging from natural selection to animal behaviour and human psychology, leaving biosemiotics at the cutting edge of the research on the fundamentals of life.

The Springer book series *Biosemiotics* draws together contributions from leading scholars in international biosemiotics, producing an unparalleled series that will appeal to all those interested in the origins and evolution of life, including molecular and evolutionary biologists, ecologists, anthropologists, psychologists, philosophers and historians of science, linguists, semioticians and researchers in artificial life, information theory and communication technology.

Kalevi Kull, Professor in biosemiotics, University of Tartu

Alexei Sharov, National Institute of Aging, Baltimore

Jana Švorcová
Editor

Organismal Agency

Biological Concepts and Their Philosophical Foundations

Springer

Editor
Jana Švorcová
Department of Philosophy and History of Science, Faculty of Science
Charles University
Prague, Czech Republic

ISSN 1875-4651 ISSN 1875-466X (electronic)
Biosemiotics
ISBN 978-3-031-53625-0 ISBN 978-3-031-53626-7 (eBook)
https://doi.org/10.1007/978-3-031-53626-7

This book was supported by the Czech Science Foundation for the project '*Contemporary Philosophy of Biology: Organism as an agent*' with grant No. 20-16633S, recipient Jana Švorcová.

This Springer imprint is published by the registered company Springer Nature Switzerland AG
The registered company address is: Gewerbestrasse 11, 6330 Cham, Switzerland

Paper in this product is recyclable.

Acknowledgments

I would like to thank all the authors who were willing to collaborate with me on this book, who trusted me and entrusted me with their wonderful contributions. I also wish to thank Anna Pilátová for her diligent language editing, translations, and valuable insights into how the texts could be improved. My heartfelt thanks go also to my colleagues at the Department of Philosophy and History of Science at Charles University, who together create an inspiring and enjoyable environment which I am happy to be a part of. Big thanks go to the Springer Publishing for publishing this book and to Alexei Sharov and Kalevi Kull for accepting the book into their series. Last but not least, I would like to thank the Czech Science Foundation, without whose funding the book would not have been possible.

Contents

Contributors

Roman Figura Department of Philosophy and History of Science, Faculty of Science, Charles University, Prague, Czech Republic

Eliška Fulínová Centre for Theoretical Study, Charles University – Czech Academy of Science, Prague, Czech Republic

Johannes Jaeger Department of Philosophy, University of Vienna, Vienna, Austria

Complexity Science Hub (CSH), Vienna, Austria

Ronin Institute, ronininstitute.org, U.S.A

Robert Kanócz Department of Philosophy and History of Science, Faculty of Science, Charles University, Prague, Czech Republic

Karel Kleisner Department of Philosophy and History of Science, Faculty of Science, Charles University, Prague, Czech Republic

Spyridon A. Koutroufinis Institute for Philosophy and History of Literature, Science, and Technology, Technical University of Berlin, Berlin, Germany

Andres Kurismaa School of Natural Sciences and Health, Tallinn University, Tallinn, Estonia

Nelly Mäekivi Department of Semiotics, University of Tartu, Tartu, Estonia

Anton Markoš Department of Philosophy and History of Science, Faculty of Science, Charles University, Prague, Czech Republic

Tina Röck Department of Philosophy, University of Dundee, Dundee, Scotland

Jana Švorcová Department of Philosophy and History of Sciences, Faculty of Science, Charles University, Prague, Czech Republic

Jan Toman Department of Philosophy and History of Science, Faculty of Science, Prague, Czech Republic

Martin Vrabec Department of Philosophy, Faculty of Humanities, Charles University, Prague, Czech Republic

Part I
Agency in Philosophy

Chapter 1
Introduction: Knowing What an Organism Is

Jana Švorcová

Abstract This book addresses various topics unified under a single overarching theme: agency. It is structured in two sections, one dedicated to philosophy, the other to biology, whereby both are intended to be equally accessible to readers interested in either field. These sections exhibit significant parallels, with the philosophical segment delving into the works of influential thinkers who have shaped our understanding of life beyond merely mechanistic views. Our goal is to comprehensively explore their philosophies, therefore we have opted for depth rather than broad surveys of philosophical trends. Instead of covering numerous philosophies superficially, we have carefully selected a few to explore in detail. The philosophical section of our book investigates the concept of organism or nature as discussed by Aristotle, Kant, Schelling, and several processualists. It explores how their ideas contributed to the development of the concept of agency. While it is impossible to cover all philosophers relevant to the study of agency, we believe our selection provides a representative overview. The biological section of our book explores various notions of agency and their relation to dissipative systems, memory, experience, phenotypic plasticity, reproduction, modularity, the development of integrated phenotypes, organismal choices, and self-representation within animal organisation. The two sections jointly offer a thorough examination of organismal agency, encompassing both its philosophical underpinnings and its biological manifestations.

The book you hold in your hands consists of contributions from biologists and philosophers who think about and investigate the nature of organisms as agents. Living things never cease to amaze us, and we believe that organisms, even the simplest single-celled ones without complex nervous systems, are a phenomenon distinct from mechanisms, machines, and the world of inanimate nature as such. The concept of agency, to which this book is dedicated, has been much in vogue in recent years – at least in certain circles of the philosophy of biology. There appeared

J. Švorcová (✉)
Department of Philosophy and History of Science, Faculty of Science, Charles University, Prague, Czech Republic
e-mail: jana.svorcova@natur.cuni.cz

J. Švorcová (ed.), *Organismal Agency*, Biosemiotics 28,
https://doi.org/10.1007/978-3-031-53626-7_1

several books consistent with ideas presented here, that is, likewise focused on trying to explain that organisms are specific in that they are agents of their own evolution (one such book was published in this series, Sharov & Tønnessen, 2021; but cf. also Walsh, 2015). Organismal agency mostly refers to the ability of living organisms to act autonomously, that is, to make decisions and take actions guided by their own goals and needs, in a way that is not determined solely by their genetic makeup, chance, natural selection, or by conditions of the external environment.

Agency can be defined through many concepts, and this book covers some of the most prominent ones. One possible way of understanding organismal agency is through the concept of *adaptation*. While traditional views of adaptation tend to emphasise the role of natural selection working on variation generated by random mutations in shaping the evolution of traits, the concept of organismal agency suggests that organisms are active participants in this process. It emphasises the concept of phenotypic plasticity, the principle of organisation, and reciprocal causality between the environment and the organism's setup. By actively seeking environmental cues, responding to them, and storing their experience with the environment, organisms can shape their evolutionary trajectories in ways that are difficult to predict or explain in purely mechanistic terms (Švorcová et al., 2018). The intellectual background to this idea is formed both by Jakob von Uexküll's concept of *Umwelt* (Uexküll 1909, 1926; also Chaps. 8 and 13) and by the idea of niche construction, originally introduced by Lewontin and further developed by Odling-Smee et al. (2003). The basic premise is that an organism does not merely adapt to the physically existing world. Rather, the environment perceived by an organism (i.e., *Umwelt*) is always a subjective, experiential realm of signs, which the organism can interpret thanks to the memory of its ancestors. Thus, the organism changes through semiotic causation (Hoffmeyer, 2008), i.e., under the guidance of interpretation of environmental signs. Similarly, we cannot imagine that organisms somehow simply fit into 'empty' ecological niches: they are active agents and constructors who transform their environment for themselves and for other organisms while maintaining their existence. It is surprising that niche construction theorists have not given much consideration to Uexküll in their work, but the Uexküllian concept of functional circuit is a beautiful example of a feedback response describing organismic activity dependent on assessment of meaning from the environment.

The uniqueness of *self-organisation* of living beings, which includes memory and heredity transferred to future generations, is essentially linked to the concept of agency. It sets apart life and non-life, organism and mechanism. Thinkers have been speculating about this subject at least since the time of Aristotle and it is thus not surprising that the theme of self-organisation appears in almost every chapter of this book. What we mean by it is the ability of living systems to organise themselves in complex ways without the need for external guidance or control. By emphasising the role of internal interactions and feedback mechanisms, the concept of organismal agency suggests that living systems are not merely the result of external forces acting on passive organisms or the result of natural selection that makes them seem purposeful. Rather, organisms are the product of complex self-organising processes which are based on organisms' inwardness and canalisation of external influences.

Nevertheless, a unified theory of self-organisation is still missing, because different – and sometimes even closely related – taxa develop with high degree of variability, which highlights the crucial role of historicity in development.

Another aspect of agency, emphasised already by Aristotle and later by for instance Shelling, is linked to having *metabolism*. Aristotle believed metabolism to be the most basic level of being but also a capacity to transform the external environment in order to produce and change its own structures. What we mean by metabolism are the chemical processes that take place in the cells. They are responsible for extracting energy from nutrients and converting it into forms that can be used for the survival and functioning of the organism. Organisms can adapt their metabolic pathways to specific environmental conditions (for example, so as to overcome a shortage of some elements of nutrition), maintain their internal balance, and thus ensure their own survival and even the survival of future generations by passing on instructions. Metabolism is thus a form of active regulation of one's own organismal state, which is the basic premise of autonomy, and hence a manifestation of organisms' agency. These ideas are associated with the influential concept of autopoiesis developed by Humberto Maturana and Francisco Varela (1980, see below), which emphasises the autonomy of living systems manifested in their self-organisation and constant self-maintenance by regulating own composition and conserving own boundaries (Brier, 2015). The continuity of organismal reproduction and regeneration is assured by manipulation of external resources (see Chaps. 2, 3, and 5).

The concept of metabolism is also related to teleology, because by reconstructing their material constitution, organisms create conditions not only for their autonomy but also for their specific teleology, that is, the production of end-states and end-directed behaviours (Hoffmeyer, 2011; Chap. 6). Teleology is not a binary concept oscillating between admitting or denying the existence of some sort of purpose in organismal behaviour. Throughout intellectual history, teleology has taken on various forms: ranging from assertions about inherently unknowable causality to claims that it is just a figure of speech and all the way to trying to understand organisms as teleological systems *per se*. Almost every chapter of this book touches upon the notion of teleological systems. It is a concept that simply cannot be bypassed in a study of agency.

The autopoietic and agential aspects of the living can also be described from the perspective of *self-reference*. Living beings not only organise themselves and produce their own structures, but also produce components of themselves, which ensure their survival and propagation. With the ability to determine its own boundaries (which define self and non-self), such self-generated beings also acquire the quality of identity and self-relation. To put it somewhat poetically: in the form of DNA, life has also written down the instructions for creating oneself. Jesper Hoffmeyer and Claus Emmeche (1991) refer to this using the term *self-reference*. A system must have the property of self-reference to be alive and self-reference is the ability to self-represent in both the *digital* and the *analogue* world. Additionally, this duality (i.e., the duality of DNA script and the body) implies that the memory of the system, represented by the DNA molecule, is simultaneously a transcription of the system itself. Nevertheless, as Hoffmeyer and Emmeche hasten to add, DNA can only be read by analogue information embedded already in the zygote. Nothing digital can work in the world on its own, just as nothing analogue on its own can undergo evolution (Emmeche, Hoffmeyer 1991; Markoš & Švorcová, 2009).

Self-reference is also manifested in organismal *self-representation*, that is, in how each life form (species) manifests itself externally through its appearance. Appearance is the self-expression of living beings: by body shape, behaviour, ornamentation, or ontogeny, and in humans also by speech, culture, etc. Organisms become meaning-carriers because other organisms attribute meaning to their appearances through the act of perception (see Chap. 14). Organisms as agents are characterised also by their inwardness, which reflects their particular experience of the world, and this experience is in turn reflected in how organisms look, behave, and interact with their environment. In the context of organismal agency, one also ought to consider other phenomena, such as sexual selection, reproduction, growth, robustness, modularity, life cycles, signalling, evolutionary novelty or life choices. This book explores all these ideas and their various forms from the perspective of several philosophers and biologists who laid down the foundations of future non-mechanistic biological concepts.

The first half of the book, consisting of historical and philosophical chapters, offers a selection of several thinkers from Aristotle to the processualists and their reflections on the living. These thinkers viewed organisms as distinct from mechanisms in the context of the themes outlined above, creating a strong foundation for contemporary non-mechanical metaphysics of biology. The non-mechanistic and non-reductionist metaphysical approaches to biology on which this book principally relies (mostly in the second part of the book) are especially *biosemiotics* (esp. Chaps. 8, 9, 11, 12, 13, and 14), *process philosophy* (Chaps. 6, 7, 9 and 10), and evolutionary concepts inspired by coming from the source of *extended evolutionary synthesis* (Chaps. 9, 11, and 12). What do these approaches have in common? In contrast to the classical substantive ontology, for instance *process philosophy* does not see organisms as mechanisms and things consisting of parts but rather as constantly self-stabilising processes (Dupré, 2021). Change pervades the entire universe continuously and is an essential precondition of stability. Rather than considering stasis as the initial state of being, we must explain why things are stable. It is by constant metabolic change, such as that mentioned above, that organisms maintain their stability, which is a state far from equilibrium. This causal continuity, i.e., stabilisation of own properties, is manifested also in self-organisation, developmental cycles (such as the constant change of structural identity in the case of amphibian development) or in the interdependence between organisms and the environment (not only abiotic surroundings but also environment formed by other organisms). These are the main arguments of contemporary process biology for not understanding organisms as machines or mechanisms. The holistic perspective of process biology, i.e., the claim that everything in nature exists in relation to everything else (Whitehead, 1978) is very close to biosemiotics. *Biosemiotics* claims that organisms are defined by semiosis, that is, by sign processes and processes of meaning attribution. As Kull, Emmeche, and Hoffmeyer put it, biosemiotics conceptualises '*living creatures not just as passively subjected to universal laws of nature, but also as active systems of sign production, sign mediation and sign interpretation, that harness the physical laws in order to live and sometimes to make a more complex living'* (Kull et al., 2011a, 1). The most common model of semiosis in the biosemiotic field is the sign relation concept introduced by Charles Sanders Peirce that consists of an irreducible triad of representamen, object, and interpretant (although the use of this concept is sometimes inconsistent). As Favareau (2015) puts it, the existence of meaning in the biosphere is not trivial: how can there be a sign relation in the physical world of matter and force, something *standing for* something or *meaning* something other than itself? The sign itself is a triadic relation that cannot be reduced to its components (see Chap. 8). Sign relations are also holistically interconnected, they connect living systems at all levels of organisation, going from individual cells to complex human communication abilities. The experiential world can thus be analysed as a dynamic realm of pure relations (Hoffmeyer, 2011).

The connection between biosemiotics and process philosophy is thus very close. The sign as a relation is never completed: it is constantly negotiated and although it can collapse into a habitus, the habitus is not permanent and can change (that is also why Charles S. Peirce is considered a processualist). Moreover, the Peircean sign triads keep on concatenating, because the interpretant of one sign relation can become a sign vehicle for the next (Favareau, 2015). Where biosemiotics works with sign relations, process philosophy considers processes. A well-stabilised process is habitus in the biosemiotic sense. Even Thomas Sebeok, the founder of zoosemiotics, was convinced that the main purpose of science is to explain stability in living systems, whereby the instrument of achieving a stable state in any living entity is semiosis (Sebeok, 1988). Life is thus a *process* based on *semiosis* and sign action (Kull et al., 2011a, b). Also, as Hoffmeyer puts it, the interpretant is never a given, once and for all, but always the result of the specific history of an organism, so that former experiences influence the interpretative process already at the earliest stages (Hoffmeyer, 2014). The emphasis on *experience* of living beings as a philosophical problem has always been present in biosemiotics (e.g., Kull et al., 2011a, b, Markoš & Švorcová, 2009, 2019), but also in process philosophy (Whitehead, 1978; Bergson 1910 [1889]) and in the works of many authors of this book (e.g., Chaps. 8, 9, 13, and 14). Nevertheless, as many scholars point out, this also places biosemiotics outside the rigid, mechanistic sciences, because subjective experience and meaning as such cannot simply be modelled and 'hard' science does not have concepts for it (Brier, 2015).

Also, the theory of autopoiesis by Maturana and Varela (1980) has much in common with biosemiotics (although Maturana and Varela try to avoid animism and therefore use a machine metaphor to describe their theory). At the core of their concept of autopoiesis is embodied information, exemplified in various biological processes such as metabolism. Embodied information goes beyond being merely a representation of the external environment: it constitutes a collaboratively formed aspect of it and implies an ongoing and reciprocal causal relationship between organisms and their environment. The physical world as such lacks meaning without organisms that actively convert the environment into functional information (Maturana and Varela endorse a semantic theory of information). In this context, such information is not a property inherent to objects but a property that emerges from interaction between objects, akin to the principles found in semiotics. The autopoietic concept of living organisms should benefit from biosemiotic metaphysics: as Brier (2015) points out, autopoiesis on its own cannot explain the subjective experience because the theory of autopoiesis lacks the necessary phaneroscopic and hermeneutical aspects. In semiotic autopoiesis, self-maintenance and self-organisation can be defined as dynamic, ongoing semiotic processes through which living systems constantly interpret and respond to their environment in order to maintain their own existence. An organism's ability to interpret signs from its environment is therefore a fundamental process of autopoiesis, a process that includes developmental, adaptive, and evolutionary aspects. The process perspective is naturally also quite isomorphic to the theory of autopoiesis: living organisms are defined by a network of processes that continuously create components necessary for their existence. Organismal self-maintenance is a constant process, a metabolic and informational flux in which and through which an organism constantly stabilises itself. Living entities are thus continuously becoming through dynamic interactions with their environment (Chap. 7).

This book will necessarily be compared to one previous successful publication on agency, also published in this series, namely the work of Alexei Sharov and Morten Tønnessen (2021). Their book offers a historical overview of the philosophy of agency, focusing mainly on the biosemiotics approach to agency. It discusses exhaustively various levels of hierarchical agency and semiotic processes, from protosemiosis to complex semiosis happening on the level of human agency and consciousness, the relation of agents to their environment, and many other related phenomena. I first learned about the book at a conference Gatherings in Biosemiotics in 2021, when our book on the same subject was already in the preparation stage. Therefore, I needed to make sure that we would not cover the same ground twice.

Our book is divided in two parts, a philosophical and biological one, both of which should be equally well accessible to readers interested in both fields. The two parts largely mirror each other. The philosophical part focuses on selected thinkers who fundamentally influenced our thinking about the living not as a mechanism. The aim here is to present their philosophies in depth and breadth: instead of an overview of many philosophical trends, I have selected only a few, which could then be presented in more detail and through particular selected topics. I have also reached out to colleagues in related fields outside of biosemiotics who work on non-mechanistic concepts of the organism to expand the discussion among several, yet through essential concepts interrelated, fields.

An Overview of Chapters

The philosophical part deals with the concept of organism or nature in the works of Aristotle, Kant, Schelling, and several processualists, and explores those of their concepts, which directly or indirectly anticipated or led to the formulation of the concept of agency. Although naturally we could not include all philosophers who would deserve to be included in a work on agency, we hope the selection is still representative. The biological part works with various concepts of agency and their relation to dissipative systems, memory, experience, and phenotypic plasticity, as well as reproduction, modularity, development of integrated phenotypes, organismal choices, or self-representation through animal organisation. All in all, this work offers a comprehensive examination of organismal agency and its philosophical and biological foundations.

Chapters two and three of this volume focus on Aristotle and his work. Aristotle is the direct or indirect source of most ideas which the study of agency investigates to this day, and his work was the main source of inspiration for thinking about the living for hundreds of years. That is why we have dedicated two chapters to his thoughts. We also wanted to investigate his ideas in greater depth; books on topics similar to ours often refer to him only in passing, superficially, but his knowledge of the living was incredibly extensive. Aristotle realised the importance of self-organisation as a principle that distinguishes the living from the non-living. The concept is emphasised alongside the significance of metabolic processes and the relationship of organisms to the environment and their clear distinction from the environment.

In the second chapter, **Eliška Fulínová** explores in detail Aristotle's doctrine of the four causes specifically in the context of biology. Aristotle believed that living beings derive their activities from within themselves, whereby the body acts as an instrument of life, so that while the father is the origin of the efficient cause, the developing organism gradually becomes an efficient cause in itself. According to Aristotle, the purpose of a living body is self-realisation and self-preservation, and the concept of a final cause reflects this. Nevertheless, Aristotle did not view purposiveness in organisms, including their ontogeny, as intentional. He believed that purposefulness is a characteristic of not the matter but rather the processes which are regularly repeated with the same outcome – and that argues against any material necessity or randomness in ontogeny. Aristotle rejected Empedocles' proto-evolutionary vision, according to which the organ parts which are functional persist, while those which are not die out, noting that this only happens due to physical interaction with the environment. Living beings actively maintain their existence through reproduction and metabolism. The fundamental goal of every being is thus its self-preservation.

The third chapter, by **Roman Figura**, investigates how Aristotle anticipated the concept of agency by emphasising the boundaries between organisms and their environment. According to Aristotle, living beings actively construct their bodies, resist external influences, and change over time, which sets them apart from non-living entities. Aristotle believed that the unifying principle among different species is ontogeny rather than evolutionary origin, and that living beings develop in the course of ontogeny from general characteristics to species-specific and individual characteristics. Aristotle's biology distinguishes living beings from the non-living environment based on concepts such as autonomy, closure, and dynamic changeability. He believed that body parts are determined mainly by their function within the whole of the organism, and that the relation between body parts and the whole organism is in its nature teleological. Aristotle's classification of living beings was based on his study of such body parts and their functional and thus ecological significance. Aristotle's thorough study of organisms, which Figura presents, established biology as a systemic science.

Robert Kanócz's chapter examines Immanuel Kant's ideas about mechanisms and teleology in nature, specifically in relation to the origin and life of organisms. Kant's concept of organism and its specific teleology has undoubtedly significantly influenced contemporary philosophy of biology. Kant was sympathetic to the idea of a purely mechanistic interpretation of nature but also believed that organisms manifest a purposeful activity regarding their coming to being and existence. Kant's transcendental philosophy tries to reconcile these two positions by analysing the powers of our mind. He argues that our cognitive abilities cannot prove natural purposiveness, we merely can – and in fact must – judge nature as if it were made by God who follows purposes and uses mechanical laws in a way that cannot be comprehended by our discursive understanding. Further, the chapter explores the reception of Kant's ideas by some of his contemporaries and by some twentieth-century thinkers who stressed the importance of aesthetic judgment. These thinkers together with Kanócz assess the usefulness of Kant's concepts for natural philosophy, since Kantian analysis is rather an analysis of our human perspectives than a study of a 'real substance'. Kanócz also invites the reader to learn to think within a perspectivist position, i.e., to give up on the search for the truth of things in themselves and universal perspectives. Instead, Kanócz says, we should reinvent reasonable combinations of perspectives that are possible given a particular situation, which he considers to be a relevant challenge for contemporary thinking about the mechanistic and teleological nature of organisms.

The fifth chapter, authored by **Martin Vrabec**, is dedicated to F. W. J. Schelling, who rejected the view that organisms and their activities are fully defined by external forces. Schelling's aim was to formulate a naturalist but not reductive, mechanistic, or deterministic (preformative) explanation of life without having to resort to positing a notion of vital force. According to Schelling, organisms can significantly modify the impact that external forces have on them. He emphasised that living entities sustain themselves by dynamically and continuously maintaining the difference between self and the external world. They are capable of metabolic activities. They do so by assimilating everything for themselves, encompassing everything within their sphere of activity. Living beings are self-constituting. Their identity is, however, not based on the identity of the matter from which their bodies are built; rather, it is based on the identity of their form and specific manner of acting with respect to themselves and the external world. They actively maintain themselves in existence through activity. Attentive readers of this chapter will see the close similarity between Schelling's philosophy and the autopoietic approach of Maturana and Varela (formulated nearly 130 years later after Schelling's death). These two perspectives are especially similar in their emphasis on organismic self-sustainability, the challenges posed by disequilibrium (thermodynamics as a discipline emerged towards the end of Schelling's life), constant maintenance of organismal boundaries, and the resulting notion of identity. Both approaches also emphasise organismal productivity (creation of products) and the cyclic nature of organismal processes.

In the sixth chapter, **Spyridon A. Koutroufinis** delves into the ideas of organismic and cellular end-state directedness, autonomy, and agency in systems biology and in the philosophy of biology. The chapter analyses the difficulties connected with comprehending these concepts and criticises the explanatory framework of the 'new mechanism' philosophy, which limits teleological explanations to a mere linguistic device with no counterpart in reality. By introducing the concept of intrinsic and extrinsic organismic factors, Koutroufinis shows how the frameworks of mechanistic and systems biology are inadequate when it comes to capturing the behaviour of actual organisms. In the last part of his contribution, Koutroufinis then proposes a process-metaphysical method based on an analysis of philosophical ideas of Whitehead and Bergson, which, he believes, should help us overcome these obstacles and do justice to the teleological nature of living beings.

The concept of identity, also discussed in philosophy since Aristotle (as noted above), is elaborated by **Tina Röck** in the following chapter in connection with the concept of organismal identity, which we need to explain phenomena such as homeostasis, symbiosis, or the ways in which organisms relate to their environment. Organisms are not just processual in their existence. They are also highly relational: collaborative, interactive, and open to and dependent on their environment. This is especially obvious in the case of organisms which exist in, for instance, mutual holobiotic interdependence. Taking inspiration from the Aristotelian concept of *tode ti* (the 'this there'), Röck distinguishes between a singular *tode ti* and a conceptual level of description, noting that the two should not be conflated. She thus formulates a scientifically useful, process-based notion of adequate qualitative identity. Röck believes that the concept of identity should be dissociated from the concepts of *individuality* and *unity*, which are traditionally considered synonymous. Röck invites the reader to determine identity through *uniqueness*. Such uniqueness emphasises distinction (*How does it relate?*) over unity (*What is it?*) and qualitative–relational continuity over numerical identity.

Anton Markoš in his chapter, which opens the biological part of this book, deals with different levels of phenomena in inanimate and animate systems. At the zero level, static phenomena are explained as systems in a state of thermodynamic equilibrium. At the first level, we can observe the role of energy gradients in the formation of dissipative structures which use the energy of the gradient to self-organise and catalyse energy dissipation, thus making the rate of entropy production more efficient. The paper then moves on to the second level, where autonomous agents (inspired by Kauffman 2000) are introduced as organised physical structures that act on their own behalf and can perform at least one work cycle – and that requires the construction of a machine that is not the product of mere self-assembly. Taking inspiration from semiotics, Markoš' contribution explores the relationships between memory, experience, signs, and meanings in the context of intersubjective and suprasubjective interactions between things and living beings. Interpreters confront incoming information with the content of their memory and experience and go on to construct a sign that better specifies the meaning of what the object is. The construction of objects (meaning attribution) is not predetermined: it should be viewed as an interpretive process that involves all living things, including plants or bacteria.

The role of memory and experience is further explored, using some concrete examples, in the following chapter by **Jana Švorcová**. She argues that epigenetic processes, such as one can observe in bacteria or eukaryotes, who are capable of internalising their lived experience with the environment, can be considered one of the attributes of agency. Through semiotic processes of meaning attribution, cells store information about the environment and by carrying instructions for future generations, organisms can be agents of their own evolution, and similar memory systems are observable at multiple levels of organisation. Such organismal memory, which stabilises the processual and plastic nature of organisms, is habitual and based on the Aristotelian conception of 'unintentional' telos and Kantian wholes as represented by first-level closures. These concepts, together with the concept of biological, organic memory, can help us formulate a concept of organismal teleology that would be acceptable for current biology.

In Chap. 10, **Johannes Jaeger** conducts a detailed analysis of one of the key topics of this book. He focuses on the relationship between ontogeny and agency (defined as organismal continuity), which he considers at the level of organism co-dependent. Jaeger emphasises the importance of organisational closure as a requirement both for ontogenesis and agency, and analyses ontogeny from the perspective of the reproducer, i.e., the minimal unit of evolution. Jaeger is convinced that if we want to understand agency, we ought to proceed via a study of cellular and developmental processes. Moreover, such study should involve a comparative approach to the investigation of underlying homologies. Rather than studying the genetic programme, we should focus on biological organisation, which imposes top-down constraints on cellular behaviour. Such constraints on cells in a developing body ensure its organisational continuity and completion of its life cycle. Organisation is what provides stability to the constant variation in organismal evolution. Jaeger concludes that in multicellular agents with multiple levels of organisation, organism-level agency functions as a constraint of ontogenesis, while cellular-level agency enables it.

The following chapter, authored by **Jan Toman**, introduces the concept of modularity, focusing on its role in evolution and its importance in both organismal organisation and agency. Exploration of the roots and evolution of modular organisation in evolving systems helps reveal its potential to facilitate transitions towards higher levels of organisation and agency. This highlights the significance of modularity and its variations as a crucial factor in the emergence of novel agents. Toman shows that while modularity exists outside the world of agents, there is no agency without modularity. Toman further argues that for a system to be self-sustaining, it is likely to require modularity, because modularity reduces its susceptibility to errors. He explains the prevalence of modularity in biological systems by pointing to the interplay between stability-based sorting and adequate macroevolutionary potential. Over time, sexual eukaryotic evolution is characterised by integration and consolidation of agency but also by reduction of possible innovations. Parcellation (i.e., multiple copying of lower-level units which are specialised on the new level) is thus the most common way of restoring macroevolutionary potential by transitioning to a higher hierarchical level. But integration, that is, the bringing together of originally independent elements into a unified whole that has a common role, is inevitable at every level of biological organisation. The processes of parcellation and integration, and their opposing effects, can thus drive the emergence of new agents on new levels and the formation of robust hierarchies.

Andres Kurismaa in his contribution discusses how the fields of life-history and allometric research could benefit from incorporating the notions of organismal agency and phenotypic plasticity, which are currently neither sufficiently well conceptualised nor modelled. To address this gap, the chapter revisits the anti-entropic approach to ontogeny developed by I.A. Arshavsky and his school, which emphasised the role of agency and phenotypic plasticity in shaping biological traits. The chapter explores how developmental stressors, such as motor work and entropy production, affect the differentiation of bioenergetic phenotypes and contribute to the physiological and morphological differences between altricial and precocial eutherians. Kurismaa also considers how the anti-entropic approach could help to frame new questions and concepts, including unresolved problems of metabolic scaling and the agential origins of prolonged gestation and reduced offspring numbers in larger/more precocial animals.

The penultimate chapter, contributed by **Nelly Mäekivi**, also addresses the agency of animals. It explores the concept of alloanimal agency in zoosemiotics, i.e., in the analysis of animal behaviour and ecological relations from the perspective of the animals themselves. Using the Umwelt theory, Mäekivi presents a case study of the reintroduction of the European mink to illustrate how changes in their environment, food resources, and social relations affect their perception and behaviour. The study uses, in addition to written sources on this subject, interviews with local people in Estonia and highlights the importance of animal agency, particularly during environmental changes. Mäekivi formulates a concept of 'Umwelt reversion' to describe a transition back to animals' natural social relations and food resources after reintroduction, indicating that meaning carriers in the environment can change due to the agency of the animal.

Karel Kleisner in the last chapter investigates the nontrivial fact of organismal appearances from unicellular to multicellular organisms. The capacity of actively appearing is what differentiates life from non-life. Kleisner describes the duality of the biological world and its evolution using a conceptual dichotomy between those who appear and those who can perceive these appearances, while emphasising that these two groups evolve in constant reciprocal interdependence. Living entities that appear are characterised by their ability to create exposed surfaces and actively influence their display through environmental affordances. We can view such external surfaces as an expression, externalisation of the internal organisation of organisms. In addition to investigating concepts such as mimicry and its evolution, Kleisner also analyses various reasons – or rather our scientific biases – why we tend to marginalise the importance of organismal surfaces in evolution.

This book is intended primarily for scholars and academics interested in the philosophy of science and philosophy of biology, as well as students interested in philosophy of biology, but it should also be of interest to evolutionary and theoretical biologists. We did our best to make the material accessible and understandable to all these groups of readers. Moreover, this anthology can also serve as a useful teaching material for courses in, for instance, the philosophy of biology or theoretical biology. We hope it will also contribute to a deeper mutual understanding between philosophy and biology in general.

References

Bergson, H. (1010 [1889]). *Time and free will: An essay on the immediate data of consciousness*. George Allen & Unwin LTD.

Brier, S. (2015). Can biosemiotics be a "science" if its purpose is to be a bridge between the natural, social and human sciences? *Progress in Biophysics and Molecular Biology, 119*(3), 576–587.

Dupré, J. (2021). *The metaphysics of biology*. Cambridge University Press.

Favareau, D. F. (2015). Creation of the relevant next: How living systems capture the power of the adjacent possible through sign use. *Progress in Biophysics and Molecular Biology, 119*(3), 588–601.

Hoffmeyer, J. (2008). *Biosemiotics: An examination into the signs of life and the life of signs*. Scranton University Press.

Hoffmeyer, J. (2011). Biology is immature biosemiotics. In C. Emmeche & K. Kull (Eds.), *Towards a semiotic biology: Life is the action of signs* (pp. 43–65). Imperial College Press.

Hoffmeyer, J. (2014). Semiotic scaffolding: A biosemiotic link between Sema and Soma. In K. Cabell & J. Valsiner (Eds.), *The catalyzing mind: Beyond models of causality* (pp. 95–110). Springer.

Hoffmeyer, J., & Emmeche, C. (1991). Code-duality and the semiotics of nature. In M. Anderson & F. Merrell (Eds.), *On semiotic modeling* (pp. 117–166). Mouton de Gruyter.

Kull, K., Emmeche, C., & Hoffmeyer, J. (2011a). Why biosemiotics? An introduction to our view on the biology of life itself. In C. Emmeche & K. Kull (Eds.), *Towards a semiotic biology: Life is the action of signs* (pp. 1–21). Imperial College Press.

Kull, K., Deacon, T., Emmeche, C., Hoffmeyer, J., & Stjernfelt, F. (2011b). Theses on biosemiotics: Prolegomena to a theoretical biology. In C. Emmeche & K. Kull (Eds.), *Towards a semiotic biology: Life is the action of signs* (pp. 25–41). Imperial College Press.

Markoš, A., & Švorcová, J. (2009). Recorded versus organic memory: Interaction of two worlds as demonstrated by the chromatin dynamics. *Biosemiotics, 2*, 131–149.

Markoš, A., & Švorcová, J. (2019). *Epigenetic processes and the evolution of life*. CRC Press, Taylor and Francis Group.

Maturana, H., & Varela, F. (1980). *Autopoiesis and cognition. The realization of the living*. Springer.

Odling-Smee, F. J., Laland, K. N., & Feldman, M. F. (2003). *Niche construction: The neglected process in evolution*. Princeton University Press.

Sebeok, T. (1988). Communication, language and speech: Evolutionary considerations. In M. Herzfeld & L. Melazzo (Eds.), *Semiotic theory and practice: Proceedings of the third international congress of the IASS Palermo* (pp. 1083–1091). Mouton de Gruyter.

Sharov, A. A., & Tønnessen, M. (2021). *Semiotic agency. Science beyond mechanism*. Springer.

Švorcová, J., Markoš, A., & Das, P. (2018). Origins of the cellular biosphere. In V. P. Sahi & F. Baluška (Eds.), *Concepts in cell biology-history and evolution* (pp. 271–290). Springer.

von Uexküll, J. (1909). *Umwelt und Innenwelt der Tiere*. Springer.

von Uexküll, J. (1926). *Theoretical biology*. Harcourt, Brace & Co.

Walsh, D. (2015). *Organisms, agency, and evolution*. Cambridge University Press.

Whitehead, A. N. (1978). *Process and reality. An Essay on Cosmology*. Free Press.

Chapter 2
Aristotle: Life as Self-Creation

Eliška Fulínová

Abstract This chapter focuses on Aristotle's theory of the four causes and the way Aristotle applies this explicative framework to living beings. Their material cause is the body parts, the functional units from which their bodies are composed. The efficient cause is identified with the father, or rather father's form (species). In the course of embryogenesis, this cause is internalised and the nascent organism itself becomes the cause of its vital movements, including the movement of self-formation. The formal cause is to be understood dynamically, as a complex of vital movements in which a particular, species-specific manner of life takes place. This form of a living body is its soul, i.e., its propre animation. Thus understood, the form merges with the final cause. Finality, or the aiming at a predefined final state, is characteristic of embryogenesis as the ontogenetic movement which (normally) leads to offspring that resemble their parents, both individually and in terms of species. The final cause of the body parts is the function they have in the body; they develop in order to exercise their vital activities. The whole organism does not serve an external purpose: its finality is to be itself, to fulfil its own form.

Keywords Self-organisation · Ontogenesis · Teleology · Goal-directedness of the living

E. Fulínová (✉)
Centre for Theoretical Study, Charles University – Czech Academy of Science, Prague, Czech Republic
e-mail: fulinova@cts.cuni.cz

J. Švorcová (ed.), *Organismal Agency*, Biosemiotics 28,
https://doi.org/10.1007/978-3-031-53626-7_2

Introduction

Aristotle is, with good reason, considered the founder of life sciences as an area of scientific knowledge.[1] Texts on living entities, especially animals, constitute about one-quarter of the corpus of Aristotle's surviving works. These texts are contained for the most part in three large treatises: *Historia animalium* (*HA*), 'Inquiries on Animals', which offers a comparative summary of a vast number of zoological facts from the areas of morphology, anatomy, and ethology of animals; *De partibus animalium* (*PA*), 'On the Parts of Animals', which summarises scientific methodology in biology and proposes a causal and functional explanation of body parts of which organisms are composed; and finally in *De generatione animalium* (*GA*), 'On the Generation of Animals', a text that deals with breeding, reproductive organs, embryology, and heredity. Along with these works, Aristotle also produced a number of smaller writings focused on partial subjects from life sciences. Aside from these, his treatise *De anima (DA),* 'On the Soul', dedicated to the subject of animation of living bodies, is also relevant. It has been noted that all in all, Aristotle in his work mentions over 500 species of animals. For comparison: folk biology of agricultural peoples tends to include on average just over 300 animal species (Berlin, 1992: 100).

Aristotle dealt with morphology and physiology, but also with the way of life and the 'ecological context' of species. We find records of dissections and on rare occasions even vivisections, although the schematic illustrations to which he on various occasions alludes unfortunately did not survive. All this is usually treated from a comparative perspective, meaning that Aristotle does not describe species after species in an encyclopaedic manner. Instead, he focuses on the shared features of larger groups of animals, eventually on unique features specific to some particular species. The corpus of Aristotle's work thus offers a monumental overview of structural and functional analogies of body parts of animals across many species, and his clear-sighted perspective – thanks to which the diverse and colourful palette of individual organisms appears as a unified domain of being and a continuum of forms – shapes our understanding of living beings to this day.

Within the context of his philosophy, Aristotle developed for the study of living entities also a theoretical, conceptual framework. His ambitions reach beyond the merely descriptive level: his aim was to *explain* the observed phenomena. According to Aristotle, knowledge amounts to more than just the grasp of facts, to knowing *that* (something is). Crucially, it includes the understanding of causes, that is, knowing *why* (something is the way it is). This is also why life sciences ought to describe not only the appearance and physical constitution of various animals, the environment they live in, how they live, and what their nature is like (*HA*, I, 1, 487a10–11). They must also explain *why* it is that way, in other words, they should try to identify the *causes* of the observed phenomena. 'It is necessary to divide off, in relation to each kind, the attributes that belong essentially to all the animals, and then to try to

[1] Aristotle of course did not call his theories about the realm of the living 'biology'. That is a modern expression.

divide off their causes.' (*PA* I, 5, 645a36–b3; cf. also *Metaph.* I, 1, 981a28–30).[2] Only knowledge which encompasses both the descriptive and the explicative approach thus leads to real understanding.

With respect to causal explanations, Aristotle distinguishes between four different types of answers to the question 'why?' (*Phys.* II, 3, esp. 194b23–195a3).[3] Let us now briefly lay out how this canonical Aristotelian teaching, familiar from countless philosophy textbooks, applies to biology. While doing so, we shall try to show that Aristotelian implementation of the structure of four possible causes to the living nature opens a still interesting perspective on what it actually means to be alive.

The Nature–Craft Analogy

Application of the theory of four causes in life sciences rests on a concept known in Aristotelian studies as the 'nature–craft analogy' (e.g., Broadie, 1987; Sedley, 2010; Witt, 2015). This analogy compares two distinct ontological domains: the area of artificial creation based on artistic skill or craftsmanship (*techne*) on the one side, and nature (*physis*) on the other: that is, the realm of what emerges naturally, as if 'on its own', without being formed by an external actor. The first broad group includes artefacts, artificial products, but also various states which are the result of human agency, such as health restored by a physician. The other area groups together living beings, i.e., plants and animals, parts of their bodies, but also the simple elemental masses (*Phys.* II, 1, 192b8–13). Aristotle proceeds by deriving the fourfold causal scheme from the area of craftsmanship, where the relevant differences are clearly distinguishable. The conceptual framework established therein, with the help of numerous examples, is then applied to the realm of natural creation and birth of living organisms (*Phys.* II, 8; Broadie, 1987).

While the area of artificial creation is epistemologically primary – it is a realm that is easier for us to understand and one where causal relations are more readily apparent – ontological primacy goes to nature. This is not just because the existence of naturally emerging living beings is a necessary condition of creation: it is above all due to the ambivalent nature of creative skills, which according to Aristotle imitate nature, develop it, and represent its culmination (*Phys.* II, 8, 199a15–18). As imitation (*mimesis*), artificial creation is dependent on nature and derived from it. At the same time, though, it can take nature further and form its culmination (*epitelein*), thus transcending it. It surpasses the abilities of nature and gives rise to something nature itself could not do. In what nature itself *can* do, it is a much more efficient creator than the creative skills are, but creative skills, in their somewhat laboured and complicated way, can create even things which are beyond the reach of nature.

---

[2] Translations from *PA* and *GA* are taken from Balme (1992); other Aristotelian treatises are quoted according to the revised Oxford translation of Aristotle, i.e., Barnes (1984) and Barnes (1985).

[3] The heterogeneity of examples listed here shows that Aristotle's aim is to formulate a universal theory of causality; causal explanation specifically in biology is addressed in *PA* I.

However, what is crucial for the possibility of applying relations that hold in the area of artificial creation to the realm of natural coming to be is the relation of *imitation*. When Aristotle claims that artificial creation imitates nature, he does not mean to say that creative actions aim at fashioning the same or similar things, although from time to time such similarities can be found. What he means is that the *process of creation* is analogical, that despite all the apparent differences it is in a sense similar to *natural coming to be*, meaning mainly the coming into existence of living organisms. In particular, the ontogenetic movement of both creating and coming into existence (birth) is causally structured in manner that is not identical but is, in a way, comparable (cf. Chap. 4).

When considering natural coming into existence in *analogy* to artificial creation, Aristotle does not want to disregard the clear differences between the two types of ontogenetic movement. The relation of analogy, which interpreters commonly use when explaining Aristotle's argumentation,[4] is viewed within Aristotle's conceptual framework as a technical term: it is a relation of similarity between different elements' relations to the wholes to which they belong. A typical Aristotelian example of an analogical relation goes as follows: what feathers are for a bird, scales are for a fish (*HA* I, 1, 486b17–22). Analogy thus *presupposes* difference. Its aim is to highlight similarity where it is not immediately apparent (cf. *Poet.* 22, 1459a5–8). In the following, we shall see what forms these differences and similarities take.

The Material and Formal Cause: Living Entities Are Hylomorphic Wholes

A question asking why a particular thing is, what it is, i.e., what causes its appearance and likeness, can be answered in two main ways: First of all, one can say what a given thing *arises from* and what persists in it. Aristotle refers to this kind of cause using the term *matter* (*hyle*), and the cause is thus traditionally known as material. The material cause of, for instance, a statue is the bronze from which it is made, while the material cause of a silver cup is silver. We can, however, also adopt a more general view and say that the material cause of both of these artefacts is metal. A different possible answer points to the *form* (*eidos*) or *model* (*paradeigma*), that is, the definition (*logos*) of the essence of the thing in question. Aristotle sometimes also uses the term *shape* (*morphe*). This cause is traditionally known as formal. In relation to the examples above, the formal cause is the shape and overall appearance of the statue or the cup; in other words, that which makes the statue a statue and the cup a cup.

---

[4] Aristotle himself does not use the term *analogy* when speaking about the two ontological domains, but – as becomes apparent below – it is a relatively accurate name for his comparative approach to them.

The more immediate similarity between the two ontological domains is that both artefacts and living entities are hylomorphic units, that is, they exist as a union of matter and form. According to Aristotle, everything in our world that has real existence has this property: every existing reality is a somehow formed matter, or perhaps rather form embodied in a particular matter, whereby these two perspectives and descriptions are fully equivalent. One should add that the form and matter are mutually related concepts: bricks and beams are the matter of a house but from the perspective of clay a brick is the form, just like wood is a matter necessary for creating the form of a beam. This holds also vice versa, so for instance, in the work of an urban planner houses are the matter from which a planned urban area is formed. To put it differently, if we call something a piece of matter, we should always know what it is a matter *of* or *with respect to what* it is matter; and the same applies to the form. It is therefore unsurprising that, according to Aristotle, a creative skill includes both familiarity with matter and knowledge of the form of the intended artefacts. In a similar way, a natural scientist must get to know the nature of investigated realities in both of the abovementioned senses, that is, both their material nature and their formal appearance.

Leaving aside heavenly bodies, how is it with the matter of living beings we can encounter in our natural surroundings? Aristotle offers the following answer: '…the matter for animals is their parts (the non-homoeomerous parts for every whole animal, the homoeomerous parts for the non-homoeomerous, and those bodies that we call elements for the homoeomerous)…' (*GA* I, 1, 715a9–11). A person who investigates nature must therefore know all the body parts, both homoiomerous ('uniform') and heteromerous ('non-uniform'), that is, in the first approximation all the organs and tissues: this is about knowledge of body *structures* (to use a later term). A natural scientist must also understand to an adequate extent simple bodies (the elements) and their interactions which co-determine the formation of body parts and thereby the living organism as a whole. The elemental masses – earth, water, air, and fire – are not, according to Aristotle, alive but neither are they the inert matter of modern science (a comparison with chemical elements would thus be misleading). The elements have an inner source of movement, albeit in comparison to living entities a rather limited one: unless something stands in their way, they necessarily move towards the cosmic locations appropriate to them, so that earth tends to the centre of the spherical cosmos, i.e., downwards, while the other elements have a tendency to accumulate around it like the layers of an onion in the sequence of water, air, and fire. From the perspective of biology this is important insofar as a living body, being a highly complex mixture of elements, exists in a sense *despite* the elements' tendency to move to their natural places. Moreover, while the elements and their interaction do form the foundation of vital processes, in virtue of their very nature their presence does not suffice for the formation of living entities. Aristotelian conception of matter is such that it does not allow for a purely material emergence of living units, and that prevents biological reductivism in the sense of reducing vital phenomena to their material causes (Balme, 1987).

To explain the appearance of a living entity, one must always consider the form (*eidos*), whereby the formal – and, as we shall see, also the final and in a somewhat different sense also the efficient – cause of each being is, according to Aristotle, the soul (*psyche*) (*DA* II, 4, 415b9–12). It is the soul as the formative principle of a living being that is responsible for the elemental masses not 'dispersing', each in its own direction. But first, let us add a few words on the concept of form as such.

It is well known that *eidos* must often be translated as 'species', especially when it appears in a pair as *eidos – genos*, i.e., 'species' and 'genus'. In contrast to our current understanding of these terms, however, for Aristotle these concepts do not denote a fixed level within the system of biological classification but merely a relatively 'lower' or 'higher' group or class (Pellegrin, 1986). If we use *eidos* in the sense of the form of an animal, one could, in first approximation, imagine it as its species-specific form. If, therefore, the material cause of a living being is its body parts and their composition, then its formal cause is primarily its character as a species, which includes appreciation of what makes, for instance, a horse a horse or a bee-eater a bee-eater. This specific individuality (current biology would call it phenotype) moreover includes the environment in which that species lives, the manner of movement characteristic of that species (which is likewise related to its environment), and consequently also for instance the manner in which the species acquires its food. For instance, what makes a duck a duck is that it is an aquatic bird and the form of its body parts – such as the length and shape of its feet or beak – cannot be understood without taking the environment into consideration.

Furthermore, Aristotle's biological writings, especially his embryology, show that the form of a particular animal includes also its sub-specific determination, at least that which can be inherited (especially via the paternal line, because according to Aristotle no matter passes from father to the offspring: the father passes on only the form; see Balme, 1987; Lennox, 1987; Salmieri, 2018). Moreover, formal determination captures not only the individual features (such as the colour or eyes or fur) but even differences between the sexes. According to Aristotle, males and females are of the same species but they not only often look different but also play a different causal role in reproduction, and that is reflected in their physical appearance. A female is therefore not just a weaker male without a penis, and the presence or absence of, for instance, the mammary glands or the uterus cannot be viewed as a merely incidental difference comparable with the difference between a straight and a crooked nose. One can thus generally state that, in Aristotle's view, the formal cause captures an animal's species-specific but also sexually and individually specific ability to survive in its own way in the world, find its bearings in it, acquire nourishment, and reproduce, that is, its ability to fulfil with its life its own essential definition and pass it on to the next generation.

Soul and Life

Within the Aristotelian conceptual paradigm, the first among what actually exists are concrete individual things. To refer to such entities, Aristotle uses the expression *ousia*, which is derived from the verb 'to be'. This is translated as *essence* or *substance* and, in biological contexts, it can sometimes be interpreted as *being*, because it denotes a living individual. Alongside this 'first' essence or being in the primary sense (*this* person, *this* horse), the term *ousia* can also be used to refer to the essential determination of those concrete particulars. In such cases, *ousia* as the 'second' essence refers to what an individual being essentially, most internally is – and it applies on various levels of generality (this particular individual is essentially a horse but also an animal). A natural body endowed with life is thus a certain essence (in the first sense of the term) but one that is *composite*, i.e., composed of matter and form. The material body which occupies certain space is then matter and a sort of underlying substrate (*hypokeimenon*), while the soul is the form and essential definition of this substrate, that is, essence in the second sense of the term: 'Hence the soul must be a substance in the sense of the form of a natural body having life potentially within it' (*DA* II, 1, 412a19–21).

We saw above that *form* can denote both the species-specific form, the species itself (a horse, a person), but also a more general determination (e.g., animal). The most general formal determination is 'living being', whereby the soul in this broadest sense is the form that makes a body alive. In other words, soul is a name for what makes a body *alive* and *animated*, what accounts for the difference between a corpse and a living individual: '…what has soul in it differs from what has not in that the former displays life' (*DA* II, 2, 413a21–22). It should be noted that this may underlie Aristotle's profound interest in various living organisms, including the most obscure or potentially repulsive ones which do not seem to deserve the attention of a philosopher and a naturalist (*PA* I, 5, 645a4–23). At least this is how Andrew Cunningham, a historian of science, explains the project of Aristotelian biology: 'The nature of the soul is the object of Aristotle's inquiry here: and it is this inquiry which leads him to look at living creatures. Aristotle looks at *animals* because he is interested in the *soul*. … He looks at animals because they are the soul-in-action' (Cunningham, 1999: 34).

The soul is the principle of life and, at the same time, life itself ('living'), which according to Aristotle amounts to a set of certain vital movements or activities aimed at ensuring the persistence of organisms in their forms. A living body on a basic level thus amounts to a self-sustaining body, because as Aristotle says, 'By life we mean self-nutrition and growth and decay' (*DA* II, 1, 412a14–15). This does not imply a reduction of all manifestations of life to the processes listed above. Rather, it is an apt definition of the necessary and sufficient conditions for calling something alive. What seems to be the basic characteristic is thus the consumption and processing of nourishment (*trophe*) when growing one's body. It includes both its sustenance and its maintenance, which manifests itself on a scale from growth and development all the way to withering away and fading. With some caution, one could perhaps use a current term and say that the core of life is, according to Aristotle, 'metabolism' (from Greek *metabole*, meaning 'change' or 'transformation') in the sense of transforming matter into one's own body structures.

The movement of bodily self-sustenance presupposes both a difference and a relation between a living organism and its environment: an ensouled being accepts something from the outside, incorporates it, and transforms it into itself. A living being thus exists as an interface between the inside and the outside: it is the movement of transformative crossing of this difference. Naturally, one could add that this movement takes place not only in the inward but also in the outward direction, whereby this bidirectional movement cannot be reduced to the consumption of nourishment and excretion of undigested leftovers. Living entities actively enter their environment, manifest themselves to the outside in some way, thus changing the contexts in which they appear. This is why Aristotle states that the study of lives of animals should include not only their body parts and – to put it in modern terms – their physiological functions, but also their different ways of life and different characters, that is, what we would nowadays call their ecology and ethology, which includes where they live, what/whom they eat, whether they are diurnal or nocturnal, solitary or living in groups, whether they live in or build their homes, whether they are more likely to be aggressive or shy, etc. (see *HA*, I, 1, 487a10–11; *HA* VIII–IX). All these manifestations of life are vital movements, that is, manifestations of the soul.

The most basic level is thus determined by the movement of what we would nowadays call metabolism. This forms the indispensable core of the life of each living entity, but ingestion and excretion, the in and out that is the cornerstone of life as such, can take the most varied and complex forms. Aristotle's view of the realm of the living is hierarchical, based on a gradual layering of vital movements, activities, and abilities. Viewed through the Aristotelian prism, living beings form a scale which ranges from plants, which were 'only alive' (*GA* I, 23, 731b5), through variously complex animals, which are not just alive but to a varying extent also endowed with perception and the ability to engage in local movement, all the way to humans, who in addition to the above also have the capacity of rational thought. These three levels, that is, plants, animals, and humans, are characterised by three different levels of life activity, which Aristotle – in accordance with tradition – describes also as three kinds or three parts of the soul (Corcilius & Gregorić, 2010): the nourishing, sensory, and rational one. The elementary level of nourishment, growth, and preservation of the body is linked to the ability to create another body of the same form, that is, to reproduce. On its own, this ability is characteristic of plants, which is also why it is often called the 'vegetative soul'. The first layer on top of this (which appears in marine sessile animals that form the link between plants and fully developed animals; see *GA* III, 11, 761a15–32) is perception, whereby different animals' senses are developed to a different degree and their sensory equipment can be variously rich. The most basic of senses, present in all animals, is touch, which is linked to the presence of flesh (*sarx*) as a body part specific for animals. Alongside perception, most animals can also engage in locomotion appropriate to their living environment. When we add to these mental activities the capacity for rational thought, we arrive at the top of the metaphorical pyramid, that is, to humans. Aristotle notes that the reasoning part of the soul (*to dianoetikon*) is more emancipated from the body than the first two levels of psychic activity, which is why it transcends the scope of biology (*DA* II, 3, 413a11–12; Corcilius & Gregorić, 2010).

In general, one can thus say that the concept of soul includes various movements which help living beings in their active efforts to remain in existence; in other words, these are movements of self-realisation via which living entities realise themselves. This is summarised in Aristotle's definition of soul in its canonical version: 'The soul is an actuality of the first kind of a natural body having life potentially in it' (*DA* II, 1, 412a27–28). This inconspicuous sentence has had a fantastic career in the history of ideas. At this point, let us just highlight what appears to be crucial: the link between thinking about the soul, and therefore also life, and actuality vs. potentiality, a key Aristotelian conceptual pair. To wit, what we call 'alive' is that which represents a set of certain potentialities which are being actualised or realised. Alive is that which *can* accomplish something, and which moreover realises (translates into action) this potential. In other words, life is the actualisation of what is possible, and the soul, as a form or the 'second' essence, is the principle of this movement of self-realisation. A living, ensouled body is alive because its movement is in its power, because it is not merely at mercy of the effects of external movers (*DA* II, 1, 412b15–17).

Regarding the body, whose form or realisation the soul is, i.e., a natural body endowed by the potential of life, Aristotle adds that this is an *organic* (*organicon*) body, i.e., one that is formed of body parts which are suitable tools or instruments (*organa*) for executing certain functions which are characteristic of living beings. This is stated in the third, final version of Aristotelian definition of the soul: 'If, then, we have to give a general formula applicable to all kinds of soul, we must describe it as an actuality of the first kind of a natural organized body' (*DA* II, 1, 412b4–6). A living body differs from non-living nature by containing the source of its own varied movement that can go in different directions, but also by executing certain specific activities (*praxeis*) thanks to its variously differentiated and specialised body parts (*mere*, *moria*), which fulfil within the whole of the body particular tasks (*erga*) (*PA* I, 5, 545b14–28). Some vital activities and functions – and the corresponding body parts – are species-specific. Many, however, are shared across different groups of living organisms, and everything that is alive must have organs for the intake and processing of nourishment. A natural body animated by a soul is thus a unified whole of functionally connected parts (Pellegrin, 2011 therefore speaks of Aristotle's biology using the term *moriology,* cf. Chap. 4). On the most general level, animation is thus tantamount to this active participation of body parts in maintaining the living whole of an organism.

It is in this sense that we should interpret Aristotle's claim that 'all natural bodies are organs of the soul' (*DA* II, 4, 415b18–19). What is meant here by the soul is by no means some sort of homunculus operating the body. The body is an instrument of the soul in the same sense in which the eye is an instrument of vision or the saw an instrument of sawing. In a similar fashion, the body is an instrument of life, it serves to maintain life. And just as, thanks to human skill, a saw is made to saw, the body of a living entity is in the course of ontogenesis naturally fashioned so as to be ensouled, that is, so as to be capable of executing vital functions that define the life of the living being in question. The active agent acting in the world is then the living being as a whole. The soul in the sense of the source or origin of vital movements, eventually the principle of life, refers to the fact that a living, i.e., ensouled, being has the source of its activities within itself.

The Efficient Cause: The Living as the Source of Its Own Vital Movements

This brings us back to the issue of causes. In addition to matter and form, what belongs to the explanation scheme of causes is also the original source (*arche*) of change or movement, that which caused the formation of the entity in question. This is traditionally called the efficient cause. Aristotle's example of this cause is an advisor who urges certain action or a father who is the efficient cause of his offspring. In general, one can say that the creator is the cause of his creation or that the originator of a change (*metabole*) is the cause of what is changing. Neither artefacts nor living entities are eternal. They must all therefore *come into being*, and that happens not all at once but gradually. In other words, both artefacts and living entities are the result of an ontogenetic movement. In neither case can the entity provide its being to itself (if, at a certain point in time, they did not yet exist, they cannot have initiated their own formation). The movement of coming into being must therefore have a beginning, a point when this movement starts.

If, together with Aristotle, we relate the question of efficient cause to the previous examples of a statue or a cup, we find a somewhat surprising claim according to which the efficient cause of a statues is not simply the sculptor as its creator but the *art of sculpting* that is at his disposal (*Phys*. II, 3, 195a5–7). Although it is true that any particular statue, for instance the Discophorus, was created by a particular sculptor, in this case Polyclitus, this is not what makes the statue a statue: it is part of the essence of a statue that it was created by a sculptor, and it is not important for that statue which particular sculptor it was. On a more general level, the efficient cause of Discophorus is not only Polyclitus but also a 'human' or an 'animal', though from the perspective of what makes a statue what it is, these constitute in Aristotle's view merely a chain of collateral, less important causes. In a similar way, we could for instance claim that the statue was created by a 'white man' or a 'scholar' (*Phys*. II, 3, 195a32–195b3). But why does Polyclitus create statues, what is it due to? It is due to the fact that he is a sculptor, thanks to the art of sculpting he mastered. The efficient cause of a statue *as a statue* is thus the art of sculpting, realised in different instances by different concrete sculptors. That is the primary substantive cause one must always try to identify (*Phys*. II, 3, 195b21–25).

The efficient cause of an artefact is thus its creator as someone who has at his disposal some art. Equivalently, one can say that the efficient cause is the *art itself* insofar as it is implemented by someone. The art which the creator must be familiar with includes both the form of the statue and the manner of embodying it in the matter.

In the case of animals, the primary (in a temporal sense) efficient cause is the genitor, that is, the father who usually uses as the instrument of creation his sperm (*GA* II, 4, 740b24–29; *GA* I, 22, 730b8–32). To be more precise, the efficient cause is the *form* which the father by his existence embodies and which is passed onto the offspring through the process of procreation. On the other hand, the material cause of procreation and the matter from which the embryo and its body parts are later constituted come from the mother, and only from her (according to Aristotle, the father's sperm does not participate materially in the formation of the embryo). Of course, the mother is likewise a realisation of the same specific form and the embryonic matter created by the female is to some extent pre-formed, i.e., predisposed to the generation of offspring of a certain species (*GA* II, 4, 740b18–25; Connell, 2016). Nevertheless, full realisation of the form in a new individual is the result of ontogenetic movement which must be initiated by the motive impulse coming from the father.[5] The analogy between natural formation and artificial creation can thus in the present context be formulated as follows: the relation between artistic skill, *techne*, and the artefact is the same as the relation between the paternal form and the offspring. In other words, the paternal form has, with respect to the begotten offspring the same causal function, namely the function of an efficient cause, as artistic skill has to the created artefact (*GA* I, 21, 729b19–21). And just like the intervention of a craftsman or artist is necessary if something is to be created by artistic skill or craftsmanship, so analogically the contribution of a male is needed for a full transfer of a form into the next generation, while the mother stands for the matter from which something is generated.

The kind of intervention in the two cases nevertheless differs, because the creators are creating something different. A builder builds houses, not other builders (and any eventual training of other builders does not amount to an exercise of the building art). In contrast, as Aristotle reminds us repeatedly, a horse begets a horse, and a human begets a human. For a causal explanation of an animal, the fact that the animal was begotten by an individual who already in advance of this act personally embodied the relevant form is utterly crucial. A person who has a certain creative skill knows the forms appropriate to it and has the knowledge requisite for its material realisation (for instance, a physician must know what health is and how to restore it in a body). Animals, however, in relation to reproduction do not have any special skills or knowledge and neither do humans: Polyclitus cannot produce an offspring in the same way in which he creates a statue. The ability to create an individual of the same species is not based on knowledge or skill – it rests in a physical realisation of the form of a given species. One could say that the physical realisation of a form is an embodied memory of the process by which the organism itself had gradually formed. This is also why only adult individuals are capable of procreation: their bodies are in a sense a living 'record' of the ontogenetic movement through which they themselves were formed. Thanks to this, each adult individual embodies a 'recipe' of how the appropriate form should be embodied in matter. A male body can, under certain circumstances, give an impulse to the movement via which this form is then realised in the next, numerically distinct individual.

[5] From our perspective, Aristotle's position may seem to underestimate the role of females, but in the context of contemporary debates it presented a conceptual compromise between, on one side, the two-seed theories (widespread among Hippocratic physicians), which downplay or neglect the differences between the male and female contribution to reproduction, and on the other side, the notion of a 'mother as a fertile furrow', according to which the embryo is begotten solely by the father (Lefebvre, 2016).

This can, however, happen only because a living body is capable of self-creation. In other words, because a living body continuously creates itself in a particular form, it can subsequently initiate the emergence of another body of the same form. This brings us to a crucial ontological difference between artefacts and living beings: in the process of formation of a living being there comes relatively soon a moment when the originally external efficient cause of the ontogenetic movement is internalised, and the emerging organism becomes its own efficient cause. From that point onwards, the source of further movement is in it. This is the point of integration of the paternal form, which the developing individual gradually ever better embodies, but from this point this is not due to an external influence but due to an internally driven movement. In the case of artefacts, the source of movement is never internalised, which is also why, unlike organisms, they are not capable of reproduction: they remain non-living. Aristotle stresses this basic difference right at the beginning of his analysis of causality, when defining natural beings as such, which unlike artefacts have within themselves the origin of their own movement and rest, be it local movement, growth and diminution, or qualitative changes (*Phys.* II, 1, 192b13–15).[6] This holds also in reverse: everything that originates in this way exists naturally, i.e., has its own nature (*Phys.* II, 1, 192b32–33). The abovementioned equivalence thus accurately defines the extent of what is 'from nature' as opposed to that which originates artificially. Even if there were no other differences, this would be sufficient to define and distinguish natural origin from artificial creation as two different ontological domains.

The Final Cause: The Living as that Which Aims at Full Realisation of Its Essence

The last way of explaining a particular thing consists in attending to its meaning, aim, or purpose (*telos*), the 'what for' (*hou heneka*) it is in the world, what good it is, what it enables. This cause is traditionally called final, and it often starts with the causal conjunctive 'in order to' (*hina*): Why do you take walks? In order to be healthy. The final cause of the walk is thus good health. Aristotle adds that the same explanation applies also to individual links in a possible causal chain. Here, good health is thus aided not only by walks – healthy weight, physical purification, and medicines or medical instruments have the same purpose. Focus on a particular purpose can thus be ascribed both to the exercise of certain activities (*erga*), which can moreover be concatenated (movement leads to healthy weight and healthy weight leads to health), and to instruments (*organa*). Final cause, which can be

[6] Paradigmatic natural entities are animals, because they are capable of all of the abovementioned movements (plants do not engage in locomotion, and the elements are not the cause of their rest: in the absence of any obstacles, they necessarily move to their proper cosmic places).

ascribed both to instruments and to activities, is then applied to investigation of the realm of the living.

Aristotle's crucial insight is that naturalists aiming to understand the observed phenomena cannot limit themselves to mapping the physical interactions and series of efficient causes. The purpose of natural processes must also be considered. There are, meanwhile, two main areas of investigation where accounting for final causality cannot be dispensed with. It is i) the issue of ontogenesis or emergence of living beings, where the final cause takes the form of purposeful movement towards a particular form clearly given in advance,[7] and ii) 'moriology' (Pellegrin, 2011), meaning the fact that body parts are fashioned so as to fulfil certain vital functions within the whole of a living body, i.e., that they are instruments serving the life of a particular organism (see Chap. 3). The purpose of a living body as a whole, meanwhile, is self-preservation or, in perhaps a more Aristotelian spirit, one should say self-realisation: all living entities aim towards life, their goal is lasting realisation of their own form, and this is also the purpose of the body structure of living beings and their varied specific manifestations of life.

Let us note the following: Regular walks lead to improved health regardless of the subjective motivation that may have inspired the walking. It does not matter whether our aim is to lose weight and be healthier or some completely different reason. Regardless of the goals we aim at, and even irrespective of our awareness of any links between walking and physical fitness, walks *are* something that tends to improve our health. This type of causal link, which is unconnected with subjective motivations and intentional efforts to achieve a particular goal, is what Aristotle calls *telos*. Aristotle developed his concept of purposefulness of ontogenetic movements so as to make it independent of intentional activity. In his *Physics*, final causality is presented in a way that makes it clear that the existence of a purpose of a particular movement does not pressupose a conscious subject acting so as to reach a goal, although this is naturally not excluded (*Phys.* II, 5, 196b17–19; see Chap. 9). Nevertheless, Aristotle expresses himself on this matter as if he anticipated that objections of this kind might be voiced: 'It is absurd to suppose that purpose is not present because we do not observe the agent deliberating. Art does not deliberate. If the ship-building art were in the wood, it would produce the same results by nature. If, therefore, purpose is present in art, it is present also in nature' (*Phys.* II, 8, 199b26–30; cf. also *Phys.* II, 8, 199a17–18).

This point is crucial, and we can see now clearly why Aristotle emphasises that in the domain of artificial creation, the efficient cause is the creative skill and not the creator. A ship as a certain form and the process of its creation are both part of the shipwright's skill. For a ship to be built, there must be someone who undertakes its realisation and builds the ship. On the other hand, this particular individual (and his or her intentions and motivation) plays no substantive causal role in the process. A particular shipwright is as if a converter or a gear between the forms which are part

[7] Cf. Gotthelf, 1987, 207: 'First, in almost every passage in which Aristotle introduces, discusses, or argues for the existence of final causality, his attention is focused on the generation and development of a living organism.'

of the shipbuilding skill and hylomorphic reality: the form of a ship takes on the causal role of the goal, while the impetus comes from shipbuilding, which is through the shipwright gradually realised in the matter. A shipwright as an expert on shipbuilding does not ponder about whether, why, and how best to build ships: the shipwright simply builds them based on the methods he or she learned.[8] It is almost as if ships were growing on their own, by natural forces, or as if shipbuilding was in a sense present in the timbers themselves. In a similar vein, Aristotle speculates what it would be like if houses were naturally born. He says that from nature, they would be born in just the same way in which they are now constructed thanks to the builder's skill. Moreover, and that is of key importance, the same holds the other way around: if we were able to artificially create what is nowadays formed by nature, it would likewise be created by our skill in just the same way as it is now formed by nature (*Phys.* II, 8, 199a12–15).

Nevertheless, if there were something we should criticise Aristotle for it would not be a subjectivation of nature but its very opposite: in an attempt to stick to the didactically effective analogy between natural formation and procreation, he does not hesitate to de-subjectivise artificial creation. This perspective turns it into something akin to an automatised generation of predetermined outputs highly similar to a spider spinning a web or a plant creating leaves to protect a developing fruit. Naturally, neither the spider nor the plant in the process of such 'creation' investigates, decides, or displays a creative skill (*Phys.* II, 8, 199a20–30). Therefore, when Aristotle speaks about artificial creation to illustrate certain types of relations which are present also in the domain of procreation of living beings, we are not encouraged to imagine for instance a master sculptor creating his chef-d'œuvre but rather the run-of-the mill production of standardised objects of daily use. Serial manufacture is, after all, much closer to biological procreation than the highly individual artistic creation is (although both fall under the Greek concept of *techne*).

Just like certain activities usually (under normal circumstances, unless hindered) lead to particular results, so certain objects are suitable tools for particular purposes. Thus, for instance, one can pour liquid into (and out of) a cup, a house gives shelter, a statue can remind us of the person it depicts. From the perspective of these artefacts' essences, it makes no difference how we (the users) handle them, what subjective motives drove us to make or acquire them, and for which *own* purposes we will use them. That is, it makes no difference to what these artefacts are whether we use a cup to drink or for libation, whether a statue is placed in the agora or, in times of hardship, melted down to make weapons. We, people, use everything, not only our own creations, as if it was *for our own sake* – but, as Aristotle points out, this *for the sake of what* carries a double meaning (*Phys.* II, 2, 194a35–36; see also *Metaph.* XII, 7, 1072b1–3 a; *DA* II, 4, 415b2–3, 20–21). It can denote the 'user' that is, the

[8] Let us note that Greek tradition ascribes the origins of craftsmanship to the gods. Moreover, individual input and originality were not (until the Modern Era) something that would be especially appreciated in art and/or production. In this view, there is little space for 'progress' or cultural evolution, which corresponds also to the fact that Aristotle does not believe in any biological evolution.

person who finds the object in question useful. This is what one has in mind when saying that artefacts serve people as tools to satisfy their particular needs and achievement of their goals.

In the context of investigation of nature, however, what is relevant is the purposeful *what for*, which one could perhaps think of as the 'inner meaning' of the thing in question, something that is its 'intrinsic end' (Witt, 2015). A walk can – but under other certain circumstances need not – benefit the person who is walking, but its inner efficient cause is health because this is the end to which this activity tends *in virtue of its own nature* (*Phys*. II, 6, 197b23–27). In a similar way, the purpose of a house as a house is to provide shelter and whether it is used by people or a family of rodents, in fact whether it is used by anyone at all is, from the perspective of what makes a house to be a house, irrelevant. Naturally, one can also use things for other purposes than those which are their intrinsic, proper ones: 'One can use an ashtray as a hammer, but it remains an ashtray; in a similar way we can use a horse to pull a plough, but it is not its natural function' (Witt, 2015: 117).

The *telos* of a particular thing, its inner meaning or proper function, is closely linked to that thing's formal determination. It is so to such an extent that Aristotle can claim, and he does so repeatedly, that the form and the purpose are the same: 'And since nature is twofold, the matter and the form, of which the latter is the end (*telos*), and since all the rest is for the sake of the end, the form must be the cause in the sense of that for the sake of which' (*Phys.* II, 8, 199a30–32; cf. also *Phys*. II, 7, 198a24–2; *Met*. VIII, 3, 1044a34–b1). If the form captures the essence, to borrow a later expression, of the thing in question, then what that thing *on its own* is good for, what it can be used for, and what it enables, is an integral part of that essence. The 'housiness' of a house does not consist in just the spatial constitution, in the organisation of material components into an overall shape. Its integral part is that a house should provide shelter. Still, typical examples of final causes in the *Physics* are not artefacts and tools that have a particular purpose, but activities and processes defined by aiming at the achievement of a particular state of affairs. In this way, walking is that which targets health, exercise leads to physical fitness, war leads to the achievement of dominance (*Phys*. II, 3, 195a9–11; II, 7, 198a20). These examples very well show the fissure between the two distinct causal roles which the form and the goal, notwithstanding the relatedness of their content, play: if war can be defined as a 'struggle to achieve superiority', then superiority is in some sense the very essence of war, but only as something that is yet to be achieved, as the inherent end goal of warring, as something that gives direction to this movement.

It is this sense of teleological causality that is of key importance to Aristotle: by going 'from art to nature', he wants to demonstrate that just like in the area of artificial creation, so too in the area of natural becoming we are dealing with purposeful processes and movements (*Phys*. II, 8, 199a8–20). In the case of artefacts, it is immediately clear that the sequence of movements used in their creation is not accidental, that it is not merely by coincidence that it leads to a result in the form of an artefact. The process of artificial creation is in all of its individual steps directed towards a particular goal, such as casting a bronze statue, building a house, or healing a patient. What happens earlier is determined by what will take shape later. The particular stages of development are comprehensible only as intermediate stages that aim towards the formation of something that emerges only at the end. At the same time, though, this final result directs the entire sequence from the beginning as its goal or desired final outcome. The goal regulates, directs the entire movement, keeps it on a certain trajectory – and this, according to Aristotle, applies also to the development of living beings and their body parts. The ontogenetic movement of formation of a living organism is, just like in the case of artefacts, purposeful, oriented at the creation of a particular outcome, in this case a being of the same form as its parents. The process is subordinated to this end, it is a sequence of steps where each developmental stage is just a little closer to the final adult form (*Phys*. II, 8, 199a8–12).

Purposefulness is, according to Aristotle, indicated by processes taking place repeatedly, regularly, usually in the same manner, and, most crucially, with the same outcome:

> For those things are natural which, by a continuous movement originated from an internal principle, arrive at some end (*telos*): the same end is not reached from every principle; nor any chance end, but always the tendency in each is towards the same end, of there is no impediment. The end (*to hou heneka*) and the means towards it may come about by chance. … But when an event takes place always, or for the most part, it is not accidental or by chance. In natural products the sequence is invariable if there is no impediment (*Phys*. II, 8, 199b15–26).

Thanks to repetition, one can come to understand what the usual and proper course of that process is, in other words, how the movement takes place and what its outcome is supposed to be. If there is a deviation from the normal course of the process, so that its outcome is different than usual, different from what one could expect, we are right to speak of an error or, in the case of biological offspring, of a monstrosity (Witt, 2015). The argument from the possibility of an error (*hamartia*) is an important element on which the Aristotelian analogy between *techne* and *physis* rests. One of the examples which Aristotle chooses to demonstrate his thought is, meanwhile, surprisingly modern: just like a scholar can make a mistake in writing, for instance swap one letter for another, so nature, too, can miss its target and form an individual who differs from the specific form of its parents and is therefore a monster (*teras*) (*Phys*. II, 8, 199a32–b7). A statue which loses its nose due to the stonemason's carelessness is no longer a proper depiction. It fails to reflect its inner sense. In short, it is no longer a proper statue. In a similar way, if for instance a fledgeling with underdeveloped wings were not to reach the goal of its development, it would not become what it is supposed to be, that is, for instance a pigeon whose inner meaning inherently involves the ability to fly.

The normative component related to repetition and regularity plays an important role in Aristotle's argumentation against naturalists who seek to explain everything in nature only by material necessity or accident. In a remarkable passage inspired by Empedocles' zoogony, Aristotle critically considers the possibility of various body parts and even larger body structures arising in this way, that is, based on purely material interactions and accidental encounters of physical components. Those that would turn out to be useful and functional (and were thus literally 'just what they would have been if they had come to be for an end') would persist, while those that would not be such, would simply perish (*Phys.* II, 8, 198b23–199a8; Cooper, 1987: 251). Darwin praised this 'proto-evolutionary' insight but did not realise that this is not Aristotle's own position but a view he rejects (Gotthelf, 1999: 8). Aristotle's counterargument is the following: Even if coincidence could lead to the emergence of a particular structure *once*, we would still have to find an explanation for why this structure emerges in the course of ontogenesis *repeatedly*, that is, always or at least in most cases, in the same way and the same form and shape. Accident or coincidence cannot, according to Aristotle, explain the *constant outcome*, which is what we see in nature in animal reproduction. What causal mechanisms ensure that a functional solution that works takes place in the course of reproduction repeatedly while suboptimal solutions do not? Aristotle offers an example of differences between the front and back teeth: even if it were solely due to the nature of their material constituents that in the course of an animal's development incisors evolve in a place where they happen to serve well the purpose of tearing off pieces of food, while molars appear in a different place where they can serve well for grinding the food, how do we explain that the next time it will happen again in this way that is fitting and functional, and not in another way? 'For teeth and all other natural things either invariably or for the most part come about in a given way; but of no one of the results of chance or spontaneity is this true' (*Phys.* II, 8, 198b34–36).

According to the train of thought which Aristotle outlines here, the original, or anachronistically, the evolutionary emergence of a suitable body structure may have been accidental: of the many realised variations, one was simply of a kind that *simulated* purposefulness. If, however, we want to explain the persistence of this variant, its repeated emergence as new individuals of the same kind are born, we must assume not merely simulated purposefulness but a real one. In other words, we must assume that the structure in question appears repeatedly in the course of ontogenesis *because* it turned out to be functional. As such, it directs the development of new individuals not accidentally but quite purposefully towards structures that turned out to function well.

It should be, however, immediately added that this consideration is within the Aristotelian conceptual framework purely hypothetical because Aristotle himself does not work with the notion of original (evolutionary) emergence of functional body structures. Quite the opposite: his theory is based on the idea that organisation of the world is eternal, which is also why the current forms of species are without a beginning or end in time. One of Aristotle's basic axioms is the symmetry of the beginning and the end: everything that emerged in time will also necessarily perish. Nothing that emerges in our sphere of the world can aspire to eternal duration, which hints at the divine nature of existing things (especially heavenly bodies). Even so, everything that is born aims at persisting: it is one of the core characteristics of the living that its goal is self-preservation. Living beings differ from non-living entities by caring for their being, i.e., by actively maintaining themselves in existence – not only by seeking and absorbing nourishment but also by constituting themselves. Reproduction, a process whereby an organism produces an offspring of the same kind, is a process in which the living accomplishes this goal within the limitations of given ontological boundaries. Eternal persistence of a numerically identical born individual is impossible, but the persistence of an identical form embodied in a changeable matter and thus in many numerically distinct individuals is very well possible (*DA* II, 4, 415a26–b7; *GA* II, 1, 731b31–a1). According to Aristotle, the goal of life is life itself. Moreover, this holds not only in general but also in particular: each particular organism aims at preserving itself in its form, which is why each organism produces offspring so to speak 'in its own image', that is, offspring that have the same form as the parent organism. This stability of form is what enables mortal living beings to transcend the boundaries of their finality and in a sense take part in eternity and persistence.

In the course of ontogenesis, we see the formation of those body parts which an animal of a certain form (which includes the environment in which it typically lives) requires for its survival and reproduction. According to Aristotle's methodological maxim, those body parts either enable life or aid survival (*GA* I, 4, 717a15–16) because 'nature creates nothing without a purpose' (*IA*, 2, 704b15–18). Body parts are thus substantively defined mainly by their vital functions in the whole of a living body and in the course of its development assume the form that enables them to efficiently fulfil such functions. We are thus arriving at a remarkable picture: on the one hand, Aristotle in his biological theory does not allow for evolution, that is, the emergence and change of specific forms in time. But still, his functional conception of teleology resonates with the biological concept of natural selection leading to the existence of species adapted to life in a particular environment. Allan Gotthelf in this context speaks of isomorphism between Darwin's and Aristotle's overall vision (see Chap. 3) and adds that a teleological perspective is at its very core:

> Animal parts, for example, come to be, and are present, *because they are needed for the life of the organism in question, or because it is better that organisms of that type have this part than not.* But as various scholars have pointed out, natural selection makes the very same thing true of animal parts in Darwinian theory. For, when they are due to natural selection, parts are present in an organism, in just the form they are, precisely *because* their presence has made survival possible or has provided for animals that have these parts a survival advantage. The difference is that, while Aristotle takes it as a basic fact of nature that animals come well-adapted, with a capacity to reproduce themselves, Darwin offers a mechanism – viz. natural selection – by which well-adaptedness is established and maintained. (Gotthelf, 1999: 23; cf. Lennox, 1993)

Let us add a final remark: unlike ontogenesis, Darwinian evolution represents a unique historical process and as such, similarly to human history, it does not have a goal, it does not aim towards some predefined outcome.[9] Aristotelian biology is non-evolutionary and the idea that evolutionary transformation took place thanks to 'errors', that is, thanks to accidental and in principle arbitrary deviations occurring during the process of reproduction (the Darwinian principle of descent with

[9] Although there are discussions, for instance, about convergent evolution of particular traits in different groups of animals, 'hard core' finalists are among current biologists rather an exception. The position according to which any possible rational beings will necessarily look like actual humans is nowadays held basically only by Conway Morris (2003).

modification), is something Aristotle would have most likely found unacceptable and perhaps even repulsive and abhorrent. This is also the direction of his thought about why in cases where animals give birth to offspring of a different form, the offspring simply *must* be sterile. He notes that one cannot have different being born from different, limitlessly and *ad infinitum*, because 'nature flees from the unlimited; for the unlimited is incomplete, but nature always seeks an end (*telos*)' (*GA* I, 1, 715b15–16). If, therefore, purely speculatively and hypothetically, Aristotle were forced to admit (against his basic intuition of a stable nature internally structured by purposefulness and effectiveness) that there *can* emerge from the deviations that occasionally take place in the course of reproduction a new generation line, then such process of emergence of new forms, an evolutionary process, would be in his view non-teleological. Ultimately, though, his thinking dictates that such process is not really possible in nature *precisely because it is non-teleological.* Darwinian evolution viewed through an Aristotelian prism (if such notion makes sense at all) would thus clearly be of a non-finalist kind.

Conclusion

This book deals with issues of organismal agency and in this chapter, we tried to show in what sense Aristotelian theory of the realm of the living could be viewed as anticipating this approach. Although Aristotle's philosophical conception is in general non-evolutionary and works with the idea of permanent, unchanging animal species, temporality does play a key role in ontogenesis. Coming to be within a particular defined form is one of the key attributes of life which according to Aristotle demands explanation. Both embryology and reproduction play an important role in Aristotelian biology. On a more general theoretical and philosophical level, ontogenesis is explained by Aristotle's theory of causality, which offers an explanation both on the level of material interactions and motive impulses, and on the level of formal and final causality. Living entities care for themselves and actively try to maintain their existence – this kind of autonomy is, by the way, listed also by Stuart Kauffman as the basic definition of agency (Kauffman, 2000, see Chaps. 5, 8, and 10). To be able to do that, organisms form in the course of their development functional body parts based on their parents' model so that they can perform the vital activities necessary for the life of the species in question under conditions proper to that species. The resulting shape of the developing organisms is thus fundamentally adapted to the environment in which those particular organisms live. In this point, Aristotle's adaptationist perspective is strongly reminiscent of Darwin. Life is conceived of as a sum of specific movements by which the organism communicates (in the broad sense of the word) with its environment: this bidirectional communication then realises the organism's nature, which is the goal proper of its existence. Aristotle's conception convincingly shows that organisms, autonomous entities adapted to their own world, which are of themselves the goal of their complex movements, cannot be fully grasped and explained purely mechanistically, that is, by physical interaction of inert particles of matter.

References

Balme, D. (1987). Aristotle's biology was not essentialist. In A. Gotthelf & J. Lennox (Eds.), *Philosophical issues in Aristotle's biology* (pp. 291–312). Cambridge University Press.

Balme, D. (1992). *Aristotle's De partibus animalium I and De generatione animalium I.* (with passages from II.1–3). Clarendon Press.

Barnes, J. (Ed.). (1984). *Complete works of Aristotle, Volume 1.* Princeton University Press.

Barnes, J. (Ed.). (1985). *Complete works of Aristotle, Volume 2.* Princeton University Press.

Berlin, B. (1992). *Ethnobiological classification: Principles of categorization of plants and animals in traditional societies.* Princeton University Press.

Broadie, S. (1987). Nature, craft and Phronesis in Aristotle. *Philosophical Topics, 15*(2), 35–50.

Connell, S. (2016). *Aristotle on female animals: A study of the generation of animals.* Cambridge University Press.

Conway Morris, S. (2003). *Life's solution: Inevitable humans in a lonely universe.* Cambridge University Press.

Cooper, J. M. (1987). Hypothetical necessity and natural teleology. In A. Gotthelf & J. Lennox (Eds.), *Philosophical issues in Aristotle's biology* (pp. 243–274). Cambridge University Press.

Corcilius, K., & Gregorić, P. (2010). Separability vs. difference: Parts and capacities of the soul in Aristotle. *Oxford Studies in Ancient Philosophy, 39*, 81–119.

Cunningham, A. (1999). Aristotle's animal books: Ethology, biology, anatomy, or philosophy? *Philosophical Topics, 27*(1), 17–41.

Gotthelf, A. (1987). Aristotle's conception of final causality. In A. Gotthelf & J. Lennox (Eds.), *Philosophical issues in Aristotle's biology* (pp. 204–242). Cambridge University Press.

Gotthelf, A. (1999). Darwin on Aristotle. *Journal of the History of Biology, 32*(1), 3–30.

Kauffman, S. (2000). *Investigations.* Oxford University Press.

Lefebvre, D. (2016). Le sperma: forme, matière ou les deux? *Philosophie Antique, 16*, 31–62.

Lennox, J. (1987). Kinds, forms of kinds, and the more and the less in Aristotle's biology. In A. Gotthelf & J. Lennox (Eds.), *Philosophical issues in Aristotle's biology* (pp. 339–359). Cambridge University Press.

Lennox, J. (1993). Darwin was a Teleologist. *Biology and Philosophy, 8*, 409–421.

Pellegrin, P. (1986). *Aristotle's classification of animals: Biology and the conceptual Unity of the Aristotelian corpus.* University of California Press.

Pellegrin, P. (2011). *Introduction. In Aristotle, Les Parties des animaux* (pp. 7–79). Flammarion.

Salmieri, G. (2018). Something(s) in the way(s) he moves: Reconsidering the embryological argument for particular forms in Aristotle. In A. Falcon & D. Lefebvre (Eds.), *Aristotle's generation of animals: A critical guide* (pp. 188–206). Cambridge University Press.

Sedley, D. (2010). Teleology, Aristotelian and platonic. In J. Lennox & R. Bolton (Eds.), *Being, nature and life in Aristotle: Essays in honor of Allan Gotthelf* (pp. 5–29). Cambridge University Press.

Witt, C. (2015). In defense of the craft analogy: Artifacts and natural teleology. In M. Leunissen (Ed.), *Aristotle's physics: A critical guide* (pp. 107–120). Cambridge University Press.

Chapter 3
Aristotle and Functional Bauplans

Roman Figura

Abstract Aristotle's monumental work on living beings belongs to the foundations of the history of biology. In the following text, we will see how Aristotle defines living organisms and how the living differs from the non-living. We are also going to examine Aristotle's classification of living organisms into larger groups. This classification is neither evolutionary nor idealistic; rather, it is based on the differences between the manner of life of the species in question. In conclusion, we will see how Aristotle approaches the subject of mutual interconnectedness of body organs.

Keywords History of biology · Vitalism · Bauplan · Aristotle · Ancient Greek science

Introduction

Aristotle's biology offers an elaborate and highly interesting method of describing and classifying living beings based on their typical features.[1] At first glance, this classification is radically different from the Linnaean system of nested categories. It is also quite unlike the evolutionary trees which, since the Darwinian Turn, have

[1] The Aristotle's works used in this chapter are as follows: Historia animalium (HA), De partibus animalium (PA), De generatione animalium (GA), De motu animalium (MA), De incessu animalium (IA), De anima (DA), De Caelo, Meteorologica (Barnes, 1984, 1985).

R. Figura (✉)
Department of Philosophy and History of Science, Faculty of Science, Charles University, Prague, Czech Republic

J. Švorcová (ed.), *Organismal Agency*, Biosemiotics 28,
https://doi.org/10.1007/978-3-031-53626-7_3

been used to model the development of life, and it bears no resemblance to the modern cladograms constructed by high-powered computers based on vast matrices of data on species-specific features. It is a method that deserves the attention of current natural scientists and in fact anyone interested in the history of science, a method that captures the world of living organisms in a nontrivial way. Further, we also present Aristotle's conception of autonomy of living beings, his approach to describing the structure of their bodies, and his view on interpreting their mutual similarities. We also plan to demonstrate that Aristotle works within a relatively clearly defined approach according to which living beings are 'autonomous agents' of their actions.

To assemble the admirable body of knowledge about living beings which he had at his disposal, Aristotle used autopsy (*HA* III, 2, 511b14–23; 3, 513a11–15) as well as observations of a vast number of various kinds of animals during different stages of their development (GA). He stressed the importance of understanding how various species reproduce. He also gathered observations from various other persons, carefully recorded all acquired information, and checked his notes when considering various hypotheses (e.g., GA I, 18, 722a16–b5). It seems that the main question he was interested in was the purpose of various parts of bodies of living beings. A classification of plants and animals was in effect a secondary output of such research. Aristotle was moreover familiar with the fact that various parts of animal bodies can serve more than one function, a typical example of such a body part being the elephants' trunk (PA II, 16, 658b30) or the cloaca of oviparous animals (GA I, 13, 719b29–720a17).[2]

Aristotle's Biosphere

First of all, let us note that some of the premises of the overall vision of the world formulated by Aristotle differ quite dramatically from what we take for granted today. Even aside from geocentrism, something we tend to find astonishing in Aristotle's scheme of the world, especially in the sublunar sphere, is the notion of its constancy (modern people are not that surprised by the idea of a relatively unchanging supralunar sphere) (Hladký et al., 2012: 118). It is not a complete immutability, because – in contrast to heavenly bodies and their movements (Cael I, 3) – events on Earth cannot be precisely predicted, but it is a state characterised by eternal duration without a beginning or an end. In a thus conceived 'weatherlessness', it makes no sense to ask after the origins and development of species or about their decline and demise (*DA* II, 4, 415a26–b7; *GA* II, 1, 731b31–a1). This nowadays odd notion has within the context of Aristotle's research an interesting position. The island of Lesbos, where Aristotle conducted his naturalist investigations

[2] Aristotle is often viewed as an advocate of monothelism, that is, a conviction that each thing serves only one purpose. But that does not seem to apply to his biological investigations.

(Leroi, 2014), is rich in fossils. Oddly enough, neither Aristotle nor his student Theophrastus, a botanist who moreover was a Lesbos native, mention these fossils in their writings, or, in the very least, no such text is known (Hladký et al., 2012: 15–19). We can only speculate why Aristotle and his students created no 'palaeontology'. Perhaps they thought the rocks merely resembled the appearance of living organisms and as such, it was a phenomenon that did not deserve any closer attention. Or perhaps such findings simply did not fit into their general theoretical framework and were ignored for that reason. Another possible explanation of this mystery is that Aristotelian biology is interested in living beings in their dynamic changeability and static fossils therefore lie outside its scope, just like a biologist is usually not interested, for instance, in statues of animals.

Because of this notion of an endlessly continuing biosphere consisting of always the same species,[3] Aristotle is often viewed as a paradigmatically anti-evolutionary thinker (Neubauer & Fiala, 2011: 308–309).[4] In his view, species do engage in dynamic interactions, but it seems they can neither become extinct nor have offspring modified into a different species. Since species are, in Aristotle's view, very well adapted to their unique lifestyles, it is also hard to imagine that their mutual competition could lead to extinction. On the other hand, Aristotle is well aware of species with a similar way of life, for instance raptors (Hall, 1991), among whom one could imagine competition although certainly not to a degree that would lead to the demise of a species.

At this point, let us also add that Aristotle's biology is not likely to be favoured by advocates of creationistic theistic conceptions: in his view, both the world and living beings are in a sense not created. That does not, however, mean that Aristotle is an atheist in the modern sense of the word, or that he tried to use his knowledge of biology to fight against traditional religion. Aristotle will also disappoint advocates of holistic ecosystemic and planetary approaches. In his view, the whole of the biosphere is given once and for all, which is why no-one and nothing has to care for and maintain its harmony and preservation. There is no need to postulate anything akin to the 'Gaia hypothesis'.

From today's perspective, how should we characterise Aristotle's 'biosphere'? According to his method, what kinds of beings are the subject of study of 'biologists'?[5] According to Aristotle's definition, living beings contain the source of motion within themselves (*Phys.* II, 1, 192b13–15; *DA* II, 1, 412a14–15). They are therefore autonomous active beings who look after their various needs and reproduction. We can thus add that they encompass both the beginning and the purpose, the goal, of their life processes (see Chap. 2). According to this definition, living beings differ from non-living entities by their ability to 'build' their bodies during

[3] Definition of a species is provided below in this text.

[4] Vast majority of natural scientists did not envisage the development of species until the end of the eighteenth century, but in Aristotle's system, even their creation by a divine being is not considered.

[5] Aristotle does not use terms 'biosphere' and 'biology'. We use these expressions merely to explain his thoughts using contemporary terms. By biosphere, we mean that whole of all living beings on Earth, by biology a science of all living entities.

ontogenesis, by having the capacity to actively resist external influences, and by having the capacity to change (grow, hatch, age, etc.) in the course of its life (GA I, 715a–17).

It seems therefore that according to Aristotle, autonomous action – which at least in some animals includes decision-making – is included in the definition of living beings. What is also important are their boundaries, the sense of separation between them and the outer environment: living beings are capable of absorbing matter from their environment and incorporating it in their bodies in a process we would nowadays call 'metabolism' (*DA* II, 1, 412a14–15; on metabolism in Schelling's philosophy see also Chap. 5). That is why they can develop and change while maintaining and preserving their identity.

Living beings in general are viewed as belonging to particular species based on their specific features and way of life. One thus cannot encounter some unspecified bird or ambivalent fish: animals always belong to a species, although in-between some more broadly defined groups Aristotle also describes certain transitional links (*GA* III, 11, 761a15–16).[6]

We could thus view the overall theoretical framework of Aristotelian biology as a sort of vitalism, although Aristotle naturally could not use this term. Living beings differ from non-living nature by certain concrete properties, whereby one that could be used to define living beings is that everything that is alive is endowed with a soul (*DA* II, 2, 413a21–22). We should not, however, think of the soul in an eschatological or transcendental sense: to Aristotle, soul expresses the animation of body, it is an autonomous principle of movement, growth, and reproduction (Thein, 2017: 126–127). In effect, we could say that in Aristotelian biology, soul is identified with 'agency'.

Transitory elements between the living and non-living realm do exist (Mete IV; *PA* IV, 5, 681a12–15) but are rather rare and in most cases, the difference between a living and non-living entity is abundantly clear. Somewhat unclear within Aristotle's thoughts on the realm of the living is his belief in spontaneous generation, where one might perhaps speak of the emergence of life from non-living matter, not from parents (e.g., *GA* I, 1, 715b25–30; *GA* III, 9). But spontaneous generation according to Aristotle takes place only under certain quite specific conditions, whereby often crucial in that process is the presence of matter that we would nowadays call organic (humus, organic substances in water, *GA* I, 16, 721a7–8). Still, the belief in spontaneous generation is, in Aristotle's biology, something of a foreign element and it makes it harder to formulate a general definition of life within his conceptual framework.

Otherwise, Aristotle's definition of living beings is clear and inspiring to current researchers. Living beings are not thought of as, for instance, complicated machines differing from artefacts made by people only by the level of complexity. Aristotle was familiar with various sophisticated mechanical devices and movable toys (Berryman, 2003). The reason why he probably would not understand mechanist

[6] Mussels have both animal and plant characteristics.

metaphors of Early Modern natural science is because machines cannot be born, and they do not undergo ontogenesis (cf. Chap. 8). In his work, efforts to understand the manner of reproduction and development of various species is of key importance. On the other hand, naturalists proceeding in their research using Aristotelian methods also do not need to address any questions regarding the origins and higher meaning of species, and do not need to ask after anything like a 'divine origin of the soul' or issues pertaining to a 'life force'. Aristotle's understanding of living beings is heuristically elegant in the sense of being able to define life based on simple and straightforward features. Aristotelian biologists do not need to deal with questions about any further forces hiding behind the manifestations of life, nor do they need to postulate any more elaborate definitions of entities they study. In short, according to Aristotle, we recognise entities as living based on how they present themselves, and for the needs and purposes of biological investigations this 'phenomenological' approach is sufficient.

Aristotle is thus not trying to explain the living realm on the basis of the non-living one, nor is he trying to reduce the living to the non-living. But he is also not an advocate of a 'panpsychic' attitude and considers neither a 'world soul', nor any kind of complex planetary or cosmic living organism. In his view, living entities are simply surrounded by a non-living environment.

According to Aristotle, all living entities – to use current terminology, the 'biosphere' – are located in the sublunar part of the world. Dividing the world in a distinct sublunar and supralunar spheres may nowadays sound like a bizarre anachronism, moreover one often viewed as hampering scientific progress in later centuries. In biological investigations, however, it does from a certain point of view make sense. Life, as Aristotle defined it and as we investigate it until today, is to the best of our knowledge indeed found on the Earth and in its atmosphere. With the exception of the visit of people to the Moon, we have no knowledge of life in either extraplanetary space or on other cosmic bodies. In a similar fashion, Aristotle does not consider life outside of what he defines as the sublunar sphere, nor does he speculate about its arrival from elsewhere. He does not believe that birds could fly to the Moon or that dragons live on the Sun. As a biologist, Aristotle investigates life within the delimited sphere of the world and does not consider its presence or origins outside it. We can thus discern three kinds of boundaries in Aristotle's biology: boundaries between living individuals and their surrounding environment, the mutual discreteness of species (without evolution and without appearance of new species via hybridisation), and a boundary or delimitation of the 'biosphere' within the universe.

It is also important to look at the activities which various members of the 'biosphere' engage in. What is the goal and purpose of their lives? Aristotle's answer is that individuals simply care for their survival and reproduction (e.g., *DA* II, 4, 415b9–12). Although in his other writings (Politics I, 8, 1256b15–22) Aristotle is not opposed to questions after the meaning of existence of various living beings in relation to humans, in biological investigations he describes living beings as existing for their own sakes, not as aiming to fulfil any 'higher' purpose. This approach is in fact rather similar to views held by modern natural scientists.

Classification

Aristotle's approach to the structure and function of bodies of living beings is something we have traditionally thought of in connection with a logical classification of organisms based on morphological and anatomical analysis. Ideally, such classification should take the form of a binary division of species according to selected features, resulting in tree-like diagrams, though naturally not evolutionary ones. This is how Aristotle's legacy has been mediated to us by the Late Classical and Medieval (both Arabic and scholastic) traditions (Hladký et al., 2012).

But when we read Aristotle's writings on biology on their own, without trying to make them conform to neo-Platonic hierarchical categories (Porphyrian trees), we realise that the texts as such simply do not contain any such categories. Aristotle does use terms equivalent to 'species' (eidos) and 'genus' (genos) but in a different context than we are used to. To him, genus is a broader category, one characterised by shared traits, while species is a narrower concept, one that within a particular context cannot be further divided. Within a species, one also cannot try to assess which of its members is ontologically more 'proper' to it, so that for instance a black horse is no more 'horsy' than a white horse (Preus, 1977; Lennox, 1980).

All fish jointly can be a genus but surprisingly also a species. The level of description depends on the issue which Aristotle is addressing at a particular point and on the level of abstraction at which he currently works. The basic element of description he uses to grasp living beings tends to be, however, indivisible species in the current sense of the term.

Let us take an example of Aristotelian contextual classification. Within the genus of sanguineous animals (nowadays we would say approximately vertebrates), fish are a species. Within the species of fish, it then makes no sense to ask whether, for instance, a catfish is more 'fishy' than a tuna. If, on the other hand, we speak of fish as a genus, we can distinguish variously defined groups of fish or individual species of fish in roughly our sense of the word. These individual species can then be compared in effect quantitatively within the fish genus, for instance with respect to the size of their fins. Aristotle adopts this contextual approach also when defining fish and cartilaginous fish. From the perspective of their locomotion and way of life they could be viewed as belonging to one group, but from the point of view of their reproduction they belong to separate groups (*HA* I, 6; *HA* II, 15, 505b26; *PA* II, 9, 655a23; *GA* I, 8, 718a35–b5).

It thus makes no sense to look in Aristotle's work for any anticipation of Linnean detailed classification or even evolutionary analyses. In contrast to Linnaeus, Aristotle does not usually go into detailed lists of representatives of a given group of organisms. Nevertheless, Aristotle's description of the biosphere is not a chaotic conglomeration of ad hoc collected information organised depending on the scholar's current mood – such characterisation would rather apply to the texts of Pliny the Elder. In fact, Aristotle organised his work according to a strict and elaborate method which, surprisingly enough, bears comparison even with some approaches used by modern science. In that case, let us now have a closer look at the criteria he relies on (Preus, 1977).

Aristotle's descriptions of living beings and their body parts are motivated by a desire to understand their functions (*PA* I, 5, 645b14–20). Organs of living bodies are understood in their mutual structural and functional connectedness within which they contribute to the preservation of life and to reproduction of the species in question. Each species' unique appearance thus enables it to live a particular lifestyle. When comparing species, Aristotle looks first of all on the manner of life that is characteristic for them. It is reflected in the appearance of their tissues, organs, and limbs – at least that is how we tend to understand Aristotle's descriptions of species nowadays: he speaks of parts being homoiomerous (uniform) and heteromerous (non-uniform) parts (*GA* I, 1, 715a9–11). In his 'classification' of biosphere, Aristotle thus compares how various species use their organs to deal with tasks connected with their life strategies. His approach is not evolutionary, but it does work with carefully collected information about morphology, anatomy, and ecology. We most definitely do not encounter in his writings any naive catch-all categories such as 'birds' applied to anything that flies, 'fish' designating all aquatic animals, 'beasts of prey' for predators generally, or 'vermin' for anything that slithers or crawls.

Particular species as well as broader groups of organisms are thus, according to Aristotle's method, characterised based on their shared traits (in current terminology their tissues, organs, and body parts). Within the genera as Aristotle defines them, individual species are then compared based on quantitative criteria, so that for example various birds differ in the size of their beaks (e.g. *PA*, I, 4, 644a16–21), eventually by the presence or absence of some organs, such as stingers or wings among the *entoma* ('segmented') organisms (a category in which we would nowadays place the insects and other terrestrial arthropods; Weiss, 1929). Between the genera, we can then compare organs based on functional, and partly also structural, analogy. In this way, we can compare for instance fish scales with birds' feathers or the scales of some terrestrial quadrupeds (nowadays we would most likely say reptiles) (*PA*, I.4, 644a17; *HA* I, 1, 486a16).

Overall diagnosis of particular basic species or larger groups of organisms, which depending on the context are labelled either as species or as genera, therefore relies on certain typical characteristics, which naturally includes internal organs. These characteristics are not arbitrary: they are the key characteristics linked to particular ways of life (*GA* V, 1, 778a32–778b11). Aristotle does not postulate, for instance, a genus of blue-eyed animals or birds with a red crest: his aim is to understand the different ways of life of the creatures he is investigating. Individual species and larger groups are determined based on deep empirical knowledge. This is also why, in his system, some typical characteristics could be absent or redundant (that is, modular). An example of this is the number of legs used for locomotion (millipedes vs. spiders), or the presence and absence of wings (spiders vs. bees) (Weiss, 1929). This is why ostriches are birds although they cannot fly; their way of life as well as some other characteristics (*PA* IV, 5, 697b15–25) such as eyelashes and feet with two 'hooves', resemble rather viviparous quadrupeds (we would say mammals). On the other hand, it makes no sense to group butterflies or dragonflies together with birds although they do have wings and excel at flying. Their overall habitus and combination of typical characteristics are so different from birds that they can be compared with them only by analogy. Similarly, by analogy we can compare cetaceans and fish or birds and bats (Lennox, 1980).

Aristotle thus places organisms within broader groups which are defined by characteristics important for the lives of animals that belong to them. On the other hand, it is quite unnecessary for all organisms included in a group to have all the listed, clearly defined characteristics. This is because the typical characteristics function rather as useful cues from whose combination one can draw conclusions regarding the manner of life of their bearers. For instance, a black swan thus probably would not confuse Aristotle because its manner of life is not substantively different from that of other swans (and moreover, the young of some European swans also have a darker colouration). In his work on biology, Aristotle is not aiming at providing a precise definition of essence, in this case 'swanness', which is why his categories remain flexible and open to new discoveries. Naturally, these categories depend on the observer who defines them, but they are not arbitrary. They are based on a thorough study and deep understanding of lives of the creatures which are being investigated and one could even conceive of a sort of transition links between them.

Bauplan

If the Aristotelian biosphere were to have an apex, and animals their Bauplan, it would clearly be humans. In human embryogenesis nature can display all its potential to the full. Human figure is erect thanks to the highest animal heat, while the forelimbs (in this case the upper limbs) and mouth can be involved in complex activities (work, speech), which are not accessible to quadrupeds. Of all animal bodies, the human figure is the best differentiated one in all three basic axes (up–down, front–back, left–right), and can thus most fully realise its potential. Other animal species could be from a certain perspective viewed as, in a sense, underdeveloped humans, with lower animal heat and other proportions of the basic elements (*IA*, 5; *PA*, IV, 10, 686b29–687a2; *PA* IV, 10, 686a25–35).

From the perspective of proportions of the elements, the human body is balanced the best, so that people do not have to crawl around on all four with head low to the earth due to an overabundance of that element (*IA* 4, 706a19–20). The human body is as if a model Bauplan, against which other organisms are measured and described. (This is heuristically useful, because it is also the body we know the best, see for instance the didactics of biology to this day.) Surprisingly, in several analogies Aristotle compares even plants to the human body, claiming that due to overabundance of the earth element they are structured in the opposite way to humans (*PA*, IV, 10, 686b29–687a2). Their mouths are thus in earth, while their reproductive organs are up in the air. In general, though, Aristotle does not define the specific Bauplans of various groups of organisms in a way that could be compared to approaches of modern scientists: the closest he comes to it is in his description of the human body. We should also bear in mind that although in its nature, the human body is the best organised in the entire animal kingdom, it does not perform the best in all abilities. The fish can certainly swim better than people do, but humans have a much wider range of activities they can engage in.

Because Aristotle does not consider the evolution of species in the sense of gradual development with modifications, it makes no sense to consider within his conceptual framework the issue of the origins of Bauplan of various groups of animals in terms of some common ancestor. According to Aristotle's biology, the structure of animal body is simply not the legacy of evolutionary history. In the same way, neither the human body nor the body of any other animal species could be the ideal target towards which the development of other species gradually aims. In short, Aristotle does not think in terms of evolutionary teleology.

According to Aristotle, the appearance of living beings is also not dictated by some ideal geometric logic. A living organism is no passive bearer of eternal types. Organisms can thus be mutually compared, and it makes no sense to view them as actual (imperfect) imitations of theoretical (ideal) Bauplans. In fact, the structure of a body is dictated by the logic of metabolic processes and needs of the organism in question. In sexually reproducing animals, it is moreover co-shaped by inheritance of some properties from parents. The overall appearance is naturally inherited from parents but now we have in mind certain individual characteristics, such as hair colour or birth marks (*GA* I, 17, 721b28–722a1). In most cases, however, such individual variations are not important for survival. In Aristotle's non-evolutionary approach, gradual changes and hereditary variations cannot lead to the appearance of new species. In that case, within Aristotelian biology can one actually speak of some theoretical conception that would be analogous to the modern Bauplan – although Aristotle himself does not offer any such concept in terms of its definition? We believe one can.

Aristotle is often criticised for his anti-evolutionary thinking, which could create the impression that he did not understand the meaning of adaptations. But that is not the case. According to Aristotle, nature does everything because it is either better or because it is necessary (*GA* I, 4, 717a15–16). The task of a researcher is then to investigate the meaning of anatomical and morphological structures of living beings for their life and procreation (*PA* I, 5, 645a36–b3; *PA* I, 1, 640b17–22; *HA*, I, 1, 487a10–11). (The teleological goal of living organisms is thus survival and reproduction, cf. Chap 2) Although Aristotle is often viewed as a thinker who would have been opposed to Darwinism, the general end result of the two approaches is surprisingly similar. Both the Darwinist and Aristotle are looking for functional explanations of various characteristics of living organisms, whereby the former can also consider the possibility of it being inherited from evolutionary ancestors (Gotthelf, 1999). It follows from the above that for instance a horse is not a horse because it represents the model of the ideal horse, nor is it a horse in order to fulfil human need for transport or food. The structure and functions of its body stem from what it needs to survive, and the two are closely intertwined. A horse thus does not have, for instance, the tentacles of an octopus or wings like a sparrow; these would not be beneficial to the horse and may be even incompatible with its survival.

Aristotle is a non-evolutionary thinker but clearly also a scholar who thought deeply about adaptations, although one could not call him a radical adaptationist in the modern sense of the word. He is well aware that various features he describes in living organisms need not be only the means of useful interaction with the animal's environment; oftentimes, they are simply given by the logic of the body structure, while others are just manifestations of natural variability, and their meaning need not be specified (e.g., the eye colour, GA V, 1, 778a32–778b11).

If the appearance of a living organism is an expression of its way of life (in modern language we would say an adaptation to particular conditions), then broader groups of organisms, which include particular species, are characterised by a shared way of life. In Aristotelian biology, demarcation of such broader groups depends on the context in which they are being defined, whereby the context is not arbitrary: it is based on a thorough study of the organisms in question. Important characteristics of defined groups of living beings include the overall body structure, their manner of locomotion, and manner of reproduction (Preus, 1977).

Within the genera, we can compare various characteristics between the species quantitatively, typically for instance the size of various organs. The possibility of such comparisons between species demonstrates that Aristotelian species are not fully discrete and incommensurable entities. In a certain sense, one could say that differences between them are a matter of degree. In such comparisons, moreover, they are always compared mutually, not against some ideal model (Lennox, 1980).

This concept of species, between which one can observe gradual 'development' of a particular characteristic and even transitory forms between larger groups, does sound familiar to modern ears. In the Aristotelian model, the unifying principle between species is naturally not their evolutionary origin but their ontogenetic one! According to Aristotle, embryogenesis is something (almost) all organisms share. The developing embryos are at first basically identical in their properties. Initially, they even lead the life of a plant (their abilities are limited to growth and 'metabolism', *GA* II, 4, 740a24–35; b8–12). In the course of embryogenesis, we can first see some indefinite animal, then an indefinite individual of a particular species, and that later acquires characteristic features of its parents (in sexual reproduction).

The content of life of a developing embryo is thus to achieve the appearance of its species and then its individual appearance. The content of life of a newborn animal is to grow and develop to the stage of an adult and to take care of itself (*GA* I, 1, 715a1–17). The purpose of these activities is to survive and reproduce. To achieve this, the individual's basic needs must be met, for instance, search and ingestion of food, dwelling in places suitable for its life, safety, communication, escaping from predators, search for a mate, reproduction, and eventually also care for its young (Gotthelf, 1987).

Aristotelian biology focuses on anatomical and morphological characteristics of living beings and related ethological and ecological investigations of organisms. Naturally, Aristotle does not define ethological concepts in a modern sense of the word, nor is he creating ecological science. He also does not speak of anything comparable with our current definitions of ecosystems, biotopes, or other key notions of ecology as we understand it. His characterisations of living organisms and their groups can, however, be viewed as analogical to modern definitions of various ecological niches. To wit, groups of organisms and individual species (defined roughly as we would understand them) are representative of unique ways of life. According to Aristotle, however, living beings did not come to be the way they are over the course of evolutionary history but via ontogenesis, a process that proceeds from general characteristics to those which are species-specific and ultimately individual.

So, what defines living beings and their groups in Aristotelian biology? With a bit of exaggeration, one could say that these concepts express the manner in which nature responds to various situations and places in the world, ways in which nature (*physis*) in effect makes itself home in the world. Organisms are tangible responses to the various opportunities for life which the world offers. According to classical Darwinism, different forms of life emerge in the course of evolution via selection and the 'struggle for life' in which living beings engage. In Aristotelian biology, we can analogically speak about the emergence of various forms of life in the course of ontogenesis. This process takes place thanks to their nature (*physis*), although this view lacks the historical dimension (e.g. *GA* I, 1, 715b25–30, 715b8–16).

The Function and Heredity of Appearance of Particular Species

Why, during ontogenesis, do such and such body parts emerge, depending on the species? This is because the organism needs those body parts, and not others, to survive, eventually also to reproduce. Those particular body parts are either necessary for survival and reproduction or else they aid and facilitate it.

This is also where we see in action two basic methodological maxims of Aristotelian biology. First, 'Nature creates nothing without a purpose, but always the best possible in each kind of living creatures' (*IA*, 2, 704b15–18), and second, 'Nature does everything either because it is necessary or because it is better' (*GA* I, 4, 717a15–16). The inner purpose of living organisms is self-preservation. Within Aristotelian conception, they are not incorporated into a higher whole of some cosmic 'superorganism' that determines a superior purpose of their existence in terms of their participation in this whole. In other words, Aristotelian biology does not offer an answer to the question why there are in the world horses, bee-eaters, daisies, or people. Aristotle probably did not, in the context of biology, consider this question relevant. Parts of bodies of organisms, on the other hand, are incorporated into the whole of a living body in such a way that they serve its survival, and their appearance is thus subordinated to the fact that they have within the body certain clearly defined functions which biologists strive to understand.

But if men and animals and their several parts are natural phenomena, then the natural philosopher must take into consideration not merely the ultimate substances of which they are made but also flesh, bone, blood, and all other homogeneous parts, as well as the heterogeneous parts such as face, hand, or foot. One must also examine how each of these comes to be what it is, and in virtue of what force (*dynamis*, *PA* I, 1, 640b17–22).

Body parts are essentially determined mainly by their function within the whole of a living body, by what they affect in it, or by what activities they are capable of performing. The material composition of body parts is important because particular functions require certain materials (for instance, a functional hand cannot be made of wood). On the other hand, as demonstrated by the example of a corpse, while properly constituted parts are a necessary condition, they are not a sufficient condition:

> And yet a dead body has exactly the same configuration as a living one; but for all that is not a man. So also no hand of bronze or wood or constituted in any but the appropriate way can possibly be a hand in more than name. For like a physician in a painting, or like a flute in a sculpture, in spite of its name it will be unable to do the office which that name implies. Precisely in the same way no part of a dead body, such I mean as its eye or its hand, is really an eye or a hand. (*PA* I, 1, 640b33–641a5)

Body parts are thus substantively determined by their vital functions and during ontogenesis, they emerge in just the manner that enables them to perform activities which are necessary or aid the survival, eventually reproduction, of the animal. In this sense, one could say that from Aristotle's point of view, 'function precedes the organ', and this is how a teleological view of the relation between body parts and the whole of an organism is implemented (cf. Chap. 4). This is very clearly summarised in an important passage from *On the parts of animals*:

> As every instrument (*organon*) and every bodily member subserves some partial end, that is to say, some special action (*praxis*), so the whole body must be destined to minister to some plenary (*polymeros*) sphere of action. Thus the saw is made for sawing, for sawing is a function (*chresis*), and not sawing for the saw. Similarly, the body too must somehow or other be made for the soul, and each part of it for some subordinate function (*erga*), to which it is adapted. (*PA* I, 5, 645b14–20)

A living body is in effect a soul at work. The activity of an ensouled body is life and individual body parts during the formation of a new individual gradually develop so as to fit the role they would eventually have within that particular living body. Growing embryos and juveniles form their body parts always in consideration of the adult form towards which they grow, and develop towards securing certain vital functions that are a necessary part of that form. For instance, a fertilised egg of a carp is supposed to lead to the emergence of a carp. The aim or purpose of embryogenesis is the formation of a carp, and that is why the embryonic matter of a carp female fertilised by a carp male gradually develops into body parts that fit or belong to the functions necessary for carp's survival (fins, gills, scales), as opposed to forming parts of, for instance, a badger or some accidental body parts which ultimately possibly would not function at all.

Systematisation of Living Organisms in Aristotelian Biology

According to Aristotle, blood is a key part of animals' bodies. Nowadays we would describe it as a central organ of metabolic transformation. According to Aristotle, it is created in organs which digest food by 'cooking' (*pepsis*), i.e., by transformation due to the effect of vital heat. Then it is carried around the body and ultimately leads, via further transformations, to the formation of various body parts. Despite this somewhat odd phrasing, the characterisation is, in the light of our current understanding, quite apt because in addition to oxygen, blood also distributes around the body nutrients extracted from food (*PA* II, 3, 650a2–32; *GA* I, 19, 726b1–11).

Aristotle takes into account the importance of blood in his classification of animals when he defines two large genera: sanguineous and bloodless animals (*HA* I, 6). The groups of animals thus defined correspond roughly to modern categories of chordates and invertebrates. Evolutionary biology nowadays naturally does not view invertebrates as a unified monophyletic taxon, but in teaching and popularisation the term is still widely used. It should be noted that Aristotle not only appreciates the importance of blood in the bodies of animals but also describes a liquid with an analogical function: one which we would nowadays probably identify with haemolymph (*GA* I, 19, 726b1–11).

Living organisms belonging to these two groups, the sanguineous and the bloodless, radically differ both in the overall structures of their bodies and in their lifestyles. Sanguineous animals are characterised by a developed internal skeleton. Bloodless animals (*HA* IV, 1, 523b1–21), on the other hand, often rely on external support and protection of their bodies (shells, carapaces, and the like), while some lack such protection altogether (e.g., the octopus). The bodies of bloodless animals are often segmented and tend to have a large number of paired walking limbs (Weiss, 1929). We can thus see that among the bloodless animals, one finds a degree of anatomical and ecological diversity which surpasses the variety found among sanguineous animals.

Within the genus of bloodless animals, Aristotle recognises four large groups, once again defined by their overall body structure and way of life (Voultsiadou et al., 2017). In his biological works, he pays a lot of attention to the *malakia*, a group we would nowadays identify with cephalopods. Aristotle describes their body structure, way of life, and reproduction in detail, displaying a considerable amount of knowledge. He also notices species with shells, such as the argonauts and nautili, which he was surprisingly familiar with (*HA* IV, 1, 523b21). Another group of the bloodless animals are the *malacostraca*, which we would identify approximately with marine crustaceans. These are thus animals equipped with an external shell and numerous legs, which inhabit the sea and seashore (*HA* IV, 2, 525a30). The group of *ostracoderma* includes what we would call bivalves, gastropods, and echinoderms. These animals are equipped with a protective shell and in many cases, their bodies have neither a differentiated head nor differentiated limbs. Many live a sessile life and engage in no local locomotion. Terrestrial gastropods are thus within this group something of an exception (*HA* IV, 4, 527b35). In the group of *entoma* we find terrestrial animals with segmented bodies, walking legs, and in some cases even wings. From our perspective, this group thus encompasses insects, arachnids, centipedes, millipedes, and terrestrial crustaceans (*HA* IV, 7, 531b20). Aristotle was also aware of animals that do not readily fit into either of the groups defined above, such as those which in some respects resemble rather plants, for instance anemones.

Within the sanguineous group, animals are usually classified based on their general body structure and consequently also their overall way of life. The manner of reproduction is another key criterion. In Aristotelian biology, embryology plays a role in the description of living bodies that is as important as evolutionary and developmental biology is for us nowadays. We use these approaches to try and uncover the origins of living beings, while Aristotle analogically learns about them from their ontogenesis (Voultsiadou et al., 2017). (1) Among the sanguineous group, Aristotle treats humans as a special category because people are terrestrial, bipedal, and viviparous. (2) Another group are terrestrial viviparous animals who move mostly on four legs. Nowadays, we would identify them with mammals. (3) Specific within the sanguineous group are animals who lay eggs. They are mostly terrestrial and move mostly using four legs. On the other hand, Aristotle includes in this group also snakes and even describes some animals belonging to this group as viviparous (or ovoviviparous). From our point of view, this group includes roughly turtles, scaly reptiles, and crocodiles. (4) A specific, well-defined group consists of birds, characterised by bipedal movement on earth, ability to fly, and laying of eggs. Aristotle is, however, also familiar with ostriches, which in some respects resemble terrestrial viviparous quadrupeds (eyelashes, manner of movement, foot soles, *PA* IV, 5, 697b15–25). On the other hand, he also notes that the wryneck shares some features with the previous group (a tongue, hissing, etc.) (Gorman, 2022). (5) Fishes and cartilaginous fishes (approximately according to current definitions) are according to some criteria, such as their way of life and manner of movement, included in one group, but according to other criteria (being oviparous vs. viviparous) belong to separate categories. (6) The group of *kete* includes aquatic, viviparous, and air-breathing animals. This group thus includes cetaceans and probably also seals (the Mediterranean monk seal) and it would have included the dugong that live in the Red Sea had Aristotle had any reports about them.

Within the group of birds, Aristotle recognises several further, more narrowly defined groups. In particular, he distinguishes (Hall, 1991): (i) Birds with curved talons, about whom he supplies many details from their lives. From our perspective, this group would include raptors, falcons, owls, and parrots. It is rather interesting to note that according to modern classification, parrots are indeed closely related to falcons. (ii) Birds living on the ground, whose flying abilities are usually poor. This group includes landfowl (Galliformes), skylarks, and the ostrich. (iii) Birds with webbed feet who inhabit aquatic environment. Nowadays we would probably include in this group the Pelecaniformes, the Anseriformes (also known as waterfowl), the gulls, grebes, and coots. (iv) Birds living along shorelines and in the wetlands. This would encompass mostly Ciconiiformes, cranes, and large waders.

This classification is once again based on key morphological, ethological, and ecological characteristics. The groups in effect express shared ecological niches and a particular way of life expressed by similarity of appearance. Although according to current evolutionary biology such taxonomy naturally does not make much sense, to field ornithologists, it feels very natural. These groups of birds indeed often share the same biotope and are similar in terms of food habits and reproduction.

Aristotle and Topology of Animal Bodies

As noted above, Aristotle pays a lot of attention to ontogenesis of animal bodies, whereby his aim is to understand the origins of their appearance. He pays a similar amount of attention to descriptions of their sexual organs, including their exact location in the body. In fact, he does not limit himself to their detailed description: he is trying to explain what he observed from a functional perspective. Although some of his theoretical conclusions are, from our current point of view, rather bizarre, Aristotle was very thorough in his explanation of empirical observations through the lens of his theories. Aristotle ascribes much importance to vital heat (*PA* II, 2), which must work adequately to secure optimal function of particular body parts. There must not be too little or too much of it, and the cooling ability is also important. More heat is usually connected with faster movement, while slower movement has a cooling tendency. He also notes that vital heat is the highest in the middle of the trunk of the animal.

Testes and the Penis

In his investigations of sanguineous animals, Aristotle concludes that not all are equipped with testes and a penis. Since he believed that nature does everything either because it is necessary or because it is better (*GA* I, 4, 717a15–16), he claims that neither the testes nor a penis are necessary for procreation (*GA* I, 5, 717b14–33). A modern naturalist would certainly agree when it comes, for instance, to the absence of a penis in fish and some birds, but the claim that testes are not necessary would strike us as most odd. On the other hand, in many species we find not round testes, such as people have, but oblong ones, which Aristotle probably mistook for vas deferens, a structure he does ascribe to all male sanguineous animals. According to Aristotle, testes are not the source of the sperm: their role is merely to slow down its progress during copulation and to cool it in that way. In his view, sperm is created in seed ducts in the lumbar area (*GA* I, 4, 717a12–b13).

Male Sexual Organs

So how does Aristotle describe the various organisation of male sexual organs in sanguineous animals? Here are several examples.

In fish, Aristotle explains the absence of penis and testes by the need for a fast expulsion of the sperm: testes and a penis would only slow this process down. The explanation, meanwhile, of why it is necessary to expel the seed quickly is quite curious. He claims that during ejaculation, animals in general hold their breath and fish could thus suffocate (probably because they do not have lungs with the requisite store of air, *GA* I, 6, 717b33–718a17).

In snakes, the explanation of absence of both testes and penis is different. Snakes have no feet and that is why they also cannot have a penis, because from the perspective of anatomy, Aristotle considers penis a lower limb. Both the testes and penis would moreover excessively cool the expelled sperm, because even prior to ejaculation the sperm has to pass through the snake's long body (*GA* I, 7, 718a17–34).

In birds, Aristotle describes testes inside their bodies: in his view, the feathered skin of birds cannot create a scrotum (*GA* I, 12, 719a30-b17). Aristotle claims that we usually do not find a penis in birds because their legs grow from the middle of their belly, which is why there is simply no place for a penis on their trunk (*GA* I, 5, 717b14–33). On the other hand, Aristotle was aware of a penis in geese (*HA* III, 1, 509b30). Aristotle also describes testes hidden in the body, for instance in hedgehogs, elephants, cetaceans, and scaly quadrupeds ('reptiles'). Its location is then explained in the same way as in birds, i.e., by claiming that the skin of those animals cannot create a scrotum (*GA* I, 12, 719a30-b17).

Aristotle finds that sexual organs are optimally structured in humans and most viviparous terrestrial quadrupeds ('mammals') that have a penis and an external scrotum with testes (*PA* II, 10, 655b37–656a13). This constellation makes sure that the sperm passing through these organs is at optimal temperature and thereby correctly 'cooked' (i.e., that conditions are right for its 'metabolic transformation'). Moreover, this setup of sexual organs is also an expression of appropriate sexual activity in humans: following his dictum of harmfulness of any excess, Aristotle views sensible engagement in all activities as the optimal expression of the nature of living organisms. And not surprisingly, humans are in many ways viewed as the 'model' of what is sensible and appropriate.

Female Reproductive Organs of Sanguineous Animals

According to Aristotle, a male is that individual which can procreate in another, while a female can create an offspring in herself. Females thus accept male seed. According to Aristotle's conception, the development of a new individual starts when the seed meets with the menstrual blood of a female (*GA* I, 2, 716a13–15; *GA* II, 4, 740b24–25). We are not going to go into the details of the complex Aristotelian theory of the role of the two sexes in reproduction (cf. Chap. 2). For the moment being, let us just focus on a functional explanation of the anatomy of internal female reproductive organs.

In fish, internal female reproductive organs are located close to the external reproductive organs. This is linked to the fact that fish lay a large number of small eggs which do not mature in the female but develop independently in water. This is why they do not have to be exposed to the vital heat of females in whom they originated (*GA* I, 8, 718a35–27).

The females of birds and terrestrial oviparous quadrupeds ('reptiles') lay eggs with a leathery or calcareous shell. The eggs must mature prior to hatching by exposure to vital heat, which is greatest close to the centre of the body. This is why the internal reproductive organs of these females reach all the way to the centre of their bodies (*GA* I, 8, 718a35–27).

Especially interesting is Aristotle's description of the function of internal reproductive organs in cartilaginous fish: he claims they are located from the centre of their trunk all the way to the external reproductive organs. Cartilaginous fish are of a cold nature, which is why they cannot create eggs with a firm shell that could be laid and avoid destruction. This is why they form soft-shelled eggs, which they carry in the centre of their bodies. These eggs then before hatching travel down the birth canal, which is where the offspring are released from the egg so that they emerge of the female's body without a shell (*GA* I, 10, 718b32–36; *GA* I, 11, 718b36–719a12).

In humans, cetaceans, and terrestrial viviparous animals ('mammals'), reproductive organs are usually located in the lower part of their trunk, close to the external sexual organs, because it would be difficult for the female to carry the heavy, developing offspring close to the centre of the body. This way, moreover, the birth canal is shorter, which makes the entire birth easier (*GA* I, 11, 719a12–30).

Aristotle in his functional descriptions of the abovementioned organs clearly works with empirical findings acquired during dissections. His theoretical explanations of what he observed do sound to our ears rather odd, but within his paradigm that worked with vital heat, they do make sense. In these explanations, Aristotle is thus consistently connecting his empirical knowledge with a theoretical model. This is one of the reasons why he can be viewed as a true predecessor of modern research in natural sciences.

Conclusion

Aristotle's biology works with several basic premises. First of all, it quite clearly distinguishes living beings from the non-living environment using the concepts of autonomy, boundary with respect to the external environment, and dynamic changeability. Living organisms are endowed with the ability to develop, to metabolise, (with the exception of plants) to perceive, and to reproduce. All living beings are located in a well-defined place within the cosmos, namely in the sublunar sphere, and they engage in dynamic mutual interaction, but they do not form anything like a complex ecosystem that would include, for instance, internal regulation mechanisms or undergo a historical development. The Aristotelian biosphere is given once and for all, its constituent members did not evolve, were not created, and cannot go extinct. The purpose of life of the various species of living beings is to provide for themselves and to reproduce. An important sign of autonomy of living beings is the ability of individuals to build, in the course of ontogenesis, their bodies from (partially already formed) matter supplied by parents.

Aristotle's conception may be non-evolutionary, but it is strongly adaptational. Aristotle is trying to explain the functional significance of key parts of animals' bodies, that is, his aim is to find out how various body parts benefit their bearers. He is not, however, a naive adaptationist who would ascribe importance to every organ.

Based on the configuration of body parts in particular organisms (and his own observations) he then postulates wider groups of animals, which in effect define certain ecological niches. The most important characteristics he pays attention to are internal organisation of the body, manner of locomotion, and the manner of reproduction. Aristotle's classification of living beings is thus a secondary product of his study of body parts and their importance.

Aristotle was aware of the existence of various sophisticated mechanisms and movable toys. However, he is most definitely not championing a mechanistic approach to biology in modern sense of the term. This is most likely because, as he knew full well, machines are not capable of reproduction, development, and growth.

For Aristotle, the model for body structure of all animals are humans, in whom body parts are differentiated optimally. This is then why people can engage in a wide range of activities, notably various kinds of work, and in speech. From this point of view, other animals are in effect underdeveloped people, although not in an evolutionary sense of the term. On the other hand, Aristotle does not claim that humans are the evolutionary model of other developmental lines, nor is he a proponent of the kind of descent theories according to which, for instance, animals are 'fallen' people. Like other organisms, humans acquire their form during ontogenesis, which, however, takes in our species its fullest course thanks to an optimal ratio of the elements and working of the vital heat (Thein, 2017: 187). At the beginning of ontogenesis, all embryos are alike in their way of life (they have metabolism and the ability to grow) but gradually, over the course of their development, they differentiate, that is, they assume the appearance of various species. In Aristotle's biology, different species are thus linked not by similar or divergent evolutionary stories but by similar or different ontogenesis.

Within the groups of living organisms defined by Aristotle, one can speak of similar body structure (sort of proto-Bauplans), which are an expression of shared way of life. The appearance of species thus does not reflect their evolutionary history, nor is it an imitation of ideal models. It is given by the 'ecological niche' within which that organism lives. It makes sense to compare organisms among each other, not to derive them from some ideal Bauplans.

Aristotle pays much attention to the internal organisation of animals, topology of body parts, and relations between the parts of their bodies, especially when reproduction is concerned. The manner of reproduction is one of the key characteristics according to which Aristotle defines broader groups of animals (Fulínová et al., 2022). Naturally, many of Aristotle's conclusions nowadays seem rather surprising and, in some instances, they even bring a smile to our face. But one cannot deny that he created biology as a systematic science based on assembling empirical observations and incorporating them into theoretical hypotheses. In the meticulous questioning and search for answers that would make sense from the perspective of the species he investigated, Aristotle most definitely equals modern natural scientists. And if we want to grasp the history of biology in general and position within it the multitude of other scholars who were inspired by him, is it crucial to understand the theories, conceptions, and concepts Aristotle worked with.

References

Barnes, J. (Ed.). (1984). *Complete works of Aristotle, Volume 1*. Princeton University Press.

Barnes, J. (Ed.). (1985). *Complete works of Aristotle, Volume 2*. Princeton University Press.

Berryman, S. (2003). Ancient automata and mechanical explanation. *Phronesis, 48*(4), 344–369.

Fulínová, E., Figura, R., Hladký, V., Liepoldová, T., & Markoš, A. (2022). *Aristotelés O plození živočichů*. Pavel Mervart.

Gorman, G. (2022). *The wryneck*. Pelagic Publishing.

Gotthelf, A. (1987). Aristotle's conception of final causality. In A. Gotthelf & J. G. Lennox (Eds.), *Philosophical issues in Aristotle's biology* (pp. 204–242). Cambridge University Press.

Gotthelf, A. (1999). Darwin on Aristotle. *Journal of the History of Biology, 32*(1), 3–30.

Hall, J. J. (1991). The classification of birds, in Aristotle and early modern naturalists (I). *History of Science, 29*(2), 111–151.

Hladký, V., Kočandrle, R., & Kratochvíl, Z. (2012). *Evoluce před Darwinem*. Pavel Mervart.

Lennox, J. G. (1980). Aristotle on genera, species, and the more and the less. *Journal of the History of Biology, 13*(2), 321–346.

Leroi, A. M. (2014). *The lagoon: How Aristotle invented science*. Penguin Group.

Neubauer, Z., & Fiala, J. (2011). *Střetnutí paradigmat a řád živé skutečnosti*. Malvern.

Preus, A. (1977). Eidos as norm in Aristotle's biology. *The Society for Ancient Greek Philosophy Newsletter, 86*.

Thein, K. (2017). *Aristotelés o lidské přirozenosti*. Filosofia.

Voultsiadou, E., Gerovasileiou, V., Vandepitte, L., Ganias, K., & Arvanitidis, C. (2017). Aristotle's scientific contributions to the classification, nomenclature and distribution of marine organisms. *Mediterranean Marine Science, 18*(3), 468–478.

Weiss, H. B. (1929). The entomology of Aristotle. *Journal of the New York Entomological Society, 37*(2), 101–109.

Chapter 4
Immanuel Kant: Mechanism, Teleology, Organism, and the Powers of Our Mind

Robert Kanócz

Abstract The chapter investigates ideas about mechanisms and teleology in nature, especially in relation to the origin and life of organisms, as Immanuel Kant formulated them in his *Critique of the Power of Judgment* (1790). Then it focuses on one specific direction in later reception of Kant's ideas. Kant, like many other authors at the end of the eighteenth century, was attracted to the idea of a purely mechanistic interpretation of nature and tried to formulate an incontestable justification of the possibility of such an interpretation. At the same time, though, he believed it unthinkable that organisms could have come to be solely through the action of blind forces: he was convinced that one must assume a purposeful activity behind their origin and existence. His transcendental philosophy makes it possible – among other things – to reconcile these two conceptual tendencies. Kant gave up on searching for the truth of things as they are in themselves; he examined only what appears to us and the powers of our mind which form these appearances. He analysed the composition of the complex system of powers and principles inherent to our mind and searched among conceptions of the world that can be produced by this system for those ultimately based only on principles which are common to all human beings, and can therefore be universally shared. Kant believed that within such theoretical framework, he managed to prove that we can, all of us and always, successfully apply a mechanistic view of nature, whereby phenomena will even confirm its correctness. At the same time, we necessarily all apply a view of organisms as created by a (presumably divine) intention, although no such thing can be empirically confirmed. In the final section, I outline the reception of Kant's ideas by some of his contemporaries, as well as by some twentieth-century thinkers. They stressed the importance of aesthetic judgment, which according to Kant seeks the appropriate degree of emphasis on the particular aspects of things without being guided by rules that can be grasped conceptually. The thinkers in question then not only gave up on a search for the truth of things in themselves but also on search for universal

R. Kanócz (✉)
Department of Philosophy and History of Science, Faculty of Science, Charles University, Prague, Czech Republic

J. Švorcová (ed.), *Organismal Agency*, Biosemiotics 28,
https://doi.org/10.1007/978-3-031-53626-7_4

perspectives valid for everyone, always, and everywhere. At that point, all one can do is to reinvent reasonable combinations of possible perspectives, always again and anew, depending on the situation. I consider this to be a relevant challenge also for contemporary thinking about the mechanistic and teleological character of organisms.

Keywords Immanuel Kant · Transcendental philosophy · Mechanism · Teleology · Regulative principles · Perspectivism · Gilles Deleuze · Pavel Kouba

Introduction

Immanuel Kant (1724–1804) holds a key position among thinkers who – after the emergence of modern natural science associated with scholars such as Galileo, Hobbes, Descartes, and especially Newton – tried to come to terms with a new conception of nature. Mechanistic natural science at least seemed to head towards the discovery of universal deterministic laws. On the other hand, it seemed hardly compatible with our experience of our own freedom – as well as with most theological and ethical conceptions of human life – that everything should be subjected to such laws. Moreover, it was unclear how one could reconcile determinism with a meaningful order of nature, especially living nature, or what role in such an order should be left to God.

We will, of course, be primarily interested in Kant's ideas regarding organisms[1] and limit ourselves to his by far most influential contribution to the topic of their relation to mechanisms, which appeared in the *Critique of the Power of Judgment* (1790[2000]). Here, Kant in a sense articulates what is specific to organisms and why they require an explanation other than a mechanistic one. But more original and more inspiring to this day is the status which Kant attributes to this subordination to a mechanism and to the 'apparent' purposefulness. This, however, is something Kant can carry out only in his system of philosophy as a whole, within philosophy that has in a unique way conceived of and related to each other reality, phenomena, and the faculties of our mind.

The gist of his original juxtaposition of the mechanistic and (specific) non-mechanistic view of the organism and nature in general is expressed in the so-called antinomy of teleological judgment, that is, in the statement of a contradiction between the principles inherent to teleological judgment and a solution thereof.

[1] Kant speaks of 'organised beings' (*organisirtes Wesen*, elsewhere also *organische[r] Körper*, *organische[s] Geschöpf*). He does not use the word 'organism' in his works published during his lifetime. In the *Opus postumum* and in Kant's private notes, the expression does appear but has a different meaning (McLaughlin, 2015: 1720). In this text, however, I will refer to animals and plants as 'organisms' when discussing Kant's ideas, as is common in Kant scholarship.

They both form the content of the *Dialectic of Teleological Judgment*, which occupies about twenty-five pages in the second half of the *Critique of the Power of Judgment* (Kant, 2000: 257–284; AA 5 385–415). I will try to summarise Kant's ideas and then, with the help of some later writers, suggest how Kant's initiative can perhaps be further developed in one specific direction.

The Antinomy

Let us start with the antinomy of teleological judgment. Kant formulates it in §70 of the *Critique of the Power of Judgment*, where he states that neither of the following can be asserted:

> Thesis: All generation of material things is possible in accordance with merely mechanical laws.
>
> Antithesis: Some generation of such things is not possible in accordance with merely mechanical laws. (Kant, 2000: 259; AA 5 387).

And yet both of the following must be asserted:

> [T]hesis: All generation of material things and their forms must be judged as possible in accordance with merely mechanical laws.
>
> [A]ntithesis: Some products of material nature cannot be judged as possible according to merely mechanical laws (judging them requires an entirely different law of causality, namely that of final causes). (Kant, 2000: 258–259; AA 5 387).

I believe that a degree of 'uncommonness' of Kant's position can at least be suspected prior to any comment. It ought to be noted, though, that to this day there are many disputes about the details of the antinomy. There is not even a full consensus about what exactly the antinomy lies in, let alone what Kant believed its solution to be: the way the *Dialectic* proceeds is very convoluted.

First, I want to recount all that must be borne in mind if we want to understand what the abovementioned distinctions and statements mean and how Kant arrived at them. Naturally, it will be necessary to explain, at least to some extent, not only the concepts and principles directly present in the antinomy, but also those which are implicitly present in Kant's system as such or at least in the *Critique of the Power of Judgment*. We shall have to keep track of what exactly (all of it) is meant by 'mechanical laws' and what is meant by the other, teleological, causality. We must also keep track of what, when, how, and why we must explain according to the mechanical and teleological principles and how these explanations do or do not relate to how things 'really are'. Then we will clarify where the contradiction seems to lie and why and how, according to Kant, one can remove the appearance of incompatibility of the sentences of the antinomy.

Mechanical Laws

Let us begin with the mechanical laws. Kant does not directly define mechanism and mechanicity in the *Critique of the Power of Judgment*, but he repeatedly characterises mechanical laws and forces as blind laws, unguided by any purpose. In his work, he distinguishes several kinds of such blind laws, and these distinctions are relevant to our present endeavour. I (tend to) agree with Ina Goy (2017: 197–205) who claims that when, both in the antinomy and elsewhere in the *Critique of the Power of Judgment*, Kant speaks about mechanical laws in general, he means all of the following types.

The first type of blind laws is necessarily introduced into all things as we know them by our 'understanding' (*Verstand*). This is certainly a strange proposition, which is why we ought to pause here and remind ourselves of the basic outline of Kant's critical philosophy. In it, Kant gave up on knowledge of things in themselves. He states that we always deal only with the way things appear and these appearances are inevitably organised at least to some extent by the faculties of our mind. Kant wants to define these faculties, examine their roles and mutual relations, and outline the limits to their legitimate use. Understanding is one such faculty. It is our 'faculty of concepts', a faculty that provides the concepts which we can safely apply to particular perceptions and thus let appear the objects which fall under these concepts. Among such concepts, Kant distinguishes between 'empirical' concepts and 'pure' or *a priori* ones, i.e., categories. All objects of experience fall under the latter, emerging by their application to perceptions.

Kant believed he discovered all the fundamental concepts which understanding has at its disposal *a priori*, that is, irrespective of any previous experience, and that he discovered them by analysing understanding as such, regardless of the particular nature of objects to which these concepts are applied. All objects of experience, both internal and external, that is, experiences with our ideas, thoughts, etc., as well as with things in the space of the world, are somehow subordinated to these concepts (Kant, 1998: 698–699; A845–847/B873–875; Kant, 2002b: 185, 188; AA 4 469–470, 474).

Kant presented them in a clearly organised table, which appeared in the *Critique of Pure Reason* (Kant, 1998: 212; A80/B106) and in the *Prolegomena* (Kant, 2002a: 96; AA 4 302–303). The category most important for our present endeavour is that of causality. How things are in the case of *inner* experience is complicated, and Kant unfortunately does not elucidate the matter.[2] He tends to focus on explaining the causal interdependence of phenomena in the *outer* experience, and so shall we. Here, the application of the category of causality to perceptions at least insures in advance that every phenomenon has a cause, something that caused it, whereby what we mean is efficient causality, not causality determined by purpose.[3] It should

[2] See Kraus (2019).

[3] Each of the two editions of the *Critique of Pure Reason* gives a slightly different formulation of this principle: 'Everything that happens (begins to be) presupposes something which it follows in

be noted that Kant uses the term 'experience' (*Erfahrung*) in a somewhat unusual way. Far from meaning all perception or even sensation, he means only 'experience-based cognition', which is formed during an interaction between perception and concepts and applies to all people. Closely related to Kant's 'experience' is his use of the expression 'nature'. What he means by it are phenomena insofar as they are subject to universally valid laws as well as these laws themselves. Until we arrive at 'experience', we cannot thus relate to 'nature', and 'nature' is 'by definition' subject to efficient, blind causality.

Let us note that the question of how, when, and on what levels categories are used along the way towards cognition of nature, i.e., in our experience, is not an easy one. There are disputes as to whether and how, according to Kant, any intuition (eventually perception) is pre-formed by categories at least in a rudimentary way. In particular, the distinction between judgments of perception and judgments of experience in §§18–21 of the *Prolegomena* seems to suggest that there are even *judgments* that do not depend on categories, which naturally implies that perception would not depend on them either. The abovementioned paragraphs are, however, rather puzzling and some scholars believe them to be a mistake on Kant's part, that is, something that is not compatible with the rest of his critical philosophy.[4] Regardless of whether, during philosophical reflection, we are driven, according to Kant, to the conclusion that categories *somehow* apply to perception and to all judgments (which I consider probable),[5] it seems that we 'fully' apply categories only in judgments in which we seek universally valid cognition of objects. In such instances, our goal is to grasp the categorial features (and thus acquire 'experience') – but that is not always the case when we speak or think.

The Powers of Our Mind: A Detour

Naturally, understanding is not the only capacity of our mind. To proceed further, we must therefore outline Kant's broader conception of our mental faculties, which is why I take the liberty of taking a little detour from the exposition of the types of mechanical laws.

First of all, Kant distinguishes three basic faculties or capacities of the soul (*Seelenvermögen oder Fähigkeiten*),[6] namely the *faculty of cognition*, the *feeling of*

accordance with a rule.' (Kant, 1998, 304; A189) vs. 'All alterations occur in accordance with the law of the connection of cause and effect.' (Kant, 1998, 304; B232).

[4] The main views of commentators writing before 1970 regarding this distinction and the challenges it entails are summarised by Prauss (1971: 139–158). A brief overview of the latest literature is found in one of the most recent texts on the subject; see Thielke (2022: 92–93).

[5] See explanation in Longuenesse (1998: 170–197).

[6] In the 'First Introduction' to the *Critique of the Power of Judgment* (Kant, 2000: 1–51; AA 20 195–251), which Kant even before publication replaced by another one, he speaks of 'powers of the mind' (*Gemütskräfte*).

pleasure and displeasure, and the *faculty of desire* (Kant, 2000: 64–65; AA 5 177–178).[7] In the course of our mental life, all these faculties of the soul continuously cooperate, and they do so in specific ways depending on the mental life area. But Kant's primary ambition is not to describe our mental life in its vast diversity: what he is interested in is whether and how any given faculty can be governed only by *a priori* principles adequate to it and thus be autonomous while relying on something that has universal validity. When a faculty is used in this way, it becomes a *higher* faculty.[8]

Second, Kant distinguishes three cognitive faculties, which are not, however, simply three branches or kinds of the cognitive faculty of our mind. What they are is a different trinity on a different level. Before listing the three cognitive faculties, let us recall that in addition to active cognitive faculties, our *a priori* cognitive apparatus includes two *a priori* forms of intuition, namely time and space. Time is a form of intuition through an inner sense, while space is a form of intuition through the five outer senses.

The three cognitive faculties are *understanding*, *reason*, and the *power of judgment*. The *power of judgment* is the ability to think of the particular as subordinate to the general, and we will return to it later. *Reason* is the capacity that seeks or 'demands' the unconditional. But nothing unconditional is given to us in any form: no intuition corresponds to the concepts of reason called 'ideas'. Kant identifies three ideas of reason which apply to the exercise of the *cognitive* faculty, namely the idea of the soul, of the all-encompassing world, and of God. These are the three aspects of something one might call the totality of all possible experiences.[9]

The *cognitive faculty of the soul* is a higher faculty, insofar as understanding – as indicated above – merely prescribes to nature *a priori* (i.e., categorial) forms and relations. In this higher exercise of the cognitive mental faculty, judgment and reason are subordinated to understanding, which supplies the *a priori* principles, that is, which works 'purely'.[10] Reason, through the ideas of the soul, the world, and

[7] Cf. the *Preface* and *Introduction* to the *Critique of the Power of Judgment*. Gilles Deleuze (1984, 2004) uses this basic framework, with only minor variations, as the basis for his unified interpretation of Kant's critical philosophy. My interpretation in the following paragraphs is also inspired by him. (For Deleuze's inspiration by the *Preface* to the *Critique of the Power of Judgment*, especially by its ending, cf. Sauvagnargues, 2021: 196.)

[8] 'In regard to the faculties of the soul in general, insofar as they are considered as higher faculties, i.e., as ones that contain an autonomy, the understanding is the one that contains the *constitutive* principles a priori for the *faculty of cognition* (the theoretical cognition of nature); for the *feeling of pleasure and displeasure* it is the power of judgment, independent of concepts and sensations that are related to the determination of the faculty of desire and could thereby be immediately practical; for the *faculty of desire* it is reason, which is practical without the mediation of any sort of pleasure, wherever it might come from, and determines for this faculty, as a higher faculty, the final end, which at the same time brings with it the pure intellectual satisfaction in the object' (Kant, 2000: 82, AA 5 196–197).

[9] For a clear and concise explanation of the nature of ideas, see Kant (2002a, §40, 119–120; AA 4 327–328). Cf. also Kant (1998: 605–623; KRV A669/B697–A704/B732).

[10] '[T]he critique of pure reason [...] consists of three parts: the critique of the pure understanding, of the pure power of judgment, and of pure reason, which faculties are called pure because they are

God, directs understanding in its construction of a systematic unity of knowledge. It guides understanding to combine knowledge as if all phenomena were affected by the fact that they are experienced by one entity, embedded in the totality of all phenomena, and created by an infinite reason, or God, although none of this can be experientially verified. The power of judgment here, of course, serves understanding by placing particular phenomena into categories.

The *faculty of desire*, that is, the ability to want something, can become a higher faculty if it does not ultimately desire anything specific because it is attracted to it. If it did so, it would be dependent on the ability to feel pleasure. It can be a higher faculty only if it pursues the ultimate, unconditional purpose. To put it differently: it is a higher faculty if it follows a universal and unconditionally valid imperative or law, but that can be provided only by reason as the requirement of the unconditional, that is, if the reason works 'purely'. The basic form of the law is: '[A]ct only in accordance with that maxim through which you can at the same time will that it become a universal law' (Kant, 1996: 73; AA 4 421). Understanding and the power of judgment are subject to reason when the faculty of desire operates at a higher level.

Later, we shall see that a specific contact with the universal can also take place in the sphere of feeling, under the guidance of the power of judgment when it operates 'purely'. A different cognitive faculty thus takes the lead depending on which mental faculty is operating at a 'higher' level.

Returning to the Mechanical Laws

We can now return to the types of mechanical laws to which nature is subjected. So far, just before starting the short detour on the system of the powers of the mind, we have mentioned one type of these laws. In particular, we spoke about the mechanical laws to which reflection in the exercise of universally valid cognition by application of categories necessarily subjects all objects, not only objects of the outer senses but also those of the inner sense, through which we acquire the contents of our mind.

Aside from this, Kant developed a special metaphysics of material things in his *Metaphysical Foundations of Natural Science*, which was published in 1786 (Kant, 2002b). In this treatise, he claims that thanks to the empirically acquired bare concept of matter (characterised as 'impenetrable lifeless extension'; Kant, 1998: 699; A848/B876), he succeeded in deducing, with the use of *a priori* principles, somewhat more specific laws which necessarily apply to all phenomena of *material objects*. This is probably the second type of mechanical laws mentioned in the antinomy. The most important for us is the Second Law of Mechanics, according to which 'Every change in matter has an external cause. (Every body persists in its state of rest or motion, in the same direction, and with the same speed, if it is not

legislative *a priori*' (Kant, 2000: 66; AA 5 179). In the *Introduction* to the *Critique of the Power of Judgment*, Kant uses the term 'critique of pure reason' for the critique of all pure cognitive faculties, i.e., reason in the narrower sense, understanding, and the power of judgment.

compelled by an external cause to leave this state)' (Kant, 2002b: 251; AA 4 543). These *a priori* laws, which govern especially the material nature, are such that wholes arise according to these laws only in the form of sums or aggregates of parts. The behaviour of the whole can be explained as the 'sum' of behaviours of the parts, but not the other way around, which is also an important feature of the 'mechanic laws' discussed in the antinomy.[11]

Thanks to the empirical knowledge delivered by natural sciences, we also know a number of other laws which phenomena follow. Kant calls them 'empirical' or 'particular' laws. They are based in some way on the general or *a priori* laws but cannot be derived from these laws without further empirical investigation. And these empirical laws, like Newton's Law of Gravitation, belong to the mechanical forces and laws which Kant apparently refers to in his formulation of antinomy as their third type.

The Beauty of Nature and the Unity of Its Particular Laws

When searching for these empirical laws, understanding is not entirely 'lawgiving', as it is when it prescribes general laws to nature. The powers of our mind cooperate here in a different way than when they subject the intuition to categories. For us, it is important to bear in mind that, according to Kant, if we are to find any laws or indeed search for them at all, we must assume a unity among the particular laws of nature. This unity then enables us to grasp the laws in a unified way by our cognitive faculties (Kant, 2000: 69–73; AA 5 182–186). This means that we must assume that to our cognitive faculties, nature is arranged as if 'purposively', although no such thing is provided by the *a priori* forms of intuition or understanding.

This assumption must be made by the power of judgment, that is, by the ability to relate the general to the particular. To be precise, we are dealing here with so-called 'reflective' judgments, which are the focus of the entire *Critique of the Power of Judgment*. Kant distinguishes them from 'determinative' judgments (Kant, 2000: 66–68; AA 5 179–181), which subordinate the particular or special to the general, if the general is given to us. Reflective judgment searches for a more general concept or rule, one that is not given, for a given individual or particular. Either the rule cannot be given or at least the reflective judgment relies on a rule that is not based on an intuition in its search for a partial higher rule that perhaps can be given. The ultimate governing rule is the abovementioned necessary assumption that there is a unity among the partial empirical rules, a unity such that we can understand it. This concept of the purposiveness of nature is an *a priori* principle of the power of judgment. The particular laws of nature are therefore mechanical, but to us they are

[11] This is highlighted and this second consequence of mechanism in the *Critique of the Power of Judgment* is emphasised by Zuckert (2007), see esp. p. 103–104.

linked to purposiveness in the sense that finding them is necessarily connected with the abovementioned assumption.

Certain purposiveness of nature is also reflected in aesthetic judgments, where, however, the judgment does not start from a particular rule or concept, nor does it seek a concept. It starts with a pleasurable feeling, which it nevertheless recognises as arising from certain suitability, an as it were purposive arrangement, of the form of the object for the basic (and thus universally shared) arrangement of one's cognitive capacities. Specifically, a form leads to a harmonious interaction between our understanding and our imagination. Guided by a reflective judgment, the ability to feel pleasure and displeasure is elevated to its higher form. This is done primarily when judging the beauty of *nature* (as opposed to art), where we do not, however, judge the forms as having been created deliberately based on their concept. Still, even works of art are ultimately created through the creator by a 'genius', that is, an 'inborn predisposition of the mind (ingenium) through which nature gives the rule to art' (Kant, 2000: 186; AA 5 307). We feel that what is truly beautiful in a work of art is not based on a (human) concept.

It is the judgment of purposiveness in nature, its various forms, and their status, that are the main theme of the *Critique of the Power of Judgment*. And, as we shall soon see, organisms have a key place in the classification of the types of purposiveness and in the *Critique of the Power of Judgment*.

Before looking in more detail at Kant's thoughts on purposiveness in nature, especially in relation to organisms, let us clarify the difference between the so-called constitutive and regulative principles. Constitutive principles are those *a priori* principles which, through determinative judgment, constitute or determine the very objects of experience. In contrast, the regulative principles and concepts (including the abovementioned ideas of reason) do not and cannot determine the objects of experience themselves. They can merely guide the way we think about them or the way we use our mental powers in general. Their validity cannot be directly verified by experience.

Another regulative principle is the systematic unity of all the special laws of nature that is comprehensible to us and whose existence we must assume. Such 'purposiveness' of nature is not an idea of reason. Rather, it is an *a priori* principle of reflective judgment, a principle which the reflective judgment, in seeking a more general law under which to subsume a special law, imposes upon itself 'and cannot derive it from anywhere else (for then it would be the determining power of judgment)' (Kant, 2000: 67; cf. a full explanation on p. 66–73; AA 5 180 and in more context at 179–186).

Purposive Causality

We have touched upon a sort of mysterious purposiveness in nature. We know that the antinomy of teleological judgment is about something like it. In contrast to the concept of mechanism, Kant explains the meaning of the notion of purposiveness in some detail. Let us first look at the various meanings of 'purposiveness', which he gradually comes to distinguish. It ought to be noted that Kant himself uses the word 'purposiveness' (*Zweckmäßigkeit*) in several senses, which are connected only by, shall we say, 'some relation to some intentions'.

Subjective purposiveness of an object means that the object is somehow arranged purposively *for the faculties of our mind.* We have already seen that nature is purposively arranged for our cognition, because it largely permits the subsumption of particular laws under more general ones. More generally, one can say that nature is purposively arranged for the faculties of our mind when we say that nature is beautiful. Importantly, though, we do not perceive this purposiveness as something that had to be applied in the creation of the things in question. It is inherent only in their relation to our mental faculties: we do not find it difficult to conceive of the origin and existence of things which we happen to find beautiful as based purely on blind mechanical causality. Kant distinguishes a subjective purposiveness from an objective one. In the latter, he distinguishes between objective *formal* and objective *material* (or *real*) purposiveness.

Objective formal purposiveness is exhibited for instance by geometric figures, which are for instance useful for solving various mathematical problems. This is due to the nature of space, which is a form of our perception. The usefulness of geometric figures for pursuing our intention to construct some other form implies no other intention (Kant, 2000: 235–239; AA 5 362–366). In this respect, this situation is akin to grass being useful to a cow, which is, according to Kant, already an *objective* and *material* (real) purposiveness but nevertheless still *external* or 'relative'. In other words, the fact that a thing is purposive for another thing is not an essential part of its concept. Here, too, we can well conceive of or think of that thing's existence and formation without this kind of purposiveness. This 'objective real external' purposiveness differs from objective formal purposiveness of geometrical figures for us in that it is not the form that is purposive but the real existence of such purposive things.

Nevertheless, objective real purposiveness can also be *internal*. This is the case when the concept of the thing itself includes the fact that it came to be intentionally, with the aim of bringing into being a thing whose concept existed beforehand (Kant, 2000: 239; AA 5 366–367). A common and uncomplicated example is all human artefacts. An equally common but highly mysterious variant is all organisms. Kant believes that we cannot think of them otherwise than as having come to be based on a prior idea of their totality: the complex ordering which they exhibit, and which is necessary for their perpetuation simply cannot have arisen by chance. And this is a mystery, because we do not see any conscious creator behind them, and nature, which seems to have created them, according to Kant 'by definition' cannot have intentions and act according to them.

In a famous passage, Kant states that it is absolutely certain that there will never be a 'Newton of the blade of grass' capable of explaining the origin of an organism, even a blade of grass, according to some blind mechanical laws.[12] He does not, however, justify this opinion.

Kant calls the things of nature which seem to have come to be by intention[13] 'natural purposes'. Like artefacts, they cannot be explained without purposive causality, that is, without an activity that pursues a goal (for comparison, see Chap. 2). But since an organism in nature has no external creators, as a thing of nature in relation to itself it must be both the cause and the effect. In short, as Kant was perhaps the first to say (Keller, 2008: 46), it must be a 'self-organising being' (*sich selbst organisirendes* Wesen; Kant, 2000: 245; AA 5 374). Kant further elaborates on this notion by saying that in a natural purpose, the individual parts of the whole not only exist for each other's sake (that is the case also, for example, in a clock), but on top of that they also *create* each other. Kant lists three phenomenal features in which this nature of the organism (he gives a tree as an example) is manifested. A tree, like essentially any organism, can (1) produce offspring of the same kind, and shape itself as an individual in both (2) growth and (3) regeneration. (Again, cf. Chap. 2; alongside procreation and growth, Aristotle does not explicitly mention regeneration).

Hannah Ginsborg (2004) speaks of the 'two kinds of mechanical inexplicability' of organisms in Kant. First, like some artefacts, organisms cannot be thought of as having come to be without reference to causality pursuing goals, and, unlike artefacts, their individual parts have created and maintain each other. Ginsborg (2004: 54–62) points out that two similar lines of reasoning about organisms can also be found in Aristotle. To wit, also according to Aristotle, organisms, like artefacts, would not have come into being solely based on the natural behaviour of the substance from which they are composed. Their existence must be therefore due to some purposive causality. At the same time, they differ from artefacts in that they

---

[12] 'For it is quite certain that we can never adequately come to know the organised beings and their internal possibility in accordance with merely mechanical principles of nature, let alone explain them; and indeed this is so certain that we can boldly say that it would be absurd for humans even to make such an attempt or to hope that there may yet arise a Newton who could make comprehensible even the generation of a blade of grass according to natural laws that no intention has ordered; rather, we must absolutely deny this insight to human beings.' (Kant, 2000: 270–271; AA 5 400).

See also Kant (2000: 250; AA 5 378): 'The internal form of a mere blade of grass can demonstrate its merely possible origin in accordance with the rule of ends in a way that is sufficient for our human faculty for judging.' or Kant (2000: 279; AA 5 409–410) '[A]bsolutely no human reason (or even any finite reason that is similar to ours in quality, no matter how much it exceeds it in degree) can ever hope to understand the generation of even a little blade of grass from merely mechanical causes.'

[13] Kant is certain that there cannot be more than two types of causality: the causality of efficient causes and that of purposive causes (Kant, 2000: 244; AA 372–373; see also Kant, 1998: 532; A532/B560)

have their own nature. In other words, the principle of their change and permanence – as well as of their formation – is inherent in themselves (cf. Chap. 2).[14]

The concept of natural purpose is an idea of reason,[15] because no operation of purposive causality in nature is given to us in an intuition.[16] But its result, or rather something that we cannot but view as its result, is given. The idea of a natural purpose is therefore different from all other ideas. Yes, we can only *judge* that a given object falls under the concept–idea, but this is a different situation from when the ideas of soul, world, and God regulate our efforts to determine by understanding more and more objects of experience. The difference between the idea of a natural purpose and ideas of the soul, the world, and God is that it is not an idea of reason for *reasoning*, but an idea of reason for *(reflective) judgment* (Kant, 2000: 274–275; AA 5 405).

As soon as reflecting upon organisms forces us to judge at least something in nature as having come to be based on purposive causality, we can view as its products (i.e., as a case of objective real internal purposiveness) those natural phenomena that cannot force us to do so on their own, such as the external purposiveness of natural beings for each other, the beauty of nature, or the unity of the particular laws of nature. Reflection upon organisms thus transforms the view of the whole of nature.[17]

One can now see that we cannot understand the presence of purposeful causality in nature (i.e., the *antithesis* of antinomy), as a constitutive principle, neither in connection with the unity of laws, nor beauty, nor organisms. But why cannot we consider the *thesis* of the antinomy, that is, the mechanical intelligibility of everything in nature, to be a constitutive principle? I believe the reason is simple: Kant realises

[14] Ginsborg (2004: 33–46; 2006: 459–463), in opposition to earlier interpreters, argues that in the antithesis of antinomy, Kant speaks only about the kind of mechanical inexplicability that organisms share with artefacts. Her suggestion is rejected, for example, by Zuckert (2007: 101–102, note 22, and 115, note 45).

[15] We noted above that the purposiveness of nature in general is an *a priori* principle of judgment. By contrast, the notion of a natural purpose – a particular natural object but one generated by causality 'striving' to realise something of which it has *a priori* notion – is really an idea of reason.

[16] This is why one can argue that '[s]ome products of material nature cannot be judged as possible according to merely mechanical laws' – yes, these 'some products' here are organisms –, but it cannot be argued that '[s]ome generation of [...] [material] things is not possible in accordance with merely mechanical laws.' (Antinomy of teleological judgment, Kant, 2000: 258–259; AA 5 387)

[17] 'Even beauty in nature, i.e., its agreement with the free play of our cognitive faculties in the apprehension and judging of its appearance, can be considered in this way as an objective purposiveness of nature in its entirety, as a system of which the human being is a member, once the teleological judging of nature by means of natural ends, which have been made evident to us by organised beings, has justified us in the idea of a great system of the ends of nature. We may consider it as a favour that nature has done for us that in addition to usefulness it has so richly distributed beauty and charms, and we can love it on that account, just as we regard it with respect because of its immeasurability, and we can feel ourselves to be ennobled in this contemplation – just as if nature had erected and decorated its magnificent stage precisely with this intention' (Kant, 2000: 251–252; AA 380).

that if we know that every single thing has a mechanical cause (which in turn has another cause, etc.), it does not yet imply knowing whether, if we knew all such causes, we would see that all states and events in nature are explained by them.

Solving the Antinomy

The line of explanation in §§71–78, which should contain the solution of antinomy, is convoluted. What is according to Kant the solution to antinomy is still a matter of controversy. For a long time, it has been generally assumed that the antinomy arises only between the two theses about constitutive principles, that is, between the theses which Kant says we cannot posit, and that it dissolves once we realise that the two principles are not constitutive but merely regulative (for an overview of such interpretations, see McLaughlin, 1990: 137–145). Such interpretation is supported by the end of §71.[18] But the name of the paragraph is '*Preparation for the resolution of the above antinomy*', which itself suggests that the solution will be somewhat more complicated. The recently prevailing view – to which I also subscribe – states that the antinomy lies between the two regulatory principles, and the subsequent considerations in §§72–78 are integral to its resolution. (See for example McLaughlin, 1990: 161–180; Ginsborg, 2006: 459–466; Guyer, 2006: 346–349; Goy, 2015; Goy, 2017, summary at 90–91).

Kant also sees an antinomy, that is, an at least apparent internal contradiction, between the two regulative principles of reflective judgment. It seems that we cannot even *judge* something to be possible and impossible at the same time. The resolution of the antinomy therefore rests in reconciling the two principles at the level of human thinking about things, not at the level of grasping the features directly given in phenomena, let alone the level of grasping the nature of things as they are in themselves. In other words, the solution is based on understanding that the two principles guiding our thinking about things do not exclude or do not necessarily exclude each other.

The key step, according to Kant, is to understand that if we are to see a purpose in the products of nature and conceive of nature itself as essentially incapable of purpose, we must think of nature as having a non-natural, supernatural basis or source, specifically a source that can pursue purposes. We must think of a being

[18] 'All appearance of an antinomy between the maxims of that kind of explanation which is genuinely physical (mechanical) and that which is teleological (technical) therefore rests on confusing a fundamental principle of the reflecting with that of the determining power of judgment, and on confusing the *autonomy* of the former (which is valid merely subjectively for the use of our reason in regard to the particular laws of experience) with the *heteronomy* of the latter, which has to conform to the laws given by the understanding (whether general or particular)' (Kant, 2000: 261; AA 389).

with infinite understanding as the author of the universe and think of nature as dependent on such a suprasensible foundation (Kant, 2000: 269; AA 5 398–399).[19]

But we cannot positively confirm the existence of an intelligent cause of the blind nature. We cannot prove it. This, too, is a regulative idea and, as such, it can only be demonstrated that we, humans, cannot think of nature except as having been made by an intelligent creator. For us, not only natural purposes but the entire world originate in it – and so do its mechanical laws. How the two principles emanating from the same source combine in one natural purpose is something we do not and cannot know. But there is a way of making the 'cooperation' between these two principles conceivable to our understanding: if we see organisms, and ultimately all of nature, as products of a divine purpose, we can view the mechanical laws as the means by which God accomplishes His purposes. In what way, we cannot understand. But we can judge the mechanical laws as mysteriously subjected to God's purposes. This does not relieve us of the obligation to use our understanding and thus to seek mechanistic explanations of natural processes 'as far as it is in our capacity' (Kant, 2000: 284; AA 5 415), but we must bear in mind that we, humans, cannot explain all forms and processes in nature by these (mechanic laws) alone. And, as noted above, even a search for those mechanistic explanations which we are able to find cannot succeed without the adoption of a teleological perspective (Kant, 2000: 280; AA 5 411).

Perhaps most remarkable is Kant's reflection on the specific features of our understanding, which enable the notion of a natural purpose and constitute the particular character of our apparatus in general from which he derives our need to distinguish between two causalities and relate them to each other, as well as the subsequent need to distinguish between nature and its intelligent cause. Kant presents it in §§76–77, where he points out that our understanding is discursive, i.e., that it works with particular, more or less general concepts; but when it comes to a further specification and existence of what falls under these concepts, it is dependent on the (sensory) intuition because the above cannot be deduced from its concepts. And if our understanding finds some sense, or unity, in what does not follow from conceptual knowledge and is therefore accidental to it, it must ultimately view it as resulting from an intention.

To appreciate that this is a specific feature of our understanding, we must contrast it with another possible understanding that would not depend on sensory judgment, that is, on 'intuitive' reasoning (implicit here is that God has such understanding; cf. Goy, 2015: 78–81). First of all, this understanding would not distinguish between the possible and the real – everything it 'sees' would be real, the notion of 'merely possible' would be out of question for it. Second, it would look at the whole and proceed from it to the parts, whereas we proceed from conceptually apprehended parts to the whole. Our understanding knows the parts but

---

[19] In 'First Introduction' to the *Critique of the Power of Judgment*, Kant speaks directly about an 'idea of nature, whose lawfulness cannot be understood without its relation to a supersensible substratum' (Kant, 2000: 46, translation modified; AA 20 247).

cannot compose them into a whole that would be superordinated to them. In this, the perspective taken by our understanding is in a sense inherently mechanistic.

An understanding that would see the whole would not necessarily perceive the distinction or opposition between mechanism and teleology. Such understanding would simply see that the whole happens as is appropriate to it. We, humans, can only understand the subordination of parts to the whole by *imagining* the whole and treating this *notion* as something that predetermines the parts (Kant, 2000: 277; AA 5 407–408). In other words, by our understanding we introduce into our comprehension of nature partial laws that govern partial processes and we define nature as subjected to these laws. Then we conclude that some phenomena cannot be explained by these laws and therefore add another, quite different type of causality to our reflections of nature. Subsequently, we try to figure out what the character of this causality is and how it relates to mechanical causality. The character of this other causality must be understood, in analogy to our own making of things, as the action of the intelligent author of nature, and we must conceive of its relation to mechanical causality as that of intentional creation to the means it uses.

Kant's Contribution and Possibilities of Its Further Development

Let us now try and see what aspects of Kant's thoughts might be inspiring for current discussions about organisms as well as mechanistic versus non-mechanistic views of the living. As suggested in the introduction, I believe that especially thoughts on the *status* of the mechanistic and teleological (or more generally, non-mechanistic) views of organisms retain their relevance. Here, in the purely philosophical sphere, lies the core of Kant's contribution.

I believe that John Zammito (2006) is right when he states that for those who simply seek a naturalistic interpretation of the generation and persistence of organisms, Kant's reflections are essentially worthless. This is because Kant, offering hardly anything in the way of an argument to support this claim, simply declared a mechanistic interpretation of organisms to be impossible and another interpretation to be valid only in a 'regulative' function. Zammito identifies other authors, both Kant's contemporaries or even earlier thinkers, such as John Locke and Georges Buffon, as precursors of current efforts to formulate a naturalistic philosophy of organisms.

The one aspect of Kant's conceptual system I want to highlight here is not far from the ideas of many twentieth-century philosophers classified as belonging to phenomenology, pragmatism, or poststructuralism. Many of these thinkers had analysed and drawn on Kant's thoughts (cf. Marino & Terzi, 2021). I want to emphasise, first, that it would be simplistic to say that Kant thought nature to be 'in fact' subjected to mechanical laws while we can, or must, view it 'as if' it were also somehow teleological. Kant also does not attribute mechanicity to things as they 'really' are in themselves. It is only guaranteed, he claims, that to all people things appear, or can appear, to be so. It is a perspective all people can apply. The 'perspective' by which we 'see' nature as realising some goals (due to being created by an infinite intellect) is simply another perspective we are also capable of, a perspective which we, as humans, according to Kant must also all apply.

Kant's topic is not the 'real substance' of the processes of nature. Rather, he analyses certain particularly important human perspectives of them and relationships between these perspectives. Kant notes that we can never know whether we know the 'real' reality and shows that, in principle, our thinking does not require its concept for anything. After showing that we, humans, must think of organisms – and ultimately the world as a whole – as if they were 'the product of a rational cause (God),' he asks:

> Now if this proposition,[20] grounded on an indispensably necessary maxim of our power of judgment, is completely sufficient for every speculative as well as practical use of our reason in every *human* respect, I would like to know what we lose by being unable to prove it valid for higher beings, on purely objective grounds (which unfortunately exceed our capacity)? (Kant, 2000: 270; AA 5 400)

Of Kant's writings, it is the *Critique of the Power of Judgment* that is of key importance for the subject of the relationship between our various faculties and the corresponding facets of the world. In fact, already some of Kant's contemporaries perceived it as such, and it is perhaps not too surprising that the most famous of them had a close relationship to art. But they did not appreciate it only for the reflection on art which it contains. For instance, Friedrich Schiller – and soon after him Friedrich Hölderlin – saw in the aesthetic judgment, or the sense of beauty, as Kant understood it, the *fundamental* principle for the conduct of one's *entire* life, the ultimate principle to which all other partial human faculties must be subordinated if one is not to be one-sided, that is, if one is to fulfil one's innermost potential. Both understanding and reason, and both cognition and moral action, must ultimately submit to the principle of beauty. One can and ought to relinquish efforts to base one's life firmly on principles that can be grasped conceptually (or purely by senses). Instead, one should be open to the interplay of all partial rules and forces, not only one's own but also those of the environment. Such interplay is not grounded in anything definite. Instead, it is anchored in a sense for the whole as it variably presents itself to us, that is, in taste. Schiller and Hölderlin argue that it is only in this way that one's life and the world as one experiences it acquire their appropriate unity: indeed, that a human being only becomes human in the true sense of the word (Schiller, 2004: 75–81; Hölderlin, 2021: 67–71).

Many twentieth-century writers praised the *Critique of the Power of Judgment* for similar reasons. Let us recall Deleuze's claim that the theory of aesthetic judgment is in a sense the basis for the whole of critical philosophy, not just a

[20] Hence '[w]e cannot conceive of the purposiveness which must be made the basis even of our cognition of the internal possibility of many things in nature and make it comprehensible except by representing them and the world in general as a product of an intelligent cause (a God)' (Kant, 2000, 270; AA 5 400).

supplementary bridge between its theoretical and practical parts (as Kant claims in the *Introduction* to the *Critique of the Power of Judgment*). In all other higher forms of mental faculties, Deleuze (2004) notes, cognitive forces are in concert under the rule of one among them, but in order for one of the cognitive forces to take charge, cooperation must already be established. This ungoverned interplay is then revealed in the perception of beauty (and, in a negative way, in the perception of the sublime; as when we experience a kind of 'hidden harmony' of reason and imagination, which surfaces in the feeling of their mutual inadequacy, in the impossibility of depicting the ideas of reason).[21]

If we were to try to follow up on Kant's ideas regarding the correspondence between our abilities and the facets of the world, which of the theses and assumptions would nowadays have to be rejected? Let us assume, as I have just indicated, that in the spirit of some of Kant's ideas, especially those from the *Critique of the Power of Judgment*, we take as the core or even as the whole of what is ultimately to be considered our actual experience, that is, the actually experienced world. This is always characterised by a conflict or interplay of several possible perspectives (ways of unifying what is experienced). In this case, it would be probably appropriate to relinquish the goal of isolating from this interplay certain partial 'pure' principles and perspectives which would then be maintained as 'higher' in the hope that they would be valid for all and always. Our thoughts arise from encounter and conflict, and the pursuit of pure autonomy and thoughts based on *a priori* universal principles is dubious and futile. We share some of our attempts at combining relevant perspectives with many people, others with just a few, and reflection is ultimately conducted by each person from their own perspective. One must then simply wait to see who may find the proposed way of discerning and matching the principles valid and who does not.

We will have to accept that there are many ways of seeing, many genres of speech, and new ones are being invented all the time; there are gradual transitions between relatively distinct types, and it is usually impossible to draw clear and sharp boundaries, certainly not once and for all. A mechanistic view of the material world, although it will certainly remain one of the important ones, will lose the status of a sort of default theoretical view of nature, which it in many ways seems to occupy in Kant's thoughts. One need not start from a universal conception of the whole of nature as being governed by mechanical laws. In fact, we can accept that our world is untidy and partly chaotic, and although we have learned in history

---

[21] '[...] *Critique of Judgment*, in its esthetic part, does not simply exist to complete the other two Critiques: in fact, it provides them with a ground. The *Critique of Judgment* uncovers the *ground* presupposed by the other two Critiques: a free agreement of the faculties. Every determinate agreement can be traced back to the free indeterminate agreement which makes the others possible in general' (Deleuze, 2004: 58).

We ought to note, though, that in comparison to the *Introduction*, Deleuze substituted the power of judgment with imagination among the three main cognitive faculties and, ultimately, questioned whether the power of judgment is a special faculty at all (Deleuze, 1984: 58–61). This reinforces the special status of the power of judgment, the core of which appears in reflective judgment (Deleuze, 1984: 60) as a fundamental 'faculty'.

lessons to present always a certain fragment of what happens in the world also mechanistically, in the world of our experience, what is seen and managed mechanically will always form just a tiny fragment. We are always exposed to different demands, we are constantly trying to decide how, when, and to what extent we can satisfy them, and we invent ways of combining the fulfilment of different imperatives in different situations.

With emphasis on the need to seek moderation when it comes to the realisation of various incompatible but interdependent possibilities and demands, Pavel Kouba (2005b)[22] tried to distinguish in Kant's thoughts the parts which are still relevant and those which failed the test of time. I consider Kouba's brief commentary to be one of the most insightful in this respect (and the perspective suggested in the previous two paragraphs is inspired by his writings as well). Kouba points out that despite the desire to escape from the actually experienced world to pure principles (i.e., to build a 'metaphysics'), Kant displays on many points an excellent awareness of the constant presence of numerous competing modes of possible unification, that is, various forms of meaning in the lifeworld. Kant often stresses the necessary and unresolvable tension between thinking and seeing, knowing and acting, the *a priori* and the *a posteriori*, the phenomenon and the thing-in-itself. Kouba suggests that one ought to try to liberate this tendency within Kant's thought from the metaphysical burden, that is, from the attempts to resolve these tensions and contradictions by escaping to the pure, *a priori* principles and their hierarchy. Like Deleuze, he states that in this way[23] an analysis of the power of judgment, and especially aesthetic judgment, in the *Critique of the Power of Judgment* would turn out to be not merely an additional bridge between the theoretical and practical parts of philosophy but 'the supporting foundation of the whole edifice' (Kouba, 2005b: 151). Kant's philosophy could then help us in our attempts to reflect upon our situation where, after abandoning metaphysics, we must again and again and without *a priori* rules try to see 'how to arrange a certain unification and how far one can go with it without losing the possibility of the other unifications' (Kouba, 2005b: 151). One might say that we must seek the limits of the legitimate use of our various abilities *depending on the situation*. According to Kouba, in order to respect the nature of meaning in general, we must not lose sight of other, often directly opposite, possible unifications when deciding on adopting a particular one, and we should be aware that in certain situations, it may well be advisable to resort to one of these other perspectives.

In other articles, Kouba (2005a, 2011) also addressed specifically science and its place in our (experienced) world. In his reflections on Anton Markoš's *Readers of the Book of Life* (2002), written at the same time as the Kantian study just cited, Kouba writes:

---

[22] Kouba's perspective has been influenced by the work of Friedrich Nietzsche, on whom he also wrote his first book (Kouba, 2001).

[23] But Deleuze does not explicitly state that his interpretive proposal is not orthodoxly Kantian. On the relation of his image of Kant to Kant himself, see, e.g., Sauvagnargues (2021).

> When trying to find one's way and decide in the tangled plots and events of our lives, no normal person looks to science for solutions. Therefore, at the most general level, our aim cannot be a discovery of the true science of life but rather an insight into *what place science has in our lives*. (Kouba, 2005a: 84)

The question of how to apply or not apply the (mechanistic or other) view of natural science reasonably, sensibly, and innovatively in different spheres of our life and how to combine it with other views will obviously be with us for a long time still. What I believe to be a reasonable and promising approach is to view the process of this search not as a conflict between understanding reality as it is on the one hand and the subjective distortions imposed by our limitations and needs on the other. Instead, we ought to treat it simply as a clash of the different powers of our mind and correlative faces or layers of the world. Immanuel Kant had significantly contributed to a formulation of this approach, and those who follow this course will surely be able to learn from him for a long time to come. In this context, Pavel Kouba writes:

> Kant's philosophy [...] can nowadays tell us many and unexpected things about what it means to understand meaning. Although his ultimate goal was to establish metaphysics, Kant has more to tell us in this regard than those who are still engaged in refuting metaphysics. (Kouba, 2005b: 152)

Conclusions

Kant, like many thinkers, especially modern ones, tried to come to terms with the fact that while the material world seems to be governed by blind mechanic laws, it is hard to believe that some phenomena, above all organisms, could have come to be by such forces alone. We feel compelled to assume some intention behind their creation and existence. Kant's transcendental philosophy as a whole allows him to resolve the contradiction in a highly original way. He proposes we ought to give up trying to find out how things really are and content ourselves with an analysis of the different ways of looking at things which all people must necessarily adopt given the composition of their mental powers. He states that both of these views of nature, a mechanistic and a teleological one, are indispensable for all humans. One essential difference between these views is that subordination to mechanic laws can indeed be claimed with respect to phenomena, while teleology is a way of thinking about phenomena which the phenomena themselves cannot directly empirically confirm.

According to Kant, given the structure of our cognitive apparatus, we necessarily link this teleology in nature and its relation to the mechanical laws we observe in it with the notion that the world is created by an infinitely intelligent and good God, who, in ways utterly incomprehensible to us, uses mechanical laws as the means of pursuing his purposes. Kant thus wants to preserve the traditionally metaphysical conception of the world as God's Creation for humanity while relegating its key theses to the realm of *regulative* principles and ideas, that is, principles and ideas which not only do not apply to things in themselves but, on top of that, cannot be verified by any intuition.

Kant's followers, however, also saw in his philosophy the possibility of focusing on the real multifaceted interplay and contention of forces and perspectives in our experience, a possibility of abandoning the metaphysical assumption of full and complete universality of certain perspectives and their hierarchy, and of focusing on inventing ways in which we could combine and balance different perspectives in order to seek the right measure of application of one or another in a particular situation. I consider this part of Kantian legacy to be most relevant for the philosophy of organisms.[24]

References

Deleuze, G. (1984). *Kant's critical philosophy. The doctrine of the faculties*. The Anthlone Press.

Deleuze, G. (2004). The idea of genesis in Kant's esthetics. In *Desert islands and other texts 1953–1974* (pp. 56–71). Semiotext(e).

Ginsborg, H. (2004). Two kinds of mechanical inexplicability in Kant and Aristotle. *Journal of the History of Philosophy, 42*(1), 33–65.

Ginsborg, H. (2006). Kant's biological teleology and its philosophical significance. In G. Bird (Ed.), *A Companion to Kant* (pp. 455–469). Blackwell.

Goy, I. (2015). The antinomy of teleological judgment. *Studi Kantiani, 28*, 65–87.

Goy, I. (2017). *Kants Theorie der Biologie. Ein Kommentar. Eine Lesart. Eine historische Einordnung*. De Gruyter.

Guyer, P. (2006). *Kant*. Routledge.

Hölderlin, F. (2021). *Hyperion, or the Hermit in Greece*. Open Book.

Kant, I. (1910–1997). *Kant's gesammelte Schriften*. ('Die Akademie-Ausgabe'). Berlin. **(Quoted as AA, followed by a volume number and a page number.)**

Kant, I. (1996). Groundwork of the metaphysics of morals. In *Practical philosophy*. Cambridge University Press.

Kant, I. (1998). *Critique of pure reason*. Cambridge University Press.

Kant, I. (2000 [1790]). *Critique of the power of judgment*. Cambridge University Press.

Kant, I. (2002a). Prolegomena to any future metaphysics that will be able to come forward as science. In *Theoretical Philosophy after 1781* (pp. 49–169). Cambridge University Press.

Kant, I. (2002b). Metaphysical foundations of natural science. In *Theoretical philosophy after 1781* (pp. 181–270). Cambridge University Press.

Keller, E. F. (2008). Organisms, machines, and thunderstorms: A history of self-organization, Part One. *Historical Studies in the Natural Sciences, 38*(1), 45–75.

Kouba, P. (2001). *Die Welt nach Nietzsche. Eine philosophische Interpretation*. Wilhelm Fink.

Kouba, P. (2005a). Hermeneutik und Wissenschaft vom Leben. In *Der Sinn der Endlichkeit* (pp. 75–85). Königshausen & Neumann.

Kouba, P. (2005b). Kant ohne das Problem der Metaphysik. In *Der Sinn der Endlichkeit* (pp. 137–152). Königshausen & Neumann.

Kouba, P. (2011). Die gelebte Objektivität. In F. Rese, D. Espinet, & M. Steinmann (Eds.), *Gegenständlichkeit und Objektivität* (pp. 235–248). Mohr Siebeck.

[24] I would like to thank Tomáš Hermann and Robin Pech for their valuable suggestions and comments on the topics discussed, and Jana Švorcová for her trust and patience.

Kraus, K. (2019). The parity and disparity between inner and outer experience in Kant. *Kantian Review, 24*(2), 171–195.

Longuenesse, B. (1998). *Kant and the capacity to judge. Sensibility and discursivity in the transcendental analytic of the critique of pure reason*. Princeton University Press.

Marino, S., & Terzi, P. (Eds.). (2021). *Kant's critique of aesthetic judgment in the twentieth century*. De Gruyter.

Markoš, A. (2002). *Readers of the book of life: Contextualizing developmental evolutionary biology*. Oxford University Press.

McLaughlin, P. (1990). *Kant's critique of teleology in biological explanation. antinomy and teleology*. Edwin Mellen.

McLaughlin, P. (2015). Organismus (Wesen, organisiertes; Körper, organisierter). In M. Willaschek, J. Stolzenberg, G. Mohr, & S. Bacin (Eds.), *Kant-Lexikon* (pp. 1720–1723). De Gruyter.

Prauss, G. (1971). *Erscheinung bei Kant. Ein Problem der „Kritik der reinen Vernunft"*. De Gruyter.

Sauvagnargues, A. (2021). The discordant accord of the faculties. Deleuzian Readings of Kant. In S. Marino & P. Terzi (Eds.), *Kant's critique of aesthetic judgment in the twentieth century* (pp. 195–206). De Gruyter.

Schiller, F. (2004). *On the aesthetic education of man*. Dover.

Thielke, P. (2022). Judgments of experience and the grammar of thought. In P. Thielke (Ed.), *Kant's prolegomena. A critical guide*. Cambridge University Press.

Zammito, J. (2006). Teleology then and now: The question of Kant's relevance for contemporary controversies over function in biology. *Studies in History and Philosophy of Biological and Biomedical Sciences, 37*(4), 748–770.

Zuckert, R. (2007). *Kant on beauty and biology. An interpretation of the critique of judgment*. Cambridge University Press.

Chapter 5
Schelling's Philosophy of Nature

Martin Vrabec

Abstract This chapter deals with the basic outlines of Schelling's conception of living nature. It had a profound influence on the development of biology, particulary in Germany, and many of its parts are highly inspirational to this day. The first part presents Schelling's general approach to explaining phenomena from living nature. His goal was to find a middle way between materialism, which tends to reduce life to mechanical and chemical processes, and vitalism, a position that assumes special principles and forces applying solely to living organisms. Schelling's position on this subject could be thought of as dynamic monism, according to which organic and inorganic nature are in effect different stages of development of one and the same principle. In the second part of this chapter, we have a look at what is, in Schelling's view, specific to the living stage of nature. We focus mainly on his theory of organism as an autopoietic, internally organised, self-sustaining whole characterised by relative autonomy and independence of its environment. The third and final part of this chapter is dedicated to Schelling's concept of evolution: he was the first thinker to use this term, which originated in the context of embryology, in the new context of evolution of species.

Keywords Schelling, F. W. J. · Autopoiesis · Self-sustainability · Relation to environment · Evolution

Introduction

Friedrich W. J. Schelling was a philosopher of a wide range of interests and although generally best known is probably his contribution to the theory of human freedom and the issue of evil, he also clearly deserves a place also in this book. This is for

M. Vrabec (✉)
Department of Philosophy, Faculty of Humanities, Charles University,
Prague, Czech Republic
e-mail: martin.vrabec@fhs.cuni.cz

J. Švorcová (ed.), *Organismal Agency*, Biosemiotics 28,
https://doi.org/10.1007/978-3-031-53626-7_5

two main reasons: first, some of his insights regarding the disputes between a mechanistic and vitalist explanation of nature and his views on the subject of natural evolution exerted a lasting influence on the development of biology, especially in Germany; second, different scholars have found in his work anticipation of various approaches and concepts found in the current philosophy of biology, for instance, the theory of processuality or theory of self-organisation and reciprocal causality. But before we embark on introducing the basic outlines of Schelling's concept of living nature, let us take a look at the contemporary context, at the situation Schelling had entered and which his views reacted to.

In Schelling's time, that is, in the late eighteenth century, biology was fast developing as a science of classification (Linné, Buffon), but its ambition was also to be an analytical science capable of offering explanations. This was happening in a situation where on the one hand, the mechanistic explanation of the functioning of organisms, which was based on an analogy between organisms and mechanical machines (Descartes, La Mettrie, etc.) and worked within a conceptual framework of inert matter and impact force, was reaching its limits, while on the other hand, the discovery of irritability by Galvani (1791), which proved a connection between electricity and physiology, and the rapid development of chemistry all seemed to offer a novel direction that would still explain organisms fully in terms of the laws and principles that operate also in inorganic nature (see Küppers, 1992: 92). But this was also a time when numerous thinkers were pointing to phenomena which, in their view, demonstrated the shortcomings of even this more broadly conceived physicochemical explanation of the existence and functioning of organisms. This pertained especially to (self)reproduction in the sense of nourishment, reproduction, and regeneration ability of some organisms, but also to organs 'taking over' the function of another, damaged organ, and finally to the fact that body parts are existentially dependent on each other and contribute to each other's continued existence (cf. Chap. 4; see also Kabeshkin, 2017: 1182). Another important direction of biological thought at that time, namely vitalism, was trying to explain these phenomena by positing that in living beings, there is at work also another type of causality and powers than merely the physicochemical ones, a 'vital force' (*Lebenskraft*) or 'formative drive' (*Bildungstrieb*). In the late eighteenth century, the most influential advocates of this approach in German-speaking lands were Georg Ernst Stahl, Joachim Dietrich Brandis, and Johann Friedrich Blumenbach.[1]

Immanuel Kant occupied a specific position in the dispute between these two approaches: he showed that life sciences cannot rely only on physical and mechanical explanations of organisms and the processes which take place in them. Moreover, he was convinced that this is not just a temporary shortcoming that could be rectified over time, but an integral feature of our discursive thinking. He noted that in biological research, one must view living beings not only as the products of commonplace efficient causality but also see them as if they were formed according to

[1] This direction of thought is presented in detail by Zammito (2018) and Richards (2002: 207–288). A distinction between the main forms of vitalism at this time is also found in Kabeshkin (2017: 1184ff.).

a plan, i.e., as if final causes were also involved. In research, this implies, among other things, a study of the purpose of particular organs and processes. In other words, it leads to functional explanations. But that does not make Kant a vitalist, because this conception of living entities as if they were natural final causes is, he adds, merely a regulative idea, a heuristic principle that guides our investigation. Importantly, it does not mean that according to Kant this is how living beings are actually formed (Kant, 2000: 247; see Chap. 4).

Kant was perceived as an authority by physiologists and physicians of his time, but they did not know how to implement his epistemological considerations in practice, because it remained unclear to what extent Kant thought that biology can be a science *sensu stricto*. This is why it was so easy for Schelling to take over this line of thought with his *Naturphilosophie*, which, at least in the beginning, presented itself as an elaboration of Kant's project, while offering a greater certainty regarding the status of biological research as such (Zammito, 2018: 323ff.). Still, as we are about to see, Schelling actually presented a separate position and one which went significantly beyond Kant's views. This is the case despite the fact that Schelling dedicated to his *Naturphiolosophie* only a relatively short time, less than a decade (starting in 1797 and ending in 1806). He published his thoughts on the subject mainly in the *Ideas for a Philosophy of Nature* (1988 [1797]), *On the World Soul* (2010 [1799]),[2] *First Outline of a System of the Philosophy of Nature* (2004 [1799]) with the later added *Introduction to the Outline* (2004 [1799]), *System der gesamten Philosophie und der Naturphilosophie insbesondere* (1804).[3]

In-Between Mechanicism and Vitalism

Let us first look at where Schelling identified the main shortcomings of both of the contemporary approaches to explaining the existence of living beings. A mechanistic explanation based on thrust and gravity is something he rejects as utterly insufficient with no further ado (*Weltseele SW* II 497f.; *First Outline* 57 = *SW* III 74). Explanation of life via chemical processes is an approach he viewed as more promising. The manner in which chemical – or, as he puts it, 'material' – physiology explains the form and function of living beings nevertheless in his view suffers from one-sidedness and incompleteness, because it seeks an explanation only in the chemical composition of living beings and in the effect of external forces. According to this approach, organisms and their activities are fully defined by external forces – and that does not correspond to the phenomena we actually encounter in the realm of the living, because organisms are able to modify, at least to some extent, the impact that external forces have on them (*First Outline* 59 = *SW* III 77). Another

[2] Only the first part of this work exists in an English translation.

[3] [*Entire System of Philosophy and of Philosophy of Nature in Particular*] This treatise is not available in an English translation.

weakness of chemical or 'material' philosophy is, according to Schelling, its failure to account for the fact that, unlike organic processes, chemical processes are not self-sustaining and self-perpetuating: a chemical process stops as soon as all initial substances have reacted and transformed into the final products (*Weltseele SW* II 500–502; *First Outline* 109f. = *SW* III 150). Organisms, however, are characterised by disbalance (see Chap. 8), and that is a precondition of metabolism and life as such.

In contrast, the other prominent contemporary explanation of life claims that organisms have a certain vital force (*Lebenskraft*), and that determines the impact – or lack of it – of external forces; it influences the effect of external factors to such an extent that it can halt or change the laws of chemical processes. Such vital force would thus transcend the laws of material nature and would not be itself material. This is why Schelling calls this explanation 'physiological immaterialism' (*First Outline* 61 = *SW* III 81). Within this theory, however, it remains unclear why the vital force affects only certain kinds of bodies and not all of them, and eventually, what are the specific empirical conditions under which it can start functioning. At the same time, it seems impossible to find any laws of functioning of this vital force, laws that would explain all phenomena of organic nature and their regularities (*First Outline* 61 note *, 111 = *SW* III 80f., 152).[4]

In his philosophy of nature, Schelling is trying to overcome the shortcomings of both of the theories described above and present a naturalist, but not reductive, explanation of life. In his view, life is something more than a physicochemical process but this 'more' cannot be an occult vital force. He does not want to assume any metaphysical forces or principles at work only in the realm of the living. His explanation aims at being monistic, i.e., such that assumes the same basic principles at work in both living and non-living nature. But that can only be achieved if the physicochemical character of nature is not assumed to be something basic and irreducible, but rather just a specific manifestation of certain more profound principles, which are at work in both living and non-living nature. We can arrive at a more concrete idea about this explanation of nature by following, one by one, Schelling's philosophical writings on nature, because Schelling elaborates his conception gradually and his theory undergoes some development.

In his first book on natural philosophy, *Ideas for a Philosophy of Nature* (1797), the theory of biology is not yet developed. The exposition ends with chemistry and Schelling touches upon the concept of organicity only briefly, in an introduction which he wrote after completing the work on the treatise (*Ideas* 30–38 = *SW* II 40–51). A theory of organicity and living nature is developed only in his next writing, *On the World Soul* (1798), although some parts of the conception remain, in comparison to his later elaboration, somewhat raw and vague. In this book, Schelling is trying to explain the entire nature not just mechanically, i.e., using the principle of thrust and gravity. Instead, his approach is guided by the basic idea of nature as

[4] For a more detailed analysis of Schelling's arguments against both chemical and immaterialist physiology, see Beiser (2002: 540ff.) and Kabeshkin (2017: 1190ff.).

an interplay of mutually opposing forces (*Weltseele SW* II 497f.). In this, he was inspired both by Kant's physical theory of repulsion and attraction and by Kielmeyer's idea that even biological processes are based on the principle of balance of opposing forces (Kielmeyer, 1993). The most adequate explanation of natural processes is thus, according to Schelling, to be found in contemporary dynamic chemistry, which works with this contraposition of forces. Still, even that theory is not quite satisfactory (see Schelling's objections above) and dynamic chemical processes themselves ought to be derived from something more general. Schelling claims that the world as a whole is organic and he defines organicity as a succession of causes and effects 'that enclosed within certain limits, flows back on itself' (*World Soul* 70 = *SW* II 349) – and thus forms a sort of loop, a feedback system. In contrast, mechanical and chemical processes that take place in the world are linear. According to Schelling, they are merely a specific, one could even say fragmentary, form of the original organic whole. In them, the initially organic forces manifest themselves only in their extremely simplified, non-complex form. For that reason, we should view a chemical process as merely an 'incomplete process of organisation' (*Weltseele SW* II 499).

What then does the initial duality of opposing forces which constitute the organic whole of the world (which, as noted above, returns to itself), and all the phenomena in it, look like? The first force is expansive: if unchecked, it would keep on expanding into infinity. This force is to be understood as the original positive principle. The second force is negating, limiting, and it is this force that creates empirical phenomena, i.e., something finite. All individual entities thus share a positive force, while their individuality and multitude are constituted by their boundaries, i.e., by restriction (*World Soul* 73f. = *SW* II 381ff.; *Weltseele SW* II 502ff.). The positive principle shared by all phenomena is not, in itself, a part of any concrete phenomenon. Schelling calls it a 'formative force' (*Bildungskraft*) or 'formative drive' (*Bildungstrieb*) (*Weltseele SW* II 565ff.),[5] noting that in antiquity, it was known as the 'common soul of nature' (*SW* II 569). The negative precondition of natural phenomena then takes the form of material and chemical processes, which in general take place as mutually antagonistic effects of various negative forces. On the level of living beings, they must reach a certain level of complexity to be able to integrate the 'vivifying' effects of the positive principle (*Weltseele SW* II 567). Still, Schelling views only the positive principle as the cause or source of life, while the negative principles constitute only its necessary conditions (*Weltseele SW* II 505). The positive principle continuously stimulates and sustains the processes of life by upsetting or disrupting the balance of powers in the body (*Weltseele SW* II, 568). We have already seen that without this disbalance, living entities would not be alive.

[5] Schelling adopts the term 'formative drive' (*Bildungstrieb*) from Blumenbach (1799), who used it to explain the reproduction of living beings in the sense of regeneration of damaged parts and propagation. Schelling generalises this conception and concludes that this force acts not only in the living nature. Cf. Richards (2002: 219).

They would sink to the level of mere chemical processes and enter a state of 'rest' (*Weltseele SW* II 514) and 'general neutralisation' (*Weltseele SW* II 493).[6]

The main problem of this approach is that the formative force or drive is transcendent, i.e., it is not part of the material world. That takes Schelling perilously close to vitalism, although – unlike the vitalists – he admits that the negative material/chemical processes are a necessary precondition of all manifestations of life. Schelling was aware of this shortcoming, which is why he tried to reformulate the conception in his following book on the subject, namely in the *First Outline of a System of the Philosophy of Nature* (1799) and in the *Introduction* (1799) which belongs to it.

The main starting point of the explanation of nature presented here is a theory of productivity. According to Schelling, we should not view nature as a mere sum of things or facts, but as constant transition and conversion between infinite productivity and its fixation in the form of products. Theories that start with the existence of fixed entities, such as atomism, must explain how and where change and development enter nature. Schelling's starting point is the opposite: in his view, nature is constant movement and change, and what is to be explained is how stable and permanent entities enter the picture (*First Outline* 202 = *SW* III 284) (see Chap. 8). In Schelling's scheme, particular entities can emerge only if, within the infinite productivity of nature, there is also the opposite tendency, one that tends to inhibit this productivity. Even that, however, merely places certain boundaries on productivity since on its own, this limiting tendency does not create individual entities. These emerge only when the initial productivity once again works against this limitation. At that point, a re-production takes place, and that fills the by now limited sphere by what we call reality. Schelling tries to elucidate this insight using the metaphor of a stream that runs into an obstacle which creates a whirl in it. Such whirl is not something fixed: it arises because new water (i.e., productivity) constantly rushes in to fill the space outlined by the obstacle. In this way, the whirl is constantly renewed (*First Outline* 18 note, 206 = *SW* III 18 Anm. 2, 289). In nature, particular things are thus not something absolutely constant and passive. Rather, they are just inhibited productivity and they come into being by an interplay of limitation and re-production. In this sense of the word, one could call Schelling a vitalist, because in his view nature is active, productive, and, indeed, vital. But it is not vitalism in the sense of the existence of some principles acting only and solely in living nature (e.g., see Kabeshkin, 2017: 1189, note 21). As Schelling puts it: '… the difference between organic and inorganic nature is only in Nature as object, and […] Nature as originally productive soars above both' (*First Outline* 232 = *SW* III 326), that is, both realms are governed by the same principles.

All natural entities and phenomena are thus just limited manifestations of one and the same productivity. According to Schelling, one can use philosophy to reconstruct a certain mutual link between them, and thus find the 'sequence of stages in

[6] Cf. Beiser (2002: 541ff.). A comparison between this Schelling's argument and recent theories of dynamic and temporary equilibrium is found in Küppers (1992: 104ff.).

nature' (*First Outline* 6 = *SW* III 6) which take the form of various degrees of development (*First Outline* 215 = *SW* III 302f.).[7] The progressive character of this development then consists in the primary opposition between infinite productivity and limitation being potentised or heightened in the sense of gradual objectification of the subjective moment. In other words, the active, productive element of nature becomes increasingly more explicit in the product itself (*First Outline* 226–232 = *SW* III 317–326). In the second and third division of the *First Outline*, Schelling thus reconstructs the development of nature from matter, through magnetism and electricity to chemical processes, and from there to the organic powers (the power of reproduction, sensibility and irritability). The higher forms of natural phenomena and processes are irreducible to the lower forms and their description requires a specific conceptual apparatus that does not make sense at the lower levels of nature. Moreover, the levels or stages are mutually linked, because they are different forms of the same primary dualism of nature: the dualism of productivity and limitation. Schelling engages in attempts to reconstruct the several degrees of nature from the foundational ontological duality also in his later works, most notably in the *System der gesammten Philosophie und der Naturphilosophie insbesondere* (1804), without, however, making further substantial changes to the ontological framework.

The Conception of Living Beings

Once Schelling leaves the general approach to nature and the place of life in it and focuses on a conception of living entities and their specific way of existing, he largely follows Kant's theory of organism (see Chap. 4). The main difference between his and Kant's approach is that Schelling does not view his theory of living beings as merely a regulatory heuristic principle but rather as a description of their actual formation. In his view, a living being is specific above all by its *organisation*, that is, by the fact that it consists of a multitude of *organs* that have different *functions* and perform different *activities*, which are nonetheless *united* (see Chap. 3 for Aristotelian view and Chap. 10 for contemporary conception). We can view this *unity* as existing in two distinct senses: according to Schelling, there exists in organisms both a reciprocal *final* and reciprocal *efficient* causality between organs (the term Schelling uses to describe this reciprocity is *Wechselbestimmung*). The latter, efficient causality rests in the fact that organs mutually create and maintain each other in existence. One could thus say that an organism is *its own efficient cause*, it exists due to itself, and undergoes spontaneous *self-organisation* (*First Outline* 51ff. = *SW* III 65–68). Nevertheless, as Michelini (2020) correctly notes, that on its own would not be a sufficient explanation of the nature of living entities, because such cyclical link between the cause and effect can be found for instance even in the

[7] For the moment being, let us leave aside the question whether Schelling understood these degrees as merely ideal types or rather as degrees of gradual development in time. We shall address this issue in the last part of this chapter.

water cycle. What is also at hand are comparisons between Schelling's theory and current theories of self-organisation (e.g., Kauffman, 1993), but if what those theories mean by self-organisation is that upon exceeding a certain threshold the system spontaneously starts establishing a certain order, it does not describe the core of Schelling's explanation of living beings. This is because at the core of Schelling's concept is the process of constitution of a *self*, a kind of active subject which aims at maintaining itself in existence and is thus its own purpose. In this context, one can therefore say that particular constitutive parts (organs) are present not only thanks to other constitutive parts (efficient reciprocal causality) but also *because of* the whole and *because of* other organs (reciprocal final causality). This conception is closer to current theories of autopoiesis and agency than to the abovementioned theories of self-organisation and theories of chaos (e.g., Maturana and Varela, 1980).

How does this self-constitution of the self which is its own purpose take place? We saw above that Schelling's aim is not to explain the emergence of movement and change in nature but the opposite: how, in the infinite stream of productivity of nature, there arises something stable and permanent, in other words, how do particular entities emerge. On the level of living beings, this constitution of entities happens in a specific manner different from what we find in inorganic nature. Only on the level of living entities can we speak about individuals *sensu stricto*, that is, entities which do not dissipate in the processes they enter. Instead, they actively sustain their current identity. A living individual maintains its identity and reproduces itself by *resisting* the effects of external physicochemical forces and by separating or singling itself out from the *external environment*. Unlike inorganic entities, living beings actively sustain themselves by dynamically and continuously maintaining the difference between the *self* (*Selbst*) and the *external world* (*First Outline* 54 = *SW* III 70). Such self-sustenance by self-delimitation vis-à-vis what is outside is, however, not 'innocent'. It amounts to more than mere resistance to external influences in the form of, for instance, thermoregulation in warm-blooded animals (*First Outline* 63 = *SW* III 83). Importantly, Schelling adds: 'No individuality in Nature can, *as such*, maintain itself, unless it begins […] to assimilate everything for itself, to encompass everything within its sphere of activity. In order that it not *be* assimilated, it must *assimilate*; in order that it not *be* organised, it must *organise*.' (*First Outline* 54 = *SW* III 70) Every living being is thus trying to maintain and sustain itself via external agency and active self-assertion (see Chap. 10) which takes the form of negating the independence of external things, and thereby also at the cost of those external things. A paradigmatic example of this is nourishment as a process of decomposition and assimilation of external substances and other living beings.

One consequence of the fact that living individuals actively keep themselves in existence is that – in contrast to non-living things – they do not merely occur. Their existence is based on activity and without activity a living individual ceases to exist. As Schelling says: 'The organic distinguishes itself from the dead simply in that the existence of the first is not an actual *being* but rather a continual *being-reproduced* (through itself).' (*First Outline* 107 = *SW* III 146)[8] Another specific feature of living individuals is that their identity is not based on the identity of the matter from which their bodies are built, because that keeps on changing through life. The identity of a living entity can thus be based only on the identity of its form, that is, its shape or likeness, and its specific manner of acting with respect to itself and the external world (e.g., *System der gesammten Philosophie*, *SW* VI 393). The abovementioned self-constitution of individuals by active setting of boundaries vis-à-vis the external environment involves a degree of self-determination and autonomy such as we do not encounter anywhere in non-living nature. Nevertheless, this autonomy does not imply any self-sufficiency of living beings or their independence of the environment. All living beings have some requirements and needs that can be met only from the *other*, in relation to what is outside (see Chap. 7). Later, Schelling, along with Hegel, called this basic feature of life *desire* (*Begierde*).[9]

So far, in relation to characteristic activities of living beings, we spoke only about self-constitution and self-reproduction. According to Schelling, however, organisms exhibit also two other basic activities: irritability and sensibility. Schelling adopts this pair of concepts from Albrecht von Haller (1771), who means by sensitivity the ability to feel for instance pain and links it to nerves. By irritability, he means reactions to stimulation, which consists in the contraction or expansion of the relevant organ and is linked to muscles.[10] Schelling adopts this distinction and tries to derive these two activities from the internal dialectics of the abovementioned process of self-constitution of organisms. Within his system, irritability is viewed as an aspect of the differentiating relation of organisms to what is external to them, i.e., as an aspect of relation to other things as other things (*System der gesammten Philosophie*, *SW* VI 408ff.). Sensibility, on the other hand, is an aspect of organisms' assimilation of what is outside. In this case, however, it is not a material assimilation, as in digestion where the other was disappearing due to organism's action. Sensibility preserves the other as other while placing it into the organism. It is thus an *ideative* assimilation, where matter is transformed into its representation, into an *immanent image* of the other (*System der gesammten Philosophie*, *SW* VI 409).

Let us now return to the primary activity of organisms, which in a certain sense includes also irritability and sensibility, that is, to self-reproduction of organisms. So far, we spoke about self-reproduction only in the sense of organisms' self-maintenance and active assertion of boundaries vis-à-vis the environment. But part of the phenomenon of self-reproduction is, according to Schelling, also reproduction, which brings us to the subject of the origin and development of individuals on the one hand and the issue of development of species on the other.

---

[8] Cf. also *System der gesammten Philosophie* (*SW* VI 387f.).

[9] Cf., for instance, '… desire, which determines the ground of every particular natural being …' (Schelling 2006: 43 = *SW* VII 376).

[10] An overview of contemporary theories and disputes regarding irritability and sensibility is presented in Richards (2002: 313–321).

Development of the Embryo and the Development of Species

Let us start with a brief outline of the chief contemporary theories of reproduction and origin of new individuals, since they form the context within which Schelling formulated his concept. On the one hand, there were the proponents of the so-called *preformism*, a theory called in Schelling's time – for us somewhat confusingly – 'evolution theory'. Preformist theories assumed that embryos are in fact fully developed miniatures of adult individuals, including their pre-prepared, albeit miniature, organs. A version of this theory formulated in antiquity usually claimed that embryos, or rather 'germs', are all around nature (e.g., Hippocrates and Heraclitus). Later, in the Modern Era, they were already situated into adult individuals: according to some into the father, according to others, into the mother. These Modern theories then also had to assume that embryos are in effect encapsulated in one another (*Einschachtelung*) *ad infinitum*, so that an adult contains an embryo, which contains another embryo, etc. The other camp consisted of proponents of *epigenesis*, who supposed that the formation of new individuals takes place by the development of organs from an unstructured embryonic substance. But this formation of organs had to be stimulated and directed by some force, which for instance Caspar F. Wolff called *vis essentialis*. Blumenbach (1799), who had an immediate influence on Schelling, then spoke about *nisus formativus* or *Bildungstrieb*, that is, a force acting towards a certain goal that is present in living nature and completely absent in non-living matter. At most, one can find some hints of it for instance in crystallisation. Dominant in the seventeenth and eighteenth century was, however, preformism, because it better met the desideratum of a mechanistic explanation of nature, while advocates of the epigenetic theory with their assumption of a vital or essential force leading to the formation of matter were in a minority. A new form of the epigenetic theory made more substantial inroads only in the nineteenth century together with the development of the cellular theory (cf. Casetta, 2020).

Schelling's contribution to this debate consisted in an effort to find some middle ground. He rejected the traditional theory of preformation, because it amounted to a mechanistic and strongly deterministic explanation incompatible with the productive character of nature (see above). At the same time, though, he was convinced that nothing in nature can arise completely anew and by accident. Schelling argued against the theory of preformation by pointing to the metamorphosis in insects, where for instance a caterpillar has a completely different organ of respiration than an adult butterfly and a completely different digestion apparatus, which corresponds to the different kind of food it seeks. Moreover, according to Schelling it does not make good sense to assume that one kind of organ of digestion is in a fully developed, though miniature, form present in another. 'All of these phenomena prove that the metamorphosis of insects does not occur by virtue of the *mere* evolution of already preformed parts, but through actual epigenesis and total transformation.' (*First Outline* 38 note = *SW* III 47 Anm.) On the other hand, epigenesis does not take place accidentally. There is a kind of predetermination, although it takes the character of preformation not of organs but rather of *directions* of the formative drive, which regulates the development of particular organs and which specifies given species (*First Outline* 37 note, 47ff. = *SW* III 46 Anm., 60ff.). It is important to note that these directions of a formative drive or natural predispositions (*natürliche Anlage*)[11] exist, according to Schelling, within a particular species always in multiple forms: only some of these predispositions are actualised, while others are not. The trigger that decides which of these tendencies will be realised takes the form of influence of the external environment (*First Outline* 44 = *SW* III 56).[12] Within a species, depending on differences in the environment, there can therefore exist various subspecies and varieties, which then persist by heredity (*First Outline* 46 = *SW* III 58). Schelling uses the term 'species' *sensu lato*, so that for instance wolves and dogs or horses and donkeys are varieties of the same species. According to Schelling, the defining feature of individuals belonging to the same species is the ability to have offspring – whether the offspring are fertile is then not relevant (*First Outline* 43, 45 = *SW* III 55, 57). But it seems he is not quite certain about absolute validity of this criterion: he notes that in many particular cases, it is difficult to tell whether there are two distinct species or just varieties of one species present, and one can decide only based on a thorough biological investigation of identity or difference of innate disposition (*First Outline* 49 = *SW* III 63). It is also important to bear in mind that Schelling's approach to species is inclusive, so that each individual embodies a different aspect of the ideal of a given species, while no individual expresses the species in full and the ideal is embodied only in the sum of all individuals belonging to it. This conception differs from the reductive concept of species, represented for instance by Linnaeus, according to whom a species is characterised by a shared set of features which fully apply to each individual belonging to the species (for more on this, see Richards, 2002: 302). Schelling's inclusive understanding of species also implies that there is no 'original' representative of a species embodying all future possibilities, a representative from which all other varieties had developed. Instead, each variety is a different, separate, and necessarily incomplete embodiment of the original predispositions of a given species whose particular realisation depends on the varying influences of the external environment.

These thoughts about changes within a species bring us to the subject of the origin of species. Schelling was the first to use the term 'evolution' – which, as we saw above, originated in embryology, where it referred to the development of preformed organs – in the sense of a particular process within the whole of nature, a process that has certain stages during which the complexity of organic forms increases (*First Outline* 42 note † = *SW* III 53 Anm. 1). Let us first revisit the basic ontological framework of Schelling's theory of nature. We saw above that, according to Schelling, nature is infinite productivity and particular living beings are merely fixed and temporary manifestations of this productivity. Nevertheless, production does not take place chaotically: in nature, we can find particular degrees or 'potences'

[11] Schelling also uses the term germ (*Keim*), without, however, implying some miniaturised preexisting forms of organs.

[12] Relevance of this consideration for current discussions about relations between a genotype, phenotype, and environmental triggering is discussed by Casetta (2020).

of this productivity and its products. They vary in the complexity at which the original opposition integral to all there is had developed. On the level of living nature, which is what we investigate now, 'organisms overall are to be seen as only *one* organism inhibited at various stages of development' (*First Outline* 43 = *SW* III 53), whereby particular stages correspond to the particular species of living beings. Schelling probably found inspiration for this conception, according to which particular species are just stages and points in the development of one general organism, in contemporary embryology. It posited that within a single individual we can distinguish various developmental stages and – as already Kielmeyer (1993) had noted, and Haeckel later famously expressed in his law of biogenetics – the stages of ontogenetic development are strikingly similar to some existing species or rather higher taxa.[13]

In our attempt to reconstruct a particular sequence of the individual degrees, however, we cannot take as our starting point the individual 'products', that is, particular existing species, because it would be very difficult to find a continuity of development at this level (*First Outline* 43 note * = *SW* III 53f. Anm. 2). Instead, we should focus on the productivity of nature and its particular stages and forms, meaning we ought to investigate the functions and activities of living beings, by which Schelling means above all the abilities of self-reproduction, irritability, and sensibility.[14] Mutual proportions and consequently also the concrete forms of these functions can be, according to Schelling, at least in broad outlines derived a priori. In particular, in this way one can derive the basic classes of living beings roughly on the level of plants, fish, amphibians, insects, mammals, etc., where for instance plants display a clear dominance of the reproductive power but lack sensibility, while in mammals, sensibility is dominant, but the reproductive potence is relatively low (*First Outline* 52f., 147f. = *SW* III 68, 203f.). A more detailed investigation should be, according to Schelling, left to comparative physiology (see Chap. 12), not to comparative anatomy, since that focuses only on the shape, which is merely an outer manifestation of organic functions (*First Outline* 43 note †, 50 = *SW* III 54 Anm. 1, 65).

What does Schelling say about the evolution of particular stages of life? In his earlier book, *On the World Soul*, Schelling admits that the development of nature might have the character of transformation (*Umgestaltung*), where one species is transformed into another and this leads to the emergence of new species, although due to the vast length of period over which such processes take place, it is very difficult to show (*On the World Soul* 69 = *SW* II 348ff.). In the *First Outline*, he abandons this notion and instead claims: 'The assumption that different organisms have really formed themselves from one another through gradual development is a misunderstanding of an idea which actually does lie in reason.' (*First Outline* 49 = *SW* III 63) Development in nature thus should not be viewed in terms of descent and

[13] For more on Kielmeyer, see Richards (2002: 244–248).

[14] Schelling himself notes that he adopts this idea from Kielmeyer (*First Outline* 141 = *SW* III 195). For more on this, see again Richards (2002: 243ff.).

transformation in the sense that all higher species evolved from a shared ancestor, some primitive kind of organism. According to Schelling, we should instead endorse a traditional view according to which a once 'fixed' species characterised by a specific mutual relation of organic functions cannot change into another. Instead, by procreation it reproduces further *ad infinitum* (*First Outline* 49 = *SW* III 62ff.). In that case, does Schelling believe at all in successive development that leads from simpler forms of life to more complex ones? How do new species appear in nature? This is somewhat unclear, and two interpretations are possible.

According to the first, all species appeared simultaneously, at a point when one and the same power of productivity stopped, as if frozen in one moment, at various degrees of development dictated by the various possible mutual relations between organic forces. If that were all there is to it, all manners of predispositions which define particular natural species would come into being at the same time. There would be an ideal-typical hierarchical order among species, but no new species could appear. They could at most die out due to geological disasters or climatic changes, about whose existence Schelling had no doubt (this interpretation is advocated, among others, by Engelhardt, 1984). In that case, any more widespread change in the composition of living beings on Earth could take place only if Schelling understood the 'species' within which variation takes place in a very broad sense (with, e.g., quadrupeds or fish forming a species), and moreover, if he did not assume that a species is defined by its members' ability to produce offspring together. This broad definition of species is ascribed to Schelling for instance by Richards (2002: 311), although he does not adduce any passage in Schelling's writings to directly support this claim.

The second, and in my view more plausible, interpretation runs as follows: according to Schelling, there appear in the history of the Earth successively new and higher species. They do not, however, emerge via a transformation of older species. Instead, 'in order to bring forth a new product, Nature would have to begin again from the start.' (*First Outline* 49 note * = *SW* III 63 Anm.)[15] By this, Schelling probably means to say that new predispositions, which consist in new mutual relations of organic functions and are specific to the nascent species, appear because after the stage of inhibition of productivity, which corresponds to one particular species of living beings, nature's productivity is renewed, and with it comes also a new and higher form of inhibition of this productivity, characterised by new mutual relations of the organic forces. This leads to the formation of a new species, which then maintains itself by reproduction, while the free productivity of nature is once again renewed. On the one hand, one can thus say about relations between species that 'one product contained the *ground* of the subsequent one. The product C could not arise before B, and this not before A had arisen.' (*First Outline* 49 note § = *SW* III 63 Anm. 2)[16] On the other hand, it is not so that a new species arises by direct trans-

[15] Cf. also Schelling's claim that 'Nature must have begun all over again with a totally new natural predisposition for each product that appears fixed to us.' (*First Outline* 49 = *SW* III 63)

[16] Cf. also *First Outline* 218 = *SW* III 307.

formation of a previous species. Individuals of the new species do not originate in individuals belonging to the previous species genealogically, and individuals of the previous species are thus not their real ancestors. Although it remains somewhat hard to grasp how, according to Schelling, individuals of a new species come into being, it seems rather clear that his thoughts proceeded in this direction, and he did actually assume the succession of new species in nature. One piece of indirect evidence can be found in the fact that he aimed the criticism contained in the *First Outline* against the theory of descent,[17] but voiced no objections against the more basic theory of successive development and formation of new species, which are both implied in the descent theory.

Conclusion

Schelling's philosophy of nature influenced nineteenth-century German biology mainly in three areas. First of all, it offered a monistic theory of nature. This, in turn, enabled contemporary biologists to abandon materialism, which reduces living nature to physicochemical reactions, without adopting dualism of the vitalist kind. At the same time, it formulated a view of nature as a productive process, thus enabling a conception of living beings as self-organising subjects who actively maintain their identity by asserting their boundaries vis-à-vis the external world while engaging in reciprocal causality among their constituent parts. This processual conception of nature eventually led to the idea of development, which on the ontological level takes place epigenetically and on the phylogenetic level takes the form of a sequence of stages, in which an order of ideal types can be distinguished. We also saw that many of these views are inspiring to this day and resonate with some of the subjects of current philosophy of biology that formed in late twentieth century and are the main topic of this book (as attested by cross-references to other chapters). Closeness to especially the theories of autopoiesis and agency is mentioned above, but Gare (2013) goes even further. He argues that Schelling, with his striving to overcome the Newtonian paradigm and find a more adequate ontology that would help us grasp the physical world in a way that makes intelligible the emergence of life, provided both direct and indirect inspiration to anti-reductionist thinking in twentieth-century mathematics and theoretical biology. He identifies and reconstructs in detail this fragmented tradition involving mathematico-physico-chemical morphology (J. Needham, C.H. Waddington et al.), biosemiotics movement (J. Hoffmeyer, K. Kull, M. Barbieri, H. Pattee et al.),[18] and relational biology focused on anticipatory systems (R. Rosen).

[17] According to Richards, the most concrete object of Schelling's criticism was the theory of Erasmus Darwin (grandfather of Charles Darwin); see Richards (2002: 300ff.).

[18] Even C. S. Peirce, one of the main inspirational sources of biosemiotics, was directly influenced by Schelling in his ontology (Ibri, 2022: 223f.).

References

Beiser, F. (2002). *German idealism: The struggle against subjectivism, 1781–1801*. Harvard University Press.

Blumenbach, J. F. (1799). *Handbuch der Naturgeschichte*. Dieterich'schen Buchhandlung.

Casetta, E. (2020). Preformation vs. Epigenesis: Inspiration and haunting within and outside contemporary philosophy of biology. In C. Emilo, I. Corriero, & G. Hamilton (Eds.), *Rethinking Schelling. Nature, myth, realism* (Vol. 74, pp. 119–138). Rivista di Estetica.

Engelhardt, D. (1984). Schellings Philosophische Grundlegung der Medizin. In H. J. Sandkühler (Ed.), *Natur und Geschichtlicher Prozess: Studien zur Naturphilosphie F. W. J. Schellings* (pp. 305–325). Suhrkamp Verlag.

Galvani, L. (1791). *De viribus electricitatis in motu musculari commentarius*. Ex Typographia Instituti Scientiarium.

Gare, A. (2013). Overcoming the Newtonian paradigm: The unfinished project of theoretical biology from a Schellingian perspective. *Progress in Biophysics and Molecular Biology, 113*, 5–24.

Haller, A. (1771). *Primae Lineae Physiologiae in usum Praelectionum Academicarum*. Grasset et Socios.

Ibri, I. A. (2022). *Semiotics and pragmatism: Theoretical interfaces*. Springer.

Kabeshkin, A. (2017). Schelling on understanding organisms. *British Journal for History of Philosophy, 25*(6), 1180–1201.

Kant, I. (2000). *Critique of power of judgement* (P. Guyer & E. Matthews Trans.). Cambridge University Press.

Kauffman, S. A. (1993). *The origin of order. Self-organization and selection in evolution*. Oxford University Press.

Kielmeyer, C. F. (1993). *Über die Verhältnisse der organischen Kräfte*. Basilisken-Presse.

Küppers, B.-O. (1992). *Natur als Organismus. Schellings frühe Naturphilosophie und ihre Bedeutung für die moderne Biologie*. Vittorio Klostermann.

Maturana, H., & Varela, F. J. (1980). *Autopoiesis and cognition. The realization of the living*. D. Reidel Publishing Company.

Michelini, F. (2020). The paradox of living: Jonas and Schelling on the Organism's autonomy. In C. Emilo, I. Corriero, & G. Hamilton (Eds.), *Rethinking Schelling. Natue, myth, realism* (Vol. 74, pp. 139–157). Rivista di Estetica.

Richards, R. J. (2002). *The romantic conception of life: Science and philosophy in the age of Goethe*. The University of Chicago Press.

Schelling, F. W. J. (1799). Von der Weltseele. In K. F. A. Schelling (Ed.), (1857). *F. W. J. Schelling Sämtliche Werke* 2 (pp. 345–583). J. G. Cotta.

Schelling, F. W. J. (1804). System der gesammten Philosophie und der Naturphilosophie insbesondere. In K. F. A. Schelling (Ed.), (1860). *F. W. J. Schelling Sämtliche Werke* 6 (pp. 131–576). J. G. Cotta.

Schelling, F. W. J. (1988). *Ideas for a philosophy of nature* (E. E. Harris & P. Heath Trans.). Cambridge University Press.

Schelling, F. W. J. (2004). *First outline of a system of the philosophy of nature* (K. R. Peterson Trans.). State University of New York Press.

Schelling, F. W. J. (2006). *Philosophical investigations into the essence of human freedom* (J. Love & J. Schmidt Trans.). State University of New York Press.

Schelling, F. W. J. (2010). *On the world soul* (I. H. Grant Trans.). In R. Mackay (Ed.), Collapse *Vol.* VI: Geo/Philosophy (pp. 66–95). Urbanomic.

Zammito, J. H. (2018). *The gestation of German biology. Philosophy and physiology from Stahl to Schelling*. The University of Chicago Press.

Chapter 6
Organismic Teleology and Agency Beyond Systems Theories: A Process-Metaphysical Perspective

Spyridon A. Koutroufinis

Abstract This chapter investigates two central biological concepts: teleological end-state directedness and agency. Its aim is to point out some key difficulties in how these concepts are understood in contemporary biology and philosophy of biology, and to show how a process-metaphysical approach to organisms can help us overcome these problems. To this purpose, we analyse the understanding of teleology and agency in contemporary biological systems theories. We focus in particular on the concept of mechanism, which plays an important role both in these theories and in modern biology and philosophy of biology. This examination shows that, first of all, mechanistic thinking ultimately reduces teleological explanations to merely a useful way of speaking that has no ontological relevance, and second, that the ability of organisms to autonomously and profoundly transform their material structure exceeds the explanatory power of mechanistic explanations. As an alternative, I propose a process-philosophical understanding of organismic end-state directedness and agency that is based on the central metaphysical and bio-philosophical concepts and ideas introduced by Alfred N. Whitehead and Henri Bergson.

Keywords Organism · Teleology · Agency · Process-philosophy · Mechanism · Whitehead · Bergson · Aristotle · Organismic memory

Introduction

The term 'teleology' was coined in 1728 by the influential German philosopher Christian Wolff. It was introduced as a name for the doctrine of purposes in nature. Etymologically, the expression stems from the Greek word 'telos', which denotes the 'final cause' ('causa finalis'), one of the four kinds of causes which Aristotle

S. A. Koutroufinis (✉)
Institute for Philosophy and History of Literature, Science, and Technology, Technical University of Berlin, Berlin, Germany
e-mail: spyridon.koutroufinis@campus.tu-berlin.de

J. Švorcová (ed.), *Organismal Agency*, Biosemiotics 28,
https://doi.org/10.1007/978-3-031-53626-7_6

identifies in nature (see Chap. 2). In biology, teleology is a central and yet problematic concept. David. L. Hull, one of the founders of contemporary Anglo-Saxon philosophy of biology, claimed that although one might get the impression that evolutionary theory has eradicated the idea of teleology, biologists continue to talk teleologically because the biological phenomena which inspired this idea still occupy our attention (Hull, 1974: 101).

In contemporary biology, the term ‘teleology’ can carry either of two meanings: *end-state directedness* or *purposiveness*. The end-state of a biological event is the final state with which it reaches its completion. Most biological phenomena, including the cell cycle, embryogenesis, growth, foraging, mating, or migration, are directed towards a particular end-state. Taking this into account, some influential philosophers of biology and biologists define teleological entities and events as those which tend to ‘approach, attain, or maintain certain preferred states’ or to oscillate around them (Hull, 1974: 107ff.). The purpose of an organic event is, in contemporary biology, identified with its function within an organism or another biological entity (e.g., an ecosystem). Both meanings can be traced back to the philosophical and biological writings of Aristotle (see Chaps. 2 and 3; cf. Chap. 4). Due to space limitations, this chapter focuses on the end-state directedness of cells and organisms and largely ignores the topic of purpose.

In the following, I shall focus on one essential characteristic of living beings that clearly distinguishes them from inorganic phenomena of all kinds: In stark contrast to artificial and natural inorganic systems, organismic and cellular processes not only reach their characteristic end-states but, moreover, autonomously create the conditions necessary for the production of these states by profoundly influencing their own material constitution (cf. Chap. 5). This essential autonomy of biological entities is subsumed under the concept of *agency*. This term emphasises the ability of biological processes to create conditions for their own preservation or to carry out an orderly transformation through their own dynamics.

The aim of this chapter is to point out some crucial difficulties in understanding teleological end-state directedness and agency in contemporary biology and philosophy of biology and to show the relevance of a process-ontological understanding of organismal dynamics. Let us therefore start by a critical introduction of the system-theoretical approach, which dominates thinking about teleology in contemporary biology.

Reduction of End-State-Directedness to Mechanisms of Self-organisation and the Degradation of Teleology to a Heuristically Useful Way of Speaking

Ever since the seventeenth century, physics and biosciences have been anchored in an abstract understanding of the term ‘system’ as a set of elements standing in rule-governed interrelations (von Bertalanffy, 1971: 55) determined by both the nature of those elements and by the conditions under which the elements interact with each other. As a result, almost all contemporary bioscientists and philosophers of biology subscribe to some version of materialistic metaphysics that is based on this conception of system.

The new discipline of systems biology is based on the assumption that the nature of organisms and cells can be, in principle, captured by the theory of self-organised dynamic systems. Leading biologists of today tend to view systems biology as supporting a renaissance of the concept of organism, because it promotes a shift of attention from DNA to an organism-centred perspective. Systems-theoretic understanding of organisms implies that their dynamics is the emergent result of interactions of their elements, none of which alone can determine the organismic dynamics. Contemporary theoretical biologists and systems biologists view organisms as energetically and materially open self-organising dynamic systems composed of nonlinearly interacting biomolecules. A 'dynamic system' is defined as an entity whose state at any given moment can be described by a limited set of time-dependent or 'state variables' $x(t) = x_1(t), x_2(t), \ldots, x_n(t)$, for which one can formulate a mathematical function F that defines a connection between states at times t and $t + \delta t$. Properties of this function reflect the causal relationships between elements of the system. Development of such system in time is described by function F, which also entails a group of control parameters. The following is an abstract expression of a dynamic system:

$$x(t+\delta t) = F(x(t), p, \delta t); p = p_1, p_2, \ldots, p_m \text{ (Ebeling and Sokolov 2005 : 40)}$$

Function F is the most central element of this model. In most cases, F is a system of coupled nonlinear differential equations, while p always denotes a set of control parameters. During the process of calculating the variables, control parameters are usually kept constant, whereby it is the model maker who determines their values. This common methodological foundation of different forms of mathematical modelling of biological dynamics (Koutroufinis, 2017: 27–31) deserves particular attention, because the control parameters are treated as constants whose values must be adjusted by the model maker so as to make the resulting values of variables fit experimentally obtained data (of the variables in question) describing real biological entities and processes.

In contemporary science, the real or hypothetical causal structures which produce or predict the course and outcome of natural or artificial phenomena are generally referred to as *mechanisms*. In current life sciences, a satisfactory explanation of a phenomenon is identified with the formulation of a mechanism. One of the most dynamically developing areas of contemporary philosophy of science is the 'New Mechanical Philosophy' or 'New Mechanism.' Its advocates have in recent decades proposed several definitions of the term 'mechanism' in the philosophy of biology (Glennan, 1996; Machamer et al., 2000; Bechtel & Abrahamson, 2005; Glennan & Illari, 2018). According to Glennan, '[a] mechanism underlying a behavior is a complex system which produces that behavior by the interaction of a number of parts' (Glennan, 1996: 52). A more recent formulation of this definition states that '[a] mechanism for a phenomenon consists of entities (or parts) whose activities and interactions are organised so as to be responsible for the phenomenon' (Glennan & Illari, 2018: 92). These characterisations indicate an implicit systems-theoretical underpinning of the term 'mechanism', because interacting entities can only be characterised as 'parts' if they are viewed as elements of a system. For Machamer, Darden, and Craver, '[m]echanisms are entities and activities organised such that they are productive of regular changes from start or set-up to finish or termination conditions' (2000, 3). The 'set-up conditions' consist of both the start conditions (relevant entities and initial conditions) and the enabling conditions (such as boundary conditions, often labelled 'constraints') of a phenomenon. By 'termination conditions', the authors mean a 'privileged endpoint, such as rest, equilibrium' or an activated state (Machamer et al., 2000: 11 f.). The set-up conditions together with the mechanism as such constitute the explanans, while the termination conditions are the explanandum or, in some cases, the prediction of a new and not yet observed phenomenon (Machamer et al., 2000: 21). In systems biology, computer simulations show how the explanandum results from a mathematical model, which is an abstract mechanism. Computer simulations of both small and large systems of equations are considered a type of mechanistic explanation (Brigandt et al., 2018: 363).

Philosopher of biology Denis M. Walsh summarises the 'mechanical worldview', noting that from the perspective of mechanistic thought 'to explain a natural phenomenon one adverts to the causal mechanisms that produce it' (Walsh, 2015: 188). The proponents of mechanistic thought consider their explanations to be '*complete* in the sense that citing causes is sufficient to account fully for the occurrence of an event' and '*exhaustive* too, in the sense that every event has a complete mechanistic explanation' (Walsh, 2015: 188, italics added). With the following assertion, Walsh joins the line of a vast majority of contemporary biologists and philosophers of biology:

> … goal-directedness is an unproblematic causal consequence of the architecture of an adaptive system. It is also an observable feature of a system's dynamics. It consists in the capacity of a system as a whole to enlist the causal capacities of its parts and direct them toward the attainment of a robustly stable end-point. That end-point is the system's goal. […] This *systems-theoretic* approach may tell us what it is for a system to have a goal. As such, it offers an account of the conditions under which teleological explanations apply. (Walsh, 2015: 195–196, italics added).

Walsh claims here that end-state directedness can be unproblematically reduced to the architecture of a sufficiently complex 'system'. From this, he concludes that teleology – and thus biological end-state directedness – can be explained by systems-theoretical mechanisms (Walsh, 2015: 186–207). It is then an inevitable consequence that, according to Walsh, 'teleological explanations are allowed, but [this] does not demonstrate that they are in any way needed' (2015: 201, addition by S.K.). Before trying to show the misguidedness of this position with respect to the alleged dispensability of teleological reasoning, we ought to take a closer look at the sense in which Walsh allows teleological explanations in biology to stand alongside mechanistic ones. Why does he claim that 'the completeness of mechanism notwithstanding, teleology has some indispensable explanatory role to play' (Walsh, 2015: 189)?

According to Walsh, the main difference between mechanistic or systems-theoretic explanations on the one hand and teleological explanations on the other is that the former focus on the *causal* conditions under which teleological explanations pertain without telling anything about the *content* of those explanations:

> [The systems-theoretic approach] offers an account of the conditions under which teleological explanations apply. But, importantly, it does not tell us about the *content* of those explanations. […] The conditions required for a teleological explanation can be given in strictly causal terms. One just has to specify what causes the goal-directed behaviour. In contrast, the *content* of a teleological explanation cannot be specified in strictly causal terms. A teleological explanation does not mention the cause of an event's occurrence. It mentions the goal to which it contributes. In a teleological explanation, the goal state figures *as a goal* in the explanans. No causal explanation does that. Teleological explanations, properly construed, are not a species of causal explanations. (2015: 196, addition by S.K.)

In this passage, the terms 'causal' and 'cause' refer to those causes that can be exhaustively grasped by the methods and terminology of contemporary physico-chemical sciences. In the context of New Mechanical Philosophy, one could call them 'mechanical causes'. They could also be viewed as 'blind causes,' where 'blind' means that the causes have no sensitivity to their effects on beings or the larger context because they lack any imaginable form of mental ability. The idea of blind mechanical causality – which is almost never explicitly mentioned – forms the hidden but omnipresent metaphysical backbone of systems-theoretical biological mechanisms. In a further explanation of his position, Walsh analyses the expressions used in mechanistic and teleological explanations, focusing on the linguistic differences between 'producing' and 'conducing' descriptions of events:

> We successfully explain an occurrence mechanistically when we demonstrate that it is *produced* by some particular mechanism. By extension, we successfully explain an occurrence teleologically when we demonstrate that it *conduces* to some particular goal. Conducing isn't just causing. […] 'Producing' and 'conducing' descriptions of an event are different. *They carry different information*. The 'producing' description specifies that the earlier event is the *mechanism* of the later. It tells us 'how' the later event occurred. The 'conducing' description signifies that the later event is a goal. In doing so it specifies 'why' the former event occurred. (Walsh, 2015: 199, italics added)

Mechanisms *produce* occurrences. In describing productive relations between mechanisms and their effects, 'we use causal dispositional concepts – like pushing, pulling, attracting, binding' – and if we 'know what pushing, pulling, or attracting is, then [we] know what happens to Y when X pushes, pulls, or attracts Y' (Walsh, 2015: 198). In teleological explanations, however, one uses an entirely different kind of descriptive vocabulary: we use terms such as 'in order to', 'for the purpose of', 'for the sake of' to signify that the effect in question is a goal, and identify means as *conducing* to their ends (Walsh, 2015: 198ff.). In other words, while specific mechanisms (which Walsh calls 'causes') explain the specific effects produced by those mechanisms, the teleological mode of thought emphasises the *role* of those effects as goals. It is obvious that Walsh's approach equates the essential difference between mechanistic and teleological explanations with a *linguistic difference* between two scientific ways of speaking. This line of reasoning, based on the typically Anglo-Saxon focus of analytic philosophy on the structure of the scientific or philosophical language used, considers teleology to be *a mode of speaking*.

Based on the organisational closedness of organismic dynamics, Schlosser presented an understanding of teleology that is close to Walsh's position (Schlosser, 1998), which has deep roots in the history of biology: In the nineteenth century, some authors viewed organisms as natural objects which are to be both mechanically explained and teleologically described. This led to the establishment of a research programme for which the term 'teleomechanism' was proposed (Lenoir, 1981: 297; Toepfer, 2011: 804 f.). One of the most influential founders of modern physiology, Claude Bernard, affiliated his understanding of organism as a microcosm of mutually sustaining elements with this programme (Toepfer, 2011: 805). Toepfer argues in favour of a position similar to Walsh's when he states that teleological and causal explanations of organisms are complementary, because the teleological approach operates on a different level than the causal one (2011: 816). To put it differently, '[o]rganismic teleology is not dichotomous opposed to mechanism, it is a complementary aspect of material systems' (Asma, 1996: 141).

The common starting point for all these thinkers is the *metaphysical* hypothesis according to which causal relationships within organisms and cells can be, in principle, fully explained by mechanisms. If, as this hypothesis suggests, mechanistic explanations of biological end-states are complete, they cannot be made more complete by adding that these states are also goals because 'that fact is strictly incidental to a mechanistic explanation' (Walsh, 2015: 199). But this position logically leads to the momentous conclusion – paradigmatically articulated by the philosopher of biology Alexander Rosenberg – that '[t]eleology is unavoidable in biology for contingent and *nonconceptual* reasons' (Rosenberg, 1985: 65, italics added). This perspective supports the conviction, widespread among contemporary biologists, that teleology has no ontological but merely heuristic and methodological relevance to biology (Toepfer, 2011: 817ff.; Toepfer, 2005: 49ff.).

In the following, I argue against this position and show that biological teleology can only be understood through a radically anti-mechanistic ontology and is therefore much more than just a way of speaking.

Organismic Teleology Beyond Mechanistic Thought

The belief that organisms can be explained mechanistically received a significant impetus from conceiving of organisms as self-organised beings whose dynamics can be, at least in principle, grasped by modern mathematical systems theories, such as the theory of dynamical systems. The latter provides the theoretical basis for systems biology, which reduces the end-state directedness of embryogenesis, cell cycle, and other final state directed phenomena to the dynamics of enormously complex systems of coupled biomolecular reactions. In the relevant equations, the time-dependent values of concentrations of these molecules are represented by the abovementioned variables x(t), whose values are calculated by solving the mathematical function F, usually consisting of a system of coupled nonlinear differential equations. As noted above, from the viewpoint of the New Mechanical Philosophy, function F is an abstract (mathematical) mechanism. An essential element of every systems-biological mechanism is a set of parameters p whose values are determined by the modeller. It is important to bear in mind that, as noted above, during the calculation of variables all parameters are kept constant, so that the dynamic of the system has no effect on them. This external fixation is justified by pointing out that these parameters canalise the development of the time-dependent variables; therefore, they are the logical presuppositions of the system's dynamic. By testing different combinations of parameters, a model builder produces the desired behaviour of the abstract mechanism (a mathematical model). In other words, calculation ends with the computation of the appropriate final values (end-states) of the variables. The (*in silico*) computed values are considered correct only if they approximately match the corresponding experimentally measured final values of real (*in vivo*) organismal processes. Once this is achieved, a model is considered successful and is used to calculate the behaviour of the system under various conditions. Based on this methodology, a model for the behaviour of bacterium *Escherichia coli* has been proposed a few years ago. The authors of the model reduced the bacterium's dynamic to ten variables, for which they employed ten coupled differential equations containing 58 parameters (Van Hoek, 2008: 18–20, 45–47). In a model of the yeast cell cycle, the dynamics of the cell was reduced to 36 state variables. For their computation, the model makers used 143 parameters (Panning et al., 2007). On average, therefore, the model makers thus used four parameters to compute one variable – and it should be noted that similar, or even larger, ratios of the number of parameters to the number of variables are quite common in systems biology models.

Systems-biological mechanisms share an essential feature: They all operate on an implicit assumption about the roles of various factors (variables, parameters, or equations) in the dynamics of biological systems. In previous publications, I have shown that in formal models used in systems biology, there are two clearly distinct *types* of factors at work (Koutroufinis, 2017: 30; 2022: 15ff.). First of all, there are the *intrinsic factors* of abstract mechanisms, that is, factors generated by the dynamics of the system itself; these are the time-dependent variables x(t). Secondly, there are the *extrinsic factors* of abstract mechanisms, which include all factors that contribute to the generation of intrinsic factors but are not influenced by the intrinsic dynamics, i.e., by values of the calculated variables. Parameters are extrinsic factors. Obviously, the terms 'intrinsic' and 'extrinsic' do not refer to any spatial boundaries of the system: they simply indicate what is within the influence of its dynamics and what is outside it. In the mechanisms (mathematical models) used by systems biologists, the most complex factors are the systems of coupled differential equations which determine the variation of variables. These systems of equations express mathematical relations between the less complex intrinsic and extrinsic factors, that is, between the variables and the parameters. In current formalisms, systems of equations are not influenced by a system's change of states: they are static, which clearly qualifies them as extrinsic. Moreover, being relations between simpler factors, they can be viewed as *second-order extrinsic factors*. Analogously, variables can thus be interpreted as *first-order intrinsic factors* and parameters as *first-order extrinsic factors*. A system of coupled differential equations is then a single indivisible second-order extrinsic factor. Since the number of parameters in systems-biological mechanisms is many times greater than the number of variables, the number of extrinsic first-order factors that must be ascribed to these mechanisms will far exceed the number of intrinsic first-order factors.

The distinction between intrinsic and extrinsic factors can be applied to organisms as well. First-order intrinsic organismic factors are all the material and energetic quantities generated by an organism that have an effect on its dynamics, such as the concentration of regulatory proteins, scleroproteins, hormones, ATP molecules, etc. Environmental factors which the organism influences to improve its living conditions (e.g., regulated atmospheric humidity and room temperature) are first-order intrinsic organismic factors as well. Every organism is constantly fine-tuning the values of these factors. First-order extrinsic organismic factors are all factors which affect the dynamics of an organism but are not affected by it. These factors include the initial embryogenetic conditions, such as parental genetic constitution and the intrauterine environment of a zygote at the time of its fertilisation, the fundamental laws of nature which determine physicochemical processes, but also environmental conditions which organismal dynamics cannot influence, such as gravitation, radioactivity, geological processes, solar activity, and the forms and quantities of available energy and matter. As we shall see shortly, the idea of second-order factors applies also to organisms.

Nonetheless, although the distinction between first-order and second-order factors on the one hand and between intrinsic and extrinsic factors on the other hand applies to organisms, there are certain essential differences between real organisms and systems-biological mechanisms. These differences, which are of the utmost importance for understanding the organismal mode of being, are related to two fundamental faculties of all living beings:

1. In sharp contrast to systems-biological mechanisms, organisms can change the value of most quantities which are in mechanistic models represented by static parameters, such as the kinetic rate and boundary conditions. Current research shows that both unicellular organisms and the cells of multicellular organisms can specifically control the course of their intracellular molecular reactions by modifying the binding affinities between single biomolecules (e.g., protein or DNA) to their ligands (e.g., drugs or inhibitors). This allows cells to regulate the kinetic rate of particular intracellular biochemical reactions. As a result, in physiological settings, these rates can vary by orders of magnitude (Giuliani, 2010: 2–4). It is also well known that by regulating the architecture of their own cytoskeleton, cells can influence intracellular enzymatic reactions (Bizzarri et al., 2013: 35). These and other recent findings demonstrate that real organisms are capable of varying – or, to be exact, of fine-tuning – the values of many quantities which are in systems-biological mechanisms represented by static parame-

ters. Consequently, in contrast to contemporary biological formalisms, *in real organisms the number of first-order intrinsic factors always far exceeds the number of the first-order extrinsic ones* (Koutroufinis, 2017: 26–31; 2022: 13–16).

2. During the growth and regeneration of unicellular and multicellular organisms and during the embryogenesis of the latter, a variety of new types of proteins are synthesised and major morphological and physiological changes take place. Cells govern their metabolic and functional processes by modifying their shape, so that radical targeted configurations of the morphology of living beings take place also on cellular level (Bizzarri et al., 2013: 5). Modifications of the system's architecture can canalise biological processes in a variety of different ways, so that 'an unexpected level of causality emerges uniquely from the specific configuration adopted by the system' (Bizzarri et al., 2013: 5). By modifying their behaviour and remodelling their shape, cells can 'exert a reciprocal influence on their microenvironment […], as well as on gene expression' (Giuliani, 2010: 36). Embryonic processes display an even more radical intrinsicality of modification of organisms' morphological, physiological, and biochemical structures. Such examples imply that the network of causal relationships within real organisms is subject to change: the material structure of an organism is the outcome of a sequence of constantly generated new relationships between its own first-order intrinsic and first-order extrinsic factors. As noted above, in current formalisms these relationships are represented by fixed systems of invariable differential equations. In contrast, in both unicellular and multicellular organisms the relationship between both kinds of first-order factors is itself an intrinsic factor. Because the relationship between an organism's intrinsic and extrinsic first-order factors is a second-order factor, if the former relationship is intrinsic, then the latter, i.e., the second-order factor, is also intrinsic. These observations about the radical morphological and other restructuring of individual cells and multicellular organisms can be summarised as follows: In contrast to contemporary biological formalisms, *in real organisms, the second-order factor is necessarily an intrinsic factor* or, in other words, *there are no second-order extrinsic factors in real organisms* (Koutroufinis, 2017: 31 f.; 2022: 18–20). It should be noted that the second-order organismic faculty has a logical priority over the first-order one: Without the ability to reorganise its own material structure (intrinsicality of the second-order factor), organisms would not be able to constantly vary and fine-tune their first-order intrinsic factors.

It logically follows from this analysis that the dynamics of real organisms cannot be captured with the current systems-theoretical mathematical mechanisms and that it is so for reasons of principle. Therefore, when Walsh claims that teleology plays an indispensable explanatory role despite the completeness of mechanism (see above, Walsh, 2015: 189), I agree with his opinion about the indispensability of teleology but not with his conclusions regarding the alleged completeness of mechanistic explanations in biology, which puts him in line with most contemporary biologists and philosophers of biology. In my opinion, contrary to the view prevalent in contemporary philosophy of biology, the concept of organismal teleology goes beyond mechanistic explanations for fundamental reasons.

Organismal Teleology and Agency from a Process-Metaphysical Perspective

The intrinsicality of the second-order factor (the intrinsic relationship between the intrinsic and extrinsic first-order factors) enables cells but also unicellular and multicellular organisms to radically transform, both qualitatively and quantitatively, their material structure as and when required. It allows them to reorganise their genetic activity and create new types of proteins. This essential ability of all living beings enables them to behave in their ever-changing environment in ways which are, in most cases, likely to lead to a satisfaction of their needs. In fact, this is one of the most essential differences between living beings and the inorganic systems of physics. While the behaviour, and in some cases also formation, of inorganic systems – including the self-organising ones – is merely a *reaction* to certain boundary conditions of their physical surroundings (Koutroufinis, 2014: 103–107; 2019: 257–271; 2022: 22–24;), living beings demonstrate a remarkable *autonomy*. In particular, the inner dynamics of every organism is capable of manipulating the internal conditions under which the processes of inner dynamics take place in such a way that the organism can act purposefully in its environment and create material and energetic boundary conditions[1] which best serve the continuation of its metabolism and in general its self-preservation as an individual entity (see Chaps. 3, 5, and 10). This essential characteristic of all organisms is captured by their designation as *agents*. Walsh offers a similar understanding of *agency*, but his definition is based on the systems-theoretical concept of an 'organizationally closed system' (Walsh, 2018: 170). Both the behaviour of an organismic agent in its environment and its biochemical and physiological dynamics regularly produce certain events which serve the agent's self-preservation and its activities, such as reproduction. These reliably, predictably, and often periodically generated events, which are either end-states or means to them, would otherwise occur only accidentally and therefore rarely. This regular end-state-directedness of organismic agents indicates that there is an inherent link between biological agency and teleology.[2]

In my opinion, any attempt to establish a theory of biological agency – and consequently of teleology – must take into account the following considerations, which summarise the ideas presented in the second section of this chapter: The radical self-modification of the material structure during the cell cycle, embryogenesis, and other biological processes cannot be explained in terms of systems-theory, because systems of equations are not flexible and, moreover, require substantially more fixed parameters than variables. In other words, systems-theoretical mechanisms fail in the face of the vast dynamicity of life: if we view organisms as systems that can be exhaustively described in purely mechanistic terms, we cannot explain how come

[1] For example, through breathing and feeding, organisms autonomously control two of their vital boundary conditions.

[2] In the words of Walsh: 'Because there are agents, there are goals, means […] and a special mode of explanation—teleology' (Walsh, 2018: 172).

they are the source of structural transformations that can be set in motion by a second-order intrinsic factor. A mechanistic system cannot radically restructure itself because of the extrinsicality of its second-order factor. As a result, any transformation of a mechanistic system that is supposed to be an organism could only come from a source external to it – but that would contradict the idea of organismic agency.

Therefore, I am convinced that biology urgently needs an understanding of teleology that radically transcends the logic of the currently formal systems-theories. This form of teleology would be incompatible with modern mechanistic thinking – it would *not* be merely a way of speaking that leaves the mechanistic approach intact.

If systems-theoretical mechanisms cannot explain the end-state directedness of organismic dynamics and if they moreover, as noted above, imply the metaphysical idea of a blind cause, we must look for an essentially different category of causes. Long before philosophers began to speculate about the nature of causality, human experience provided the basis for distinguishing between two essentially different categories of causes: In the macroscopic world of our everyday life, we encounter processes caused either by entities that do not experience anything (all kinds of inanimate materiality and non-animal organisms) and consequently cannot have intentions, or by entities that have either complex or elementary experiences (animal and human agents) that enable them to pursue goals. The first kind of causation gave rise to the notion of 'blind cause.' The latter inspired scientists of earlier times and philosophers to speculate about causes that could be described as 'non-blind'. If, therefore, there is good reason to believe that blind causes cannot explain organismic teleology and agency, we must assume the efficacy of *non-blind causes* in living beings (cf. Chap. 4).

The introduction of some form of non-blind causality into biology is supported by the observation that many biological processes reach a functional end-state even under extreme conditions and via many detours (plasticity, radical restructuring of the genome, etc.). Such ability was in the past often associated with purposeful animal and human action. But, as we shall see below, the notion of non-blind causality does not necessarily imply a conscious or intentional pursuit of goals, such as is typical of humans and higher animals. In contrast to blind causality, non-blind causes are ascribed sensitivity to reaching end-states, and thus a *mental* aspect (Koutroufinis, 2014: 121–126; 2019: 575–586, 590 f.). As we shall shortly see, this idea corresponds well with one of the main pillars of Whitehead's process-metaphysics.

Alfred North Whitehead (1861–1947) is often regarded as one of the most innovative thinkers of twentieth-century philosophy of nature and metaphysics. Whitehead's metaphysics builds on the assumption that the most basic actual beings that make up the enduring micro-, meso-, and macroscopic entities of the physical universe are indivisible *processes* called 'actual occasions' or 'actual entities'. They are short-lived, indivisible events whose spatiotemporal expansion covers a wide range from microcosmic quantum phenomena to macroscopic conscious processes such as occur in highly developed mammal brains. Whitehead created his metaphysics against the background of various influential versions of mechanistic thought in the first half of the twentieth century. This is clearly evidenced by the three main metaphysical assumptions on which he based his concept of actual occasion: First, each actual occasion is to a certain degree a creative entity directed towards self-determination, and thus exhibiting end-state directedness (Whitehead, 1978: 25). All actual occasions are teleologically self-constituting entities whose final constitution is not pre-existent in them but emerges gradually during and through their self-determination. Through their self-constitution, these elementary processes determine their 'real internal constitution' (Whitehead, 1978: 43), in other words, their essence or nature. Once this determination of essence is completed, the new actual occasion appears as a short-lived spatiotemporal datum in the physical universe. In other words, the self-determination (determination of essence) of an emerging actual occasion does not take place in space and time. Actual occasions are flashes of existence which disappear after a fleeting presence (lasting from a fraction of a second to a few seconds) in the physical world. Secondly, actual occasions are indissolubly related to their physical environment (Whitehead, 1978: 73, 77, 80). This form of relatedness is aptly captured by the metaphysical concept of 'internal relationship' – meaning relations without which the related entities could neither emerge nor exist.[3] Thirdly, actual occasions are conceived of as inherently mental-physical entities: in other words, all elementary entities of the physical universe have a mental pole alongside a physical one (Whitehead, 1978: 108). Due to this mental pole, actual occasions possess an elementary subjectivity – and that is why one can view them as non-blind causes. Nevertheless, it ought to be noted that Whitehead specifically warns against conflating mental activity with consciousness (Whitehead, 1958: 16, 32; 1978: 25, 53, 56, 139, 280). Only the highest forms of mental activity (the richest forms of subjectivity) exhibit consciousness (Whitehead, 1958: 32; 1967: 180; 1978: 53), which is characterised as a 'second-order type of

[3] Whitehead calls the internal relationships that exist between actual occasions 'prehensions'. They must not be equated with physical forces, because they do not take place in physical space and time. A prehending actual occasion is not present in space because its self-determination is not yet complete. Only a prehended actual occasion whose self-determination is thus complete is in space for a short time before vanishing forever, but during that time it can be prehended by new, self-constituting actual occasions. The self-constituting prehending actual occasion participates in the *essence* of the prehended object through its prehensions. Although Whitehead calls the prehensions between actual occasions 'physical prehensions', they are not physical (if 'physical' is taken to mean 'spatiotemporal' or 'according to the laws of physics'). Since the emerging elementary process (actual occasion) is not a spatiotemporally localised entity during its self-constitution, its prehensions are not observable actions mediated by physical forces, particles, or fields that traverse the physical (spatiotemporal) universe. The concept of prehension is a purely metaphysical one in the literary sense of the term 'metaphysics.' In Whitehead's view, all basic concepts of physics (space, time, energy, force, field, elementary particles, causality, etc.) must be based on the metaphysical concept of actual occasion, which contains, among other things, the idea of prehension. Therefore, in Whiteheadian metaphysics, the concept of prehension logically underlies all kinds of physical entities and events. In Whitehead's metaphysics, there are not only 'physical prehensions', but also prehensions that take place between actual occasions and other real or ideal entities. But what all types of prehensions have in common is that the prehending entity is *always* an actual occasion.

mentality' (Whitehead, 1958: 33). During their fleeting manifestation as physical facts in space and time, actual occasions exert a causal influence on the course of material events in inanimate and living beings (Koutroufinis, 2014: 121–126).

From the perspective of the anti-mechanistic critique developed in this chapter, the key characteristic of the organismal way of being is that every living being is capable of creating its form and sustaining itself because its structure is a second-order intrinsic factor which fine-tunes the relationships between its material components. According to Whitehead's metaphysics, everything that endures in time consists of actual occasions. Consequently, living beings can be understood as coherent groups of internally related mental-physical processes (actual occasions), which are teleologically coordinated elementary subjects (Whitehead, 1978: 103). From a Whiteheadian perspective, therefore, their second-order intrinsicality, which is the most essential feature of all living beings, as well as all the biological abilities that arise from it, must be seen as the result of relationships between organismic actual occasions. The end-state of a macroscopic organismic event should then be viewed as the result of coordination of a vast number of actual occasions through their internal relationships: the teleological actions of microscopic actual occasions are tuned into one another in such a way that they give rise to harmonious macroscopic forms, such as the morphology of a cell at the end of certain phase of the cell cycle or of a multicellular organism at the end of its embryogenesis.

In my opinion, the three pillars of Whitehead's metaphysics (self-determination, internal relationality, and elementary mental capacity of actual occasions) can crucially contribute to an adequate understanding of second-order intrinsicality, which, according to my criticism above, cannot be achieved by a systems-theoretic analysis of the web of biomolecular reactions that compose an organism. From a Whiteheadian perspective, during a physicochemical event in the organism (e.g., an enzyme reaction), the electronic and protonic actual occasions which constitute atoms of the interacting molecules occur again and again in mutual interdependence (internal relationality) and disappear at a fast rate. The mutual relatedness of actual occasions is not mediated by any physical events; rather, it is so fundamental that underlies all kinds of physical events,[4] even the most elementary microcosmic ones. Therefore, it can only be grasped metaphysically. That, however, means that the internal relationality between actual occasions is a much more elementary form of relationship than one mediated by physicochemical interactions between molecules. Like all actual occasions, those which constitute living beings are not timeless and static Cartesian substances with a fixed essence. Instead, they should be understood as processes of a microchronic lifespan which determine their own essence in mutual interdependence. As an organismic actual occasion builds itself, it makes certain features of its immediate environment – which consists of other, mostly intracellular, elementary processes – particularly relevant, while making other features of its surroundings irrelevant (Whitehead, 1978: 73, 77, 80). In this way, the essence of an emerging actual occasion becomes inextricably linked to certain characteristics of

[4] See footnote 3.

its biochemical environment, which are *selected* by the new entity and appropriately included in the process of its self-determination. Because of this selective activity, organismic actual occasions are capable of determining their essence so as to be able to influence biochemical reactions inside and outside cells, thus steering the entire dynamics of the organism towards a specific pathway which produces biologically beneficial end-states. Logically, the coordination of these microscopic agents' teleological influence on the macroscopic physicochemical dynamics of the organism is greatly assisted by their mental side. As entities of experience and thus non-blindly acting elementary agents, actual occasions are sensitive to their impact on the overall dynamics of the organism in ways in which blind causes could never be.

Due to space limitations, I was not able to address here in detail the multifaceted issue of coordination of a vast number of microscopic actual occasions towards the end-state of a single macroscopic biological event. The following must suffice: The inner relationality and subjectivity of the elementary processes, as well as their activity in determining their essence, are necessary but not sufficient conditions for their mutual coordination. As I have shown elsewhere, such coordination requires the activity of a single indivisible factor that has a memory-like quality and acts throughout the organism (Koutroufinis, 2023: 239–242; 2019: 384–394). Using Henri Bergson's theory of memory, I have discussed the hypothesis that every multicellular organism has an embryogenetic and immunological memory (Koutroufinis, 2023: 240–242; 2019: 384–394, 628, 659 f.). The latter is primarily individual, because it is rooted in the organism's unique ontogenetic history (Koutroufinis, 2019: 390 f.), while the former is supraindividual, because each individual embryogenesis is rooted in the evolution of a particular species (Koutroufinis, 2019: 392–394). Gernot and Renate Falkner, Austrian biologists and Whiteheadians, spent many years studying the nature of memory in unicellular organisms (Falkner & Falkner, 2014). Gernot Falkner had recently formulated interesting philosophical arguments in support of the claim that individual organismic memory is rooted in species-specific memory (Falkner, 2023). In my own process-metaphysically grounded approach to organismic memory, I came to the following two conclusions: First, the short-lived actual occasions which constitute a living being are coordinated by a single, *long*-lived organismic subject (Koutroufinis, 2023: 240–242), which is both their source and the carrier of organismic memory (Koutroufinis, 2019: 664–666). Secondly, this long-lived organismic agent is a single, indivisible, and self-determining process with its own inwardness (Koutroufinis, 2019: 380–383; see also Chap. 9).

Finally, we should also address the nature of mental activity and the inwardness of cellular and organismic agents: even since its appearance on the scene, the teleological mode of explanation has been burdened with the assumption that it – at least implicitly – presupposes agents with psychologically complex intentionality and thus with consciousness. This notion, often accused of anthropomorphism, prompts philosophers of biology who consider themselves naturalists to emphasise the non-psychological status of agency and teleology.[5] In my opinion, we find in all biological agents, from the simplest bacterium to the blue whale, processes that *strive* to achieve different kinds of goals. While most of these goals have to do with the intra-organismic biochemical and physiological dynamics, very few, only the most complex, are related to the behaviour of the organism as a whole in its environment. But in both cases, certain end-states correspond to goals. Only in few and exceptional cases can such organismic acts of striving contain conscious representations of anticipated future situations (Koutroufinis, 2019: 652ff.; 2014: 118ff., 125, 127). These conscious processes often control the behaviour of the organism as a whole in its environment. The corresponding representations are elements of the subjective (mental) aspect of macroscopic actual occasions which manifest themselves for at most few seconds in certain parts of the brain of humans or other highly evolved mammals (Whitehead, 1978: 109). In all other cases, organismic and cellular agency and teleology are based not on a consciously reflected but merely an experienced *feeling* of a state of an organismic or cellular dynamic that has not yet been reached. This state is aimed for as an end-state because it is *positively* felt by the relevant agent. As Whitehead put it:

> In its lowest form, mental experience is canalised into slavish conformity. It is merely the appetition towards, or from, whatever in fact already is. The slavish thirst in a desert is mere urge from intolerable dryness. This lowest form of slavish conformity pervades all nature. It is rather a capacity for mentality, than mentality itself. But it *is* mentality' (1958, 33).

Conclusion

The systems theories used in current biology are based on the assumption that organismic teleological end-state directedness and agency can, in principle, be fully explained by mechanisms. This opinion is shared by many philosophers. This gave rise to the assumption – widespread in the current philosophy of biology – that final-state directedness can in principle be reduced to the architecture of a sufficiently complex 'system' and teleology can thus be explained, at least in principle, mechanistically. Teleological explanations are thus treated as heuristically and methodologically useful supplements with no ontological relevance. They are reduced to a different way of scientific speaking.

In this chapter, I have shown that for reasons of principle mechanisms cannot explain the organismic dynamics and therefore the teleology and agency of living beings. In terms of understanding organismic processuality, I believe a viable alternative to mechanistic thought is to be found in the process philosophy of Alfred N. Whitehead. His metaphysics enables a formulation of a non-anthropomorphic

[5] According to Walsh's understanding of agency and teleology, which reflects the dominant approach to these issues in contemporary philosophy of biology, 'if agency is a kind of observable activity and goals are its end states, then the natural, *non-psychological* status of agency […] is as unimpeachable as that of fluidity and viscosity' (2018: 173; italics added).

theory of organismic agency and teleology. Starting from Whiteheadian metaphysics, one can formulate a theory in which organismic processes that exhibit elementary mental activity can steer the entire dynamics of an organism towards specific pathways which produce biologically beneficial end-states. This theory must, however, be supplemented by the assumption of an organismic memory that acts throughout the living body and in Henri Bergson's theory of memory, we can find a good philosophical basis for developing the notion of organismic memory.

References

Asma, T. S. (1996). *Following form and function. A philosophical archaeology of life science.* Northwestern University Press.

Bechtel, W., & Abrahamson, A. (2005). Explanation: A mechanist alternative. *Studies in History and Philosophy of Biological and Biomedical Sciences, 36*(2), 421–441.

Bizzarri, M., Palombo, A., & Cucina, A. (2013). Theoretical aspects of systems biology. *Progress in Biophysics and Molecular Biology, 112*(1–2), 33–43.

Brigandt, I., Green, S., & O'Malley, M. A. (2018). Systems biology and mechanistic explanation. In S. Glennan & P. Illari (Eds.), *The Routledge handbook of mechanisms and mechanical philosophy* (pp. 362–374). Routledge.

Ebeling, W., & Sokolov, I. (2005). *Statistical thermodynamics and stochastic theory of nonequilibrium systems.* World Scientific Publishing.

Falkner, G. (2023). The creative power of the individual memory and the species-specific memory in the development and the evolution of living beings. In S. A. Koutroufinis & A. Araujo (Eds.), *Process-philosophical perspectives on biology: Intuiting life* (pp. 52–76). Cambridge Scholars Publishing.

Falkner, G., & Falkner, R. (2014). The experience of environmental phosphate fluctuations by cyanobacteria. In S. A. Koutroufinis (Ed.), *Life and process. Towards a new biophilosophy* (pp. 73–97). De Gruyter.

Giuliani, A. (2010). Collective motions and specific effectors: A statistical mechanics perspective on biological regulation. *BMC Genomics, 11*(1), 1–13.

Glennan, S. (1996). Mechanisms and the nature of causation. *Erkenntnis, 44*, 49–71.

Glennan, S., & Illari, P. (2018). Varieties of mechanisms. In S. Glennan & P. Illari (Eds.), *The Routledge handbook of mechanisms and mechanical philosophy* (pp. 91–103). Routledge.

Hull, D. L. (1974). *Philosophy of biological science*. Prentice Hall.

Koutroufinis, S. A. (2014). Beyond system theoretical explanations of an Organism's becoming: A process philosophical approach. In S. A. Koutroufinis (Ed.), *Life and process. Towards a new biophilosophy* (pp. 99–132). De Gruyter.

Koutroufinis, S. A. (2017). Organism, process, machine. Towards a process ontology for organismic dynamics. *Organisms. Journal for Biological Sciences, 1*(1), 23–44.

Koutroufinis, S. A. (2019). *Organismus als Prozess. Begründung einer neuen Biophilosophie.* Karl Alber.

Koutroufinis, S. A. (2022). Toward a logic of the organism: A process philosophical consideration. *Entropy, 24*(1), 1–32.

Koutroufinis, S. A. (2023). The *flowing bridge*: on the processual teleology and agency of living beings. In S. A. Koutroufinis & A. Araujo (Eds.), *Process-philosophical perspectives on biology: Intuiting life* (pp. 203–250). Cambridge Scholars Publishing.

Lenoir, T. (1981). Teleology without regrets. The transformation of physiology in Germany 1790–1847. *Studies in History and Philosophy of Science, 12*(4), 293–354.

Machamer, P., Darden, L., & Craver, C. F. (2000). Thinking about mechanisms. *Philosophy of Science, 67*(1), 1–25.
Panning, T. D., Watson, L. T., Tyson, J. J., & Shaffer, C. A. (2007). A mathematical programming formulation for the budding yeast cell cycle. *SIMULATION, 83*(8), 497–514.
Rosenberg, A. (1985). *The structure of biological science*. Cambridge University Press.
Schlosser, G. (1998). Self-re-production and functionality. A systems-theoretical approach to teleological explanation. *Synthese, 116*(3), 303–354.
Toepfer, G. (2005). Teleologie. In U. Krohs & G. Toepfer (Eds.), *Philosophie der Biologie. Eine Einführung* (pp. 36–52). Suhrkamp.
Toepfer, G. (2011). Zweckmäßigkeit. In G. Toepfer (Ed.), *Historisches Wörterbuch der Biologie. Geschichte und Theorie der biologischen Grundbegriffe* (pp. 786–834). J. B. Metzler.
Van Hoek, M. J. (2008). *Evolutionary dynamics of metabolic adaptation*. (PhD Thesis). University of Utrecht.
von Bertalanffy, L. (1971). *General system theory*. Allen Lane The Penguin Press.
Walsh, D. M. (2015). *Organisms, agency, and evolution*. Cambridge University Press.
Walsh, D. M. (2018). Objectcy and agency. Towards a methodological vitalism. In D. J. Nicholson & J. Dupré (Eds.), *Everything flows. Towards a processual philosophy of biology* (pp. 167–185). Oxford University Press.
Whitehead, A. N. (1958). *The function of reason*. Beacon Press.
Whitehead, A. N. (1967). *Adventures of ideas*. Free Press.
Whitehead, A. N. (1978). *Process and reality. An essay in cosmology*. Free Press.

Chapter 7
The Becoming of Identity: A Process-Ontological View on the Relational Co-existence of Biological Beings

Tina Röck

Abstract A fundamental issue when thinking about the agency of organisms is the question of their identity. How could we talk of a biological being's ongoing engagement with the environment, its continued maintenance of homeostasis, its ability to collaborate with other beings and the like, if it did not persist over time, if it did not have a stable identity? I use the term 'biological being' to refer to life at any level of complexity, any level of organisation, any level of integration, i.e., micro- as well as macroscopic nested structures characterising life like cells, organelles, or genes, as well as beings in symbiotic relationships, colonies, multicellular organisms, or even the biosphere. I use the expression 'biological being' instead of 'biological entity' to denote the processual character (being as a gerund implies a process, a 'going on'). Finally, I avoid using the term 'organism' to steer clear of any preconceived notions as to how 'identity' should be defined and what should be the paradigmatic entity, level of organisation, or level of integration from which we can depart to develop the concept of 'identity' in biological beings. The identity of biological beings is important not only in determining their sameness over time (diachronic identity), it is also the basis for distinguishing biological beings from each other at one point in time (synchronic identity).

While some progress has been made in addressing the processual nature of organisms and their diachronic identity, their synchronic identity, or their relational nature, has received less attention. Organisms are not only processual in their existence but also highly relational; they are collaborative, interactive, as well as open and dependent on their environment. This collaborative, open, and relational factor is especially obvious when it comes to beings persisting in mutual holobiotic interdependence, but it applies, to different degrees, to all forms of life. It is this question of how to best conceive of synchronic as well as diachronic identity of biological beings (conceived as processes) that I will address in this contribution. In what

T. Röck (✉)
Department of Philosophy, University of Dundee, Dundee, Scotland
e-mail: t.rock@dundee.ac.uk

J. Švorcová (ed.), *Organismal Agency*, Biosemiotics 28,
https://doi.org/10.1007/978-3-031-53626-7_7

follows, I begin by looking at the philosophical concept of identity, then present a critique of the prevalent metaphysical understanding of 'identity' in the context of biology, and finally introduce a process-based idea of adequate qualitative identity.

Keywords Process philosophy · Identity · Individuation · Uniqueness

Introduction

Considering the insights gained by contemporary biology, the conclusion that organisms persist via change, be it morphological or metabolic (Dupré & Nicholson, 2018; DiFrisco, 2018), seems indubitable. But this view is not entirely novel. The idea of organisms as stable, essentially unchanging entities has long faced considerable challenges. Consider, for example, this quote from Waddington, who described animal life[1] in terms of four layers of processuality in the 1950s:

> An animal functions from minute to minute or from hour to hour, in feeding, digesting, respiring, using its muscles, nerves, glands and so on. These processes of *physiological* functioning may be repeated within periods of time which are short in comparison with the lifetime of an individual animal. But there is an equally important set of processes, of a slower tempo, which require appreciable fractions of the life-history and are repeated only a few times, if at all, during one life cycle; these constitute *development*. Still longer-term processes are those of heredity, which can only be realised during the passage of at least a few generations, and which form the province of *genetics*. And finally, no full picture of an animal can be given without taking account of the still slower processes of *evolution*, which unfold themselves only in the course of many lifetimes *[my emphasis]* (Waddington, 1956: 3–4)

This growing acknowledgement of the dynamic character of life has led to more dynamic and processual forms of understanding living beings (see also Chap. 6). One first big step towards this dynamic reconceptualization of life can be traced back to systems thinking. Biological beings were conceived as complex, self-organising, and self-sustaining biological systems (see Rosslenbroich, 2014; Moreno & Mossio, 2015; Walsh, 2015) that maintain their continued existence by producing both their own parts and their boundaries. This dynamic reading of life relied on a formal or structural account of processuality as expressed in concepts such as 'self-regulation' or 'autopoiesis' (Varela, 1979; Maturana & Varela, 1980). This conceptual shift led to a characterisation of organisms as autonomous and stable self-constructing or self-maintaining wholes constituted through activities, which therefore cannot be fully understood through reductive approaches. However, this approach falls short of a fully processual account, because it also implies a certain structural stability at the heart of biological change, which is tied to continued identity and maintained by systemic closure.

[1] What Waddington describes in terms of the processual nature of animals holds for most, if not all, biological beings at least to a degree.

In a second wave of reconceptualization, this idea of a 'self-contained autonomy' which implies an ultimately contained self-maintaining stable structure was criticised from a process perspective. These critics tend to argue that organisms should not be thought as fundamentally stable auto-poietic systems, but as merely temporarily stabilised and integrated processes that persist through time (Dupré & Nicholson, 2018): 'As process ontologists see it, enduring things are never more than patterns of stability in a sea of process.' (Rescher, 2006: 14) With this further emphasis of the processual aspects and its view of biological beings as merely temporary stabilised patterns, the organisational closure implied by the concept of autopoiesis was problematised. The idea of full autonomy of biological beings was further undermined by research into the fundamental embeddedness of biological beings in their environment and their collaborative integration with other biological beings.

Let me spell out these last claims in more detail. From within the auto-poietic paradigm, any exchange between an organism and its environment tends to be seen as a form of structural coupling between a closed, autonomous, self-maintaining system and its environment:

> This is to say that they [autonomous self-maintaining systems] have the capacity to promote their own existence and to maintain their own structural and functional integrity across a range of internal and external conditions (Thompson, 2007: 44). Because autonomous systems persist by exchanging matter and energy with their environments, they must be coupled with their environments. Coupling is the ability of the system to engage in the kind of reciprocal interactions with its environment that result in its continued viability. (Walsh, 2018: 170)

Here 'coupling' is presented as the ability to collaborate that has no impact on the autonomy of the biological being in question, because there is no structural change involved in this energy exchange or collaboration. The argument is that if a biological being remains structurally unchanged (and thus self-identical), it can be considered autonomous. This reduction of a being's identity to absence of a structural change is, however, highly questionable. There are forms of exchange and of impact that do not show on a structural level but none the less affect the system's ability to maintain itself and that are thus an integral part of the system's continued survival. To say that the system is structurally autonomous from its environment, i.e., that it can in some fundamental way continue to exist without its environment, is to disregard biological reality. Consider that organisms subsist not merely by exchanging matter and energy with their environments, but through metabolism. Metabolism is the process by which an organism synthesizes the materials of which it is made. Through the exchange of matter and energy, the organism builds order internally while decreasing it in its environments (see Chaps. 5, 6, and 8). Such exchange is necessary for them to resist thermodynamic decay (Walsh, 2018: 169).

Through this 'metabolic struggle against its thermodynamic predicament an organism creates and individuates itself' (Walsh, 2018: 169). Note that in this context, the term 'environment' does not denote the 'objective world' but the affordances present for living beings to engage with (Lewontin, 1978). This ultimately means that there is no organism and no individuation without an environment,[2] which renders the environment essential for the identity and continued existence of organisms. This approach is congruent with the idea of *Umwelt*, a term coined by the Baltic German biologist Jakob von Uexkül (1909, 1926). *Umwelt* denotes the world as it is subjectively experienced. It is thus a world shaped by the sensory organs of the perceiving organism, its physiological makeup, and its constant interactions with the environment (Chaps. 8 and 13). According to Uexküll, each species has its own characteristic *Umwelt*, which determines how it perceives and responds to its environment. This individualised *Umwelt* co-creates the individuation of organisms. Biological individuation is thus an ongoing process that implies more than a simple coupling between an independent, organisationally closed entity and its material environment. In Uexküll's view, organisms are not passive recipients of environmental stimuli but active participants in constructing their subjective reality. The concept of *Umwelt* expresses the idea that the world is not some clearcut, unrelated, objective reality. Instead, it is full of subjective, individual worlds shaped by the unique characteristics and history of each organism. This fact, i.e., that more or less stable constellations of biological processes are inherently open to and dependent on their environment, is explicitly acknowledged for instance in certain strands of systems biology (Noble, 2006, 2017), especially where it is influenced by Rosen's relational biology (Rosen, 1984).[3]

Biological individuation is thus a process that implies an inherent openness to and exchange with the environment, not only regarding the physical or material environment but also the biological environment, as evolutionary research continues to reveal. Contemporary evolutionary researchers tend to conceptualise organisms as heterogeneous assemblages of elements that are to different degrees collaboratively integrated (Niklas & Newman, 2020). A wide variety of studies on the coevolution of different species (Wisz et al., 2013), multi-cellular organisms (Niklas & Newman, 2016), and symbionts (Gilbert et al., 2012), have provided further evidence for the all-pervasive presence of relational biological existence and the need for an updated relational and open understanding of biological identity. And to be sure, one can find innumerable examples of integrated (i.e., fundamentally related) coexistence in mutual holobiotic interdependence (rhizosphere, mycorrhizae, insects and bacteria, corrals, algae, and so on) in all types of taxa, from plants to vertebrates (Markoš & Švorcová, 2019).

This fundamental role of relationality is often (implicitly) acknowledged in process biological accounts. Consider the following definition:

> We understand living things to be most fundamentally the consequences of numerous interweaving (occasionally nested) processes. Although it is common to describe the domain of

---

[2] Just for the sake of completeness, it is also the case that there is no environment without an organism. They are in a co-constitutive relation to each other.

[3] While Rosen's approach is very interesting, it is abstractive. The aim is to model all relational structures constituting an organism as a whole (Rosen, 1984). As will become clear in this paper, this is a problematic set of presuppositions, especially the question of what constitutes one organism as one.

> biology as consisting of things, for example organisms, cells, genes, and so on, we understand even these as ultimately processual. (Bapteste & Dupré, 2013: 380)

If biological beings are processes engaged in continuously maintaining homeostasis *in constant communication, connection, and exchange* with their internal as well as external environment, as the above expression 'interweaving (occasionally nested)' suggests, then to consider organisms as individual, unrelated agents, or as completely separate identities in the traditional sense is, strictly speaking, impossible. In this contribution, I will take these discoveries into account to develop a concept of 'identity' adequate to grasp the identity of biological beings that are 'more or less stable [processual] constellation[s] with fuzzy and fluctuating boundaries' (Meinke & Dupré, 2020: 3).

Biological Identity

In their contribution *A Symbiotic View of Life: We Have Never Been Individuals,* Gilbert, Sapp, and Tauber summarise the problem of biological individuality or identity[4] in the following words:

> There are many ways in which the term 'individual' is used in biology. Individuals can be defined anatomically, embryologically, physiologically, immunologically, genetically, or evolutionarily (see Geddes and Mitchell 1911; Clarke 2010; Nyhart and Lidgard 2011). These conceptions, though, are not wholly independent of one another. Nor have these definitions of individuality often been explicitly articulated as such. Indeed, even in biology today there is a dearth of definition in what constitutes the individual organism. Still, definitions are implied, and each stems from the common tenet of genomic individuality: one genome/one organism. As such, all classical conceptions of individuality are called into question by evidence of all-pervading symbiosis. (Gilbert et al., 2012: 327)

I would argue that this diagnosis, *mutatis mutandis*, is still applicable to contemporary biological discourse.

Let me thus begin this investigation into 'identity' by introducing Leibniz' famous definition of identity as identity of indiscernibles: there cannot be two numerically different entities that have all properties in common. This is usually interpreted to mean that in order to claim that y and x are numerically identical, all properties would need to be the same. This framework presupposes a distinction between numerical and qualitative identity. It is important here to note that qualitative identity allows for degrees of sameness, while numerical identity implies absolute qualitative identity (identity of indiscernibles), because absolute qualitative identity amounts to numerical identity.[5] A being can be in a sense qualitatively the

[4] While I will introduce a distinction between individuality and identity when I develop my proposal for a processual conception of identity below, for now I will use these terms interchangeably.

[5] Strictly speaking numerical identity can only hold between a physical entity and itself at one point in time. Abstract objects, being a-temporal, are not limited by such a temporally localised form of

same but numerically different, while something is numerically identical (i.e., it can be counted as one) if and only if there is just one entity.

From this account alone, it obvious that it is difficult to apply this understanding of identity as being indiscernible rigorously when referring spatiotemporally located entities in general, since at least their temporal location changes and renders the entity at an earlier point in time discernible from the entity at a later point in time. This is even more problematic in the case of biological beings whose persistence is constituted via change, either morphological or metabolic. To ameliorate this difficulty, this stance is generally supplemented by the neo-Aristotelian claim that there is an unchanging ontological basis or ground that retains all properties and is thus numerically identical, even when the biological being is apparently changing. This basis could be a substance, a structure, a form, a set of properties, or something similar that remains unaltered through all changes. This additional postulate allows for a metaphysically rigorous account of identity but renders all observable changes merely apparent or insubstantial. This elegant solution to the problem of identity and change thus comes at a high cost: it leads to the conclusion that the phenomena we can observe do not ultimately reveal the actual identity of any biological being. In fact, only in abstracting from what we can observe can we infer the being's unchanging and conceptual identity.

Another possible way to understand identity in the context of biology is to take it to refer to what allows us to re-identify something as that same entity: 'the identity of X is what accounts for X remaining X across time, since "identity", in our ordinary discourse, includes identity through time' (Huneman, 2021: 42). Here identity is not necessarily tied to an unchanging basis or core but to the epistemic procedure of determining reliable identity criteria that allow for re-identification. While this might sound quite straight forward, in biology this procedure is not as easily applied as it might seem. It is not always clear what the countable unit at hand is and thus what can or should be taken as the basis for re-identification:

> Seeing many dandelions in a field, as Janzen (1977) famously described, would lead a lay-person to identify dozens of individuals there, individualised by the body of each visible plant: however, their genetic sameness entitles some biologists to say that there is a single individual here. Clearly, the identity conditions for each of those dandelions would differ according to those views. (Huneman, 2021: 40)

This problem of determining the appropriate criterion or level to introduce countability or numerical identity is exacerbated by the levelled or nested nature of life: 'genes assemble within cells, cells assemble into multicellular organisms, organisms into demes (if one considers the relations of interactions) or into species (if one considers their relations of genealogy; Eldredge 1985)' (Huneman, 2021: 43). Thus, the question of identity becomes the question of what level of assemblage or what degree of integration can or should be counted as a unit to provide the basis for re-identification. Should 'identity' as countability in this sense be developed in

---

self-identity.

departing from the level of genomes, cells, organs, organisms, symbiotic unities, or even fields?

These short comments show how complex the concept of 'identity' in the context of biology is. These difficulties become only more pronounced if we do not look toward common individual organisms like butterflies, dandelions, or lions as paradigmatic examples for identification but try to account for the most prolific forms of biological existence, namely multi-organism assemblages such as colonies, symbionts, mutualistic associations, communities, or biofilms. These forms of life seem to form stable but temporary networks that allow for interactions and confer an evolutionary advantage. While these sorts of temporary forms of stabilisation can be well accounted for in the process perspective that I am taking, it is difficult to ascribe any form of individuality to them. What allows us to count a colony or its members as one or to distinguish symbionts?

With these more complex but all-pervasive forms of life in mind, a further potential answer to the problem of identity has been proposed, namely to ground biological identity in functional approaches (Bouchard, 2009, 2010). Many contemporary functional approaches to identity are rooted in Wilson and Sober's notion of biological individuality, which is couched in terms of functional integration and common fate (Wilson & Sober, 1989). Bouchard paraphrases this type of account as follows: 'A biological individual is a functionally integrated entity whose integration is linked to the common fate of the system when faced with selective pressures from the environment' (Bouchard, 2018: 188). Here identity is rooted in a processual account and is thought to be the result of continuous selection through evolutionary processes. Nevertheless, 'functional integration and common fate can be achieved at various levels of organisation, one can have overlapping individuals operating at different temporal scales and with different levels of transiency and continuity' (Bouchard, 2018: 190). Which leads us back to the problem of the stratified nature of biological beings we encountered above.

One possible answer to this overwhelming complexity when it comes to biological identity might be that within the realm of biology, there simply are different kinds of individuals, identifiable by different identity criteria. It may be thus impossible to formulate one overarching definition that would hold for all kinds of entities, on all levels, and for every type of investigation.

How should we determine biological beings and their individuality or identity in that case? One solution could be that the relevant identity criterion is determined by the specific research interest shaping the investigation:

> Consider a lichen. This consists of a fungus (a multicellular organism) and a very large number of photosynthetic bacteria. Presumably all of these, the fungus, all of the bacteria, and the lichen itself are individuals. What about the population of bacteria? Or subpopulations with a particular mutation? Or what about the lichen-covered tree of which this particular lichen is part? (Dupré, 2021: 39)

As this example demonstrates, identity can be seen as a function of the investigation itself. Different accounts of what is to be counted as an 'individual', or to be considered identical, become relevant or adequate depending on the object or issue investigated. If the object of investigation is the evolution of bacteria within a lichen, the individuality of the bacteria might be the more relevant unit. If, however, the object of investigation is the interaction between tree and lichen, it might be more adequate to disregard the identity of the bacteria and focus on the tree and the lichen as two individuals. And if, finally, the question is how the lichen-tree reacts to environmental changes, the lichen and the tree might most adequately be considered as a single individual. John Dupré refers to this position, which allows us to integrate or distinguish different sets of processes (and thus allows different ways of spelling out 'identity') depending on our investigative aim and further pragmatic considerations, 'promiscuous individualism'. He states that 'the point is not that there are no boundaries suitable for delineating individuals, but that there are too many' (Dupré, 2021: 39). While I hold this to be the case, this promiscuous answer to defining 'identity' in biology is an answer many philosophers of biology and especially metaphysicians interested in biology would find insufficient.

Consider the following quote that clearly attempts to propose an objective level on which identity (i.e., the organism) should be situated and draws an objective distinction between what belongs to a unit (its identity) and what does not (as mere part of the unit), irrespective of the issue investigated:

> For, at least on the traditional understanding of these things, to be a part of an organism is precisely *not* to be an organism, to be an organism is *not* to be a collective, and so on. My gut bacteria and I live in a symbiotic relationship, but it would be plain bad science to say that the bacteria were literally *part* of me – organs like my kidneys or liver, or parts like my skin or blood cells. A lion might belong to a pride but is not *itself* a pride, nor a member of a school of fish itself a school, nor is any organism literally *composed* of other organisms, and so on. (Oderberg, 2020: 23)

To address this kind of objection to the idea of 'promiscuous identity', I will qualify my support for it by introducing a further distinction. This is the distinction between concrete, actual living beings (*tode ti*) on the one hand, and concepts, categories, substances, and the like, that are used to refer to or to identify living beings as organisms or individuals, on the other hand. The argument will be that it makes a difference whether we investigate the relation between the concrete singular *me* and *my* gut bacteria, or a 'lion' and its 'pride' on a general or conceptual level. Here, too, the aim of the investigation has an impact: the shift from investigating a concrete living being's embodied relation (investigating me and my gut) to a logical relationship between concepts (investigating the concept 'is part of', and determining that in this context it is to be read as denoting a part–whole relationship and not as denoting a constitutive element of a symbiotic whole) has an impact on what answers to the question of identity are relevant, or even possible.

In what follows, I will introduce this additional distinction between an investigation focused on the concrete singular *tode ti* and an investigation focused on the general or conceptual relations using Christopher Austin's arguments against a processual approach to biological identity (Austin, 2020). I use Austin's arguments as a paradigmatic example of a confusion between these levels of investigation not because he is alone in confusing them, but because in his paper this distinction is hinted at but not actually considered in the discussion.

Can Process Ontology Provide a Robust Notion of Identity?

From a process-ontological perspective, the world appears to be fundamentally related, be it through causal links, evolutionary stages, or via other processes: 'What makes an egg, a caterpillar, a pupa and a butterfly stages in the career of a single organism are the causal links connecting the stages, not an intrinsic property they share in common' (Dupré, 2021: 45). Within this framework, it makes sense to tie identity to this processuality itself: a living beings' diachronic identity is then its development over time, whereby the 'what it is now' is the result or a product of what it was before plus the changes it underwent, while its synchronic identity is defined through every single relation it stands in. Austin argues that this is not a metaphysically robust notion of identity. He contends that simply describing a living being's relations and its development over time – as proposed by processual approaches – is not sufficient to answer *what that living being is fundamentally*. To him 'the identity of an organism is determined by (I) a set of properties it could not fail to possess as long as it exists which (II) define and delimit its range of possible developmental trajectories and determine its characteristic mereological-cum-morphological features' (Austin, 2020: 12).

He characterises process ontology as defining an organism as 'a life-cycle-process – that is, an autopoietic activity which produces a four-dimensional pattern of characteristic property possession' (Austin, 2020: 12). But he argues that this is not sufficient, that process ontologies, POs) fall short of being able to account for *what it is to be* such a life-cycle-process. While, as he admits, the identity of a singular empirical phenomenon, a concrete individual process, can be described well using the process-ontological paradigm (as this specific lifecycle), the underlying structure, substance, or essential properties cannot be accounted for without a substantialist (SO) reconceptualization:

> …while PO's answer to the organism question [i.e. determining its identity in terms of a life-cycle-process] is capable of saving the phenomena – of sufficiently cataloguing the natural world by sufficiently distinguishing this organism from that one – it is incapable of explaining the phenomena – of producing that catalogue by its own means. As I illustrate in what follows, PO can only successfully perform that important latter task by appealing to the central principles of SO. (Austin, 2020: 13)

Austin's claim is thus that process ontology can account for the phenomena but not for the conceptual structures that shape our scientific, philosophical, and linguistic approaches. Let me investigate this issue in more detail by looking at an Aristotelian distinction at the heart of this discussion. Aristotle famously proposes different potential answers to the question *to ti hen einei* (what it was to be this), and therefore to what can ground identity in a metaphysically robust way. The most relevant candidates he discusses are the substance/essence (*ousia*), substance/subject (*hypokeimenon*), the universal (*kath'holou*), matter (*hyle*) as well as a composite of matter and form (*hyle* and *morphe,* see also Chap. 2), i.e., the concrete 'this' (*tode ti*). In what follows, I will look at three main options relevant for this paper, namely *ousia* (substance as essence), *hypokeimenon* (substance as subject), and the *tode ti* (this there).[6] The first concept, *ousia*, represents the kind of ontological essential property, structure, or feature Austin is looking for when he searches for a metaphysically robust grounding for identity. The second one, the *hypokeimenon,* is generally used by Aristotle in the context of predication (it is defined as that which cannot be predicated of another but of which properties can be predicated). In other words, this is the substance as a (grammatical) subject. During his argumentation, Austin sometimes combines these first two aspects of ontology and language (and he is in great company doing so: many interpreters have tried to unify these accounts), by assuming an analogy between the way we *conceptualise* and *talk* about something (in this case 'identity') with its *metaphysical grounding in reality* (i.e., the *ousia*). The third aspect I want to draw attention to here is the *tode ti* (the 'this there' or 'a this'). This term denoted here is the material process, the object of an ostensive definition, the being we can pick out by pointing towards it. This aspect does also feature in Austin's account: it appears in the form of the experientable phenomenon.[7]

The merging of all three layers shapes most (neo-Aristotelian) metaphysical discussions around a processual and relational view of biological identity, because within most substantialist frameworks the substance (both as essence and as subject) is what is thought to guarantee the identity of the *tode ti*, the actual concrete particular. Aristotle enabled this confusion by arguing that the substance (in either interpretation) is *per definitionem* also a concrete particular. That this is actually the case, i.e., that a substance (or a substantial definition) is as concrete as the actual concrete being, remains doubtful. In any case, this position *de facto* implies that knowing a being's substance is sufficient to fully know all relevant aspects of the being in question, any information not included in its substantial definition is *ex definitione* irrelevant. The *tode ti*, all the concrete material evolving and changing aspects that are the biological being, cannot provide more *relevant* or *important* information about its (self-)identity than is already contained in its conceptual identity, i.e., its substance. Process-ontological approaches show how limited this thinking, which merges or confuses an investigation of the *substance* with an investigation of the *tode ti,* is. They point out that this confusion leads to an overemphasis of the role of the abstract and the conceptual when it comes to accounting for identity in concrete (biological) beings. It is this fusion of layers of investigation that Whitehead terms the *fallacy of misplaced concreteness*. This fallacy emerges when conceptual accounts are investigated as if they were the concrete entities, that is, as if concepts, however well formed, could stand in for the concrete individual currently under investigation without loss of relevant information or potentially relevant complexity.[8]

[6] To be sure Aristotle uses *tode ti* to characterise substance in places as separable and a this there (*chôriston kai tode ti*), but generally it is used to point towards a concrete phenomenon a this there (i.e., the morpho-hyletic compositum).

[7] For a detailed discussion of various modes of interpreting the *tode ti* and its use by Aristotle see Corkum (2019).

[8] A similar account can be found in Markoš and Švorcová (2009).

But what if we did not commit the fallacy of misplaced concreteness and did abstract from the concrete being when it comes to the question of identity? If all information and complexity present with the continuously evolving and processual *tode ti* were considered relevant to our account of individuality, we would need to conclude that a concrete organism's identity is determined by every single relation it stands in, including relations to its past becoming, its environment, what the currently evolving being encounters, what it incorporates, and so on. Not a single one of these factors can be disregarded when it comes to determining this particular biological being's identity as a whole, this organismal *tode ti*. While this level of precision might not yield a scientifically useful concept of identity, or a metaphysically robust one, it describes the actual identity of any living being.

Naturally, this level of concrete precision in our concept of identity explodes the concept and thus renders it useless – for certain purposes. Just like a map that includes every single geographic detail, down to the most minute ones, becomes a copy instead of a tool, such concrete description of identity loses its categorical usefulness or explanatory power, just like Austin argued. However, acknowledging the reality of identity on the level of the *tode ti* does not necessarily affect the more conceptual substantialist approaches via *ousia* or *hypokeimenon*. Since these approaches abstract from some of the complexity present in the concrete dynamic and relational nature of *tode ti*,[9] they remain *true enough*. It merely affects the certainty with which we can claim to have arrived at an absolute or final definition of 'identity' in the realm of biology. Furthermore, while I grant Austin that substantialist approaches are much more powerful in enabling scientific engagement and conceptual ordering, I would also argue that this is not the only possible investigative interest we might have. We might also want to investigate how concrete reality exists and how this can help us creatively expand the knowledge we already have. To confuse the scientifically sound and useful descriptions with actual reality (and the potential creative advance its descriptions can initiate) is again to commit the fallacy of misplaced concreteness, because one is not a substitute for the other – and this holds both ways.

Therefore, it seems sensible to adopt John Dupré's concept of promiscuous individualism for the purpose of scientific discovery, while keeping in mind that the 'identities' generated in this manner are the result of a selection of identity criteria or identity categories that is guided by specific investigative aims. The results of the selection might thus be highly useful and true enough, but they are not necessarily congruent with actual reality or fundamental in any way. At this point, I would like to return to Bouchard's functional definition of identity introduced above in order to provide an (implicit) example of such an approach. Bouchard argues that this definition has a dual aim. It is intended both as an operational definition that helps us make sense of certain scientific projects as well as a 'metaphysical claim about what inhabits the universe independently of our needs and uses for certain types of

[9] As Aristotle already argued in Metaphysics Z, while substances are definable (and thus knowable at least to a degree), it is impossible to define the *tode ti*, i.e., *it is impossible to fully conceptualise it.*

scientific explanations' (Bouchard, 2018: 188). In this quote, he distinguishes two relevant levels of investigation, namely the level of scientific knowledge and actual metaphysical reality, and he considers his answer to be relevant to both. However, at a later stage in the same text he argues that operational definitions can diverge from metaphysical truth without conflict: it might be the case that 'individuals are real enough [for the purposes of scientific research] and should retain a prominent role in our explanations' (Bouchard, 2018: 188). I would argue that for many research questions, a full process-ontological account would involve far too much uncertainty, complexity, ambiguity, interrelatedness, and dynamicity to provide useful general categories for our scientific investigations. Therefore, there is no need to eliminate more substantial conceptualisations, where those concepts are appropriate for the issue investigated, because certain definitions of identity, individuality, and what it means to be an organism, are *real enough* to be helpful in scientific discourse.

However, I would also argue that the metaphysician should remain concerned with both, that is, with what is actually real and with understanding why and in what regard certain conceptions are real enough to be helpful or useful to scientific discovery. Furthermore, it is the metaphysician's job to keep developing new concepts and constructs that allow for further (potentially more) adequate renderings of the *tode ti*, concepts that might be helpful for novel scientific questions while acknowledging that it is ultimately impossible to reach the full reality of the *tode ti*. The metaphysician's contribution to scientific understanding is thus to provide a critical reflective interrogation of modes of conceptualising that are real enough, to aid in interrogating whether the concepts are fit for a specific investigative aim or purpose, and to continuously develop concepts that are potentially more adequate than those currently available.

Developing a Useful Concept of Concrete Identity

Let me now take on this role of the metaphysician and try to develop the concept of 'identity' further in an attempt to develop a concept of 'identity' that is scientifically useful and while providing a more accurate description of the identity of actual biological beings as merely stabilised (potentially collaborative) processes with fuzzy borders, fundamentally open to their environment.

Traditionally 'identity' has been conceived as that which renders an entity this specific entity (ontologically) and as a marker that allows us to differentiate one entity from another (conceptually/epistemologically). Furthermore, 'individuality', which fundamentally names a countable unit of some kind (Lowe, 2009), and 'identity' have been considered largely synonymous. However, this association between individuality and identity is not necessary:

> Not every entity that does or could exist is an individual object of some kind, although every entity is necessarily self-identical. An individual object is an entity which, quite literally, counts as one entity of some kind and for that to be the case, it must have unity. Only unitary entities can qualify as individual or single objects, capable in principle of being enumerated

> along with other such objects. There are or could be entities which lack unity, which we might call 'dividuals', as opposed to individuals. Putative examples of such entities are quantities of homogeneous and infinitely divisible matter. (Lowe, 2003: 335)[10]

Such 'dividuals' would be characterised by identity (at the very least by self-identity) without implying a fundamental unity. Lowe's putative example has two characteristic features, namely uniformity (homogeneous) and the assumption that there is no ultimate basic unity (infinitely divisible). This can be transposed to the realm of biology in terms of 'promiscuous individuality' to provide a second example: biological beings are fundamentally interrelated processes that can be grouped into individuals in various (potentially infinite) ways without any ultimate base unit that the various forms of individuality could be reduced to.

How does this change our concept of identity? 'Identity' is often defined as a predicate that *identifies* one being and *differentiates* it from others. The conceptual shift implied by a disassociation of identity and unity allows for a view of identity as naming what makes this one being unique irrespective of unity. All the qualities that make one concrete being unique are what grounds its identity. The dissociation of 'identity' from apparent cognates such as 'unity' and 'individuality' thus renders a different, more processual, and relational view of identity possible, because it moves our focus from numerical identity to degrees of qualitative identity. In this reading, identity is then qualitatively determined as those properties or relations that make a being unique. These qualities can change, be exchanged, or transform without affecting its uniqueness. 'Uniqueness' here should not be read as an additional quality, but rather as a placeholder for all the relations a concrete being stands in and thus for what actually constitutes the unique character, i.e., the identity of the *tode ti*, over time. Read in this way, 'uniqueness' emphasises distinction (How does it relate?) over unity (What is it?) and qualitative–relational continuity over numerical identity. The concept of 'qualitative identity' thus is an attempt to emphasise the fundamentally relational nature of identity, which is often disregarded in unitarian/numerical definitions of identity because it denotes i) how this being relates to itself (self-identity), i.e., what makes it unique, and at the same time ii) how it relates to others, i.e., how it differs in its uniqueness. This very basic definition already shows how intimately qualitative identity is connected to relationality.

Even just on a minimal level, by virtue of being physical, of being spatiotemporally located (whether that location is extended or not), any actual biological being is necessarily unique. Even a cloned being is distinct from its clone: at any given time, it has at least a different material composition and therefore a different spatial location. If we thus focus on 'qualitative identity', the term can still be used to refer to or determine what an entity is (what makes it unique) by focusing on how exactly it differs from other entities. Consequently, 'qualitative identity' can still account for i) what it is and ii) how it differs from others, which is why it should suffice as a

[10] Consider also: '[…] a principle of individuation, we might say, is not so much a criterion of *identity* as a principle of *unity*: countable items are singled out from others of their kind in a distinctive way that is determined by the sortal concept under which they fall [...]' (Lowe, 2001: 33).

basis for iii) synchronic qualitative identity and iv) diachronic qualitative identity. To paraphrase the famous Hegelian dictum: qualitative identity is the identity of self-identity (uniqueness) and non-identity (difference).

Conclusion

The concept of 'qualitative identity', if divorced from numerical identity or unity, can provide a more adequate concept for scientific research focused on uncovering the essential role of collaboration, co-evolution, and symbiotic relationships than more traditional approaches to organismal identity. Twin studies are a good example for a further possible application of this tension between identity and uniqueness; identical twins have the same genome but the respective epigenetic changes that result from the twins' exchange with concrete environmental influences create uniqueness beyond genomic identity. Still, these possibilities do not imply that other concepts of identity should be eliminated in favour of this qualitative understanding of uniqueness. They remain valuable for investigations that presuppose and rest on clear and final distinctions between individuals or between essential structures and accidental presentation.

References

Austin, C. J. (2020). Organisms, activity, and being: On the substance of process ontology. *European Journal for Philosophy of Science, 10*, 1–21.

Bapteste, E., & Dupré, J. (2013). Towards a processual microbial ontology. *Biology & Philosophy, 28*, 379–404.

Bouchard, F. (2009). Understanding colonial traits using symbiosis research and ecosystem ecology. *Biological Theory, 4*(3), 240–246.

Bouchard, F. (2010). Symbiosis, lateral function transfer and the (many) saplings of life. *Biology & Philosophy, 25*(4), 623–641.

Bouchard, F. (2018). Symbiosis, transient biological individuality, and evolutionary processes. In D. Nicholson & J. Dupré (Eds.), *Everything flows: Towards a processual philosophy of biology* (pp. 186–198). Oxford University Press.

Corkum, P. (2019). This. *Ancient Philosophy Today, 1*(1), 38–63.

DiFrisco, J. (2018). Biological processes. Criteria of identity and persistence. In D. Nicholson & J. Dupré (Eds.), *Everything flows. Toward a processual philosophy of biology* (pp. 76–95). Oxford University Press.

Dupré, J., & Nicholson, D. (2018). A manifesto for a Processual philosophy of biology. In D. Nicholson & J. Dupré (Eds.), Everything flows: Towards a Processual philosophy of biology (pp. 3–45). Oxford University Press.

Dupré, J. (2021). *Metaphysics of biology*. Cambridge University Press.

Gilbert, S. F., Sapp, J., & Tauber, A. I. (2012). A symbiotic view of life: We have never been individuals. *The Quarterly Review of Biology, 87*(4), 325–341.

Huneman, P. (2021). Biological individuals as "weak individuals" and their identity. Exploring a radical hypothesis in the metaphysics of science. In A. S. Meincke & J. Dupré (Eds.), *Biological identity. Perspectives from metaphysics and the philosophy of biology* (pp. 40–62). Routledge.

Lewontin, R. C. (1978). Adaptation. *Scientific American, 239*(3), 212–230.
Lowe, E. J. (2001). *The possibility of metaphysics. Substance, identity and time*. Clarendon Press.
Lowe, E. J. (2003). Identity, individuality, and unity. *Philosophy, 78*(305), 321–336.
Lowe, E. J. (2009). *More kinds of being: A further study of individuation, identity, and the logic of Sortal terms*. Wiley-Blackwell.
Markoš, A., & Švorcová, J. (2009). Recorded versus organic memory: Interaction of two worlds as demonstrated by the chromatin dynamics. *Biosemiotics, 2*, 131–149.
Markoš, A., & Švorcová, J. (2019). *Epigenetic processes and the evolution of life*. Taylor & Francis.
Maturana, H., & Varela, F. (1980). *Autopoiesis and cognition: The realization of the living*. D. Reidel Publishing Company.
Meinke, A. S., & Dupré, J. (2020). Biological identity. Why metaphysicians and philosophers of biology should talk to one another. In I. A. S. Meinke & J. Dupré (Eds.), *Biological identity: Perspectives from metaphysics and the philosophy of biology* (pp. 1–21). Routledge.
Moreno, A., & Mossio, M. (2015). *Biological autonomy. A philosophical and theoretical enquiry*. Springer.
Niklas, K. J., & Newman, S. A. (2016). *Multicellularity. Origins and evolution*. The MIT Press.
Niklas, K. J., & Newman, S. A. (2020). The many roads to and from multicellularity. *Journal of Experimental Botany, 71*(11), 3247–3253.
Noble, D. (2006). *The music of life. Biology beyond genes*. Oxford University Press.
Noble, D. (2017). *Dance to the tune of life: Biological relativity*. Cambridge University Press.
Oderberg, D. S. (2020). Siphonophores: A metaphysical case study. In A. S. Meinke & J. Dupré (Eds.), *Biological identity. Perspectives from metaphysics and the philosophy of biology* (pp. 22–39). Routledge.
Rescher, N. (2006). *Process ontological deliberations*. Ontos.
Rosen, R. (1984). Relational biology and the origin of life. In K. Matuno, K. Dose, K. Harada, & D. L. Rohlfing (Eds.), *Molecular evolution and protobiology* (pp. 421–431). Plenum Press.
Rosslenbroich, B. (2014). *On the origin of autonomy. A new look at the major transitions in evolution*. Springer.
Thompson, E. (2007). *Mind in life: Biology, phenomenology, and the sciences of mind*. Harvard University Press.
Varela, F. J. (1979). *Principles of biological autonomy*. North-Holland.
von Uexküll, J. (1909). *Umwelt und Innenwelt der Tiere*. Springer.
von Uexküll, J. (1926). *Theoretical biology*. Harcourt, Brace & Co.
Waddington, C. H. (1956). *Principles of embryology*. George Allen and Unwind.
Walsh, D. M. (2015). *Organisms, agency, and evolution*. Cambridge University Press.
Walsh, D. M. (2018). Objectcy and agency: Towards a methodological vitalism. In D. Nicholson & J. Dupré (Eds.), *Everything flows: Towards a processual philosophy of biology* (pp. 167–185). Oxford University Press.
Wilson, D. S., & Sober, E. (1989). Reviving the superorganism. *Journal of Theoretical Biology, 136*(3), 337–356.
Wisz, M. S., Pottier, J., Kissling, W. D., et al. (2013). The role of biotic interactions in shaping distributions and realised assemblages of species: Implications for species distribution modelling. *Biological Reviews, 88*(1), 15–30.

[illegible] (20[illegible]). [illegible] Naturalis [illegible] 22(2), 1–8.

[illegible]

[illegible]

[illegible]

[illegible] (2018). [illegible]. *[illegible]*, [illegible] (13), 1–18.

[illegible] (19[illegible]). [illegible]

[illegible]

[illegible] (2019). [illegible]

Marone, A. & Mayo, [illegible] (20[illegible]). [illegible]. Springer.

[illegible] (20[illegible]). [illegible] Press.

[illegible] (20[illegible]). [illegible]

[illegible]

[illegible]

[illegible] (20[illegible]). [illegible]

[illegible] (2015). [illegible]

[illegible] (20[illegible]). [illegible]

[illegible]

Thompson, [illegible] (20[illegible]). [illegible]

[illegible]

[illegible]

[illegible]

Walsh, D. M. (20[illegible]). *Organisms, agency, and evolution*. Cambridge University Press.

Walsh, D. M. (20[illegible]). [illegible] and agency: Towards [illegible]. In D. Nicholson & [illegible] (Eds.), [illegible] (pp. [illegible]). Oxford University Press.

Williams, L. [illegible] (20[illegible]). Revealing [illegible]. *[illegible]*, [illegible].

Wisz, M. S., Pottier, J., Kissling, W. D., et al. (2013). The role of biotic interactions in shaping distributions and realised assemblages of species: Implications for species distribution modelling. *Biological Reviews, 88*(1), 15–30.

Part II
Agency in Biology

Chapter 8
(Bio)Semiosis as Life-Specific Form of Agency

Anton Markoš

What absorbs me in these birds, beyond their beautiful whiteness, their astounding numbers, the great vigor of their lives, is how adroitly each bird joins the larger flock or departs from it. And how each bird while it is a part of the flock seems part of something larger than itself. Another animal. [...] I never saw two birds so much as brush wingtips in the air, though surely they must. They roll up into a headwind together in a seamless movement that brings thousands of them gently to the ground like falling leaves in but a few seconds.
Barry Lopez, 1986: 154–155

Call a system able to act on its own behalf in an environment an "autonomous agent." All free-living cells and organisms are autonomous agents. But a bacterium is "just" a physical system. In its Kantian form, my core question became, What must a physical system be such that it can act on its own behalf?
Stuart Kauffman, 2000, x

Abstract This text is an attempt to bring into harmony various models aimed at characterising the realm of the living, from purely physical notions such as dissipative structures, through autonomous agents, up to analogies with human cultures. Life is characterised by innerness and the ability to extract meaning from both the surrounding world and mutual understanding of different life-forms in the biosphere, while drawing upon the memory and experience of individuals and/or communities. Individuals as well as entire communities interpret all such inputs and decide upon their ways of living in the present, as well as in the nearest future.

A. Markoš (✉)
Department of Philosophy and History of Science, Faculty of Science, Charles University, Prague, Czech Republic
e-mail: anton.markos@natur.cuni.cz

J. Švorcová (ed.), *Organismal Agency*, Biosemiotics 28,
https://doi.org/10.1007/978-3-031-53626-7_8

Keywords Autonomous agent · *Umwelt* · Innerness · Semiosis and evolution

Introduction

From time to time, a catchy concept – such as 'system', 'self-assembly', or 'hologram' – grips the scientific community and for a moment it seems to work like magic, capturing and naming as if the most general, holistic principle in the universe. 'Agency' used to be very much in vogue, only to recently give way to 'active matter'. The latter is a good example of such concepts: it concerns systems consisting of high numbers of constituents. By consuming and dissipating incoming energy, systems maintain themselves far from an equilibrium and moreover, they behave like a *coherent* whole. Their principal characteristic is *cooperativity*, that is, the ability to react as a whole (see Chap. 4). This feature can be found at many levels of description, from (bio)polymers to cell assemblies (multicellular bodies, colonies), all the way up to flocks of animals, schools of fish, and perhaps even entire ecosystems. What all such systems have in common is that they can be modelled using a small set of common rules: it is as if they followed shared 'physical laws'.

In the following, I first discuss some of such collective domains of description and/or organisation. At first glance, one can intuitively distinguish two groups within such cooperative entities: (1) physical systems, including cooperative biopolymers (e.g., proteins such as haemoglobin tetramer), and (2) living beings. Nonliving systems are *dissipative*. Life, on the other hand, is characterised by *innerness* (see Chap. 9) that allows it to be in power of deciding about the resulting collective behaviour, such as when a school of fish consisting of thousands of individuals abruptly changes the direction of its movement. Later in the text, I compare these two groups of systems but first, let us take a closer look at the state of null agency.

Level Zero: Static Phenomena

In our analysis, a 'null system' is a system in a state of thermodynamic equilibrium. The probability of leaving such a state spontaneously is extremely low despite the fact that the system may contain substantial amounts of *internal* (or *latent*) *energy*. For example, undisturbed atmosphere (at 1000 hPa, 27 °C) contains about 2.8×10^5 J/m^3 in the form of kinetic energy of its molecules. This corresponds (in each cubic meter of air!) to the kinetic energy of a 1,000 kg car moving at the speed of 85 km/h. In comparison with this, the kinetic energy of winds, hurricanes, raising gas columns, and similar collective phenomena (as discussed in the following section) is negligible, comprising about 0.5% of overall energy content of the atmosphere (Seinfeld & Pandis, 1998).

Typical speed of a nitrogen molecule in an undisturbed atmosphere as defined above is about 728 m/s (2600 km/h).[1] Molecules, however, move randomly and engage in mutual collisions every 10^{-10} s, meaning that this is also how often they change their direction and speed. As a result, the average speed of a molecule in gas is zero.[2] Despite this feverish whirling of zillions of constituents (internal energy), *nothing happens* in gas when it is in a state of equilibrium. A change of parameters can be imposed only from the outside, by adding or removing energy. Unless this happens, its microscopic (speed of molecules) and macroscopic (temperature, pressure) parameters remain constant. In other words, the gas 'does nothing'. In solid objects, such as stones, the situation is similar. As in gasses, the molecular dynamics on the microscopic level has no impact on the macroscopic level.

Nevertheless, we do actually observe many apparently spontaneous dynamic macroscopic phenomena in non-living nature, phenomena such as the growth of crystals and emergence of hurricanes, or flames. This brings us to dissipative systems.

Agency Level 1: Dynamics Imposed by Energy Gradients

If a gas or a liquid is exposed to a steep gradient of energy (either external, such as radiation, or internally created, as in the case of burning), the random movement of its molecules may become *organised into macroscopic currents*, such as winds, Bénard disequilibrium, gyres, flames, or stars. Here, singularities – that is, bunches of several molecules moving occasionally in parallel in the same direction – may become amplified up to the level of becoming macroscopic. In such case, the energy flow triggered by the gradient is canalised and maintains the overall macroscopic assembly. The system uses the energy of the gradient to organise itself and, at the same time, it catalyses a *dissipation* of energy (hence dissipative structures; Prigogine & Stengers, 1985). In other words, it makes the rate of dispersion of the same gradient (e.g., entropy production in its environment) more efficient.[3] Dissipative structures develop their macroscopic appearance at the expense of such energy flows: once the energy gradient had dissipated and the energy flow-through ceases, the structure vanishes (e.g., hurricanes) or leaves behind a 'frozen', ordinary, 'zero-level' structure (e.g., crystals). The rate of entropy production is thus the sole measure for comparing different systems (e.g., flames with gyres).

---

[1] Compare this with meteorological data: wind of 30 m/s (108 km/h) is viewed as accounting for rather windy weather.

[2] Meaning that a spontaneous appearance of collective behaviour of molecules is highly improbable.

[3] For example, the decay of a wooden hut releases the same amount of energy if it burns down within 2 h or rots away over the course of a century. On the other hand, the *rate* of entropy production in its environs (aided by the dissipative structure of fire) will be vastly different for the two events.

Box 8.1 Active Matter

The phenomenon of *synchronous* behaviour of an entire system can be considered a special case of a dissipative structure that coordinates (binds) its constituents in a way that upon a specific cue they act *as a whole*. But this description does not always allow for the supposition of passive behaviour of the constituents of such a system. For example, a flock of birds (which has also been presented as a case of active matter) is coordinated in a different way than proteins in a contracting muscle or the self-propelling system of polymers. This brings us to the concept of an agent.

Agency Level 2: 'On Their Own Behalf'

Stuart Kauffman (2000) made a substantial leap from ordinary dissipative structures by introducing the concept of an *autonomous agent* (see Box 8.2), an organised physical structure that *acts on its own behalf* in its environment, that is, it is not only pushed and pulled by energy gradients. Autonomous agents 'manipulate' the universe around us but in a different way than dissipative structures do. The initial definition of an autonomous agent may sound trivial: it is a system that can perform at least one work cycle[4] and thanks to this, it can *grow and multiply* in such a way that it produces entities similar to itself. At this point, what we have in mind is an extremely simplified model of a cell or cell assemblages such as multicellular bodies, bacterial mats, or meadows. If, however, we take this makeshift definition further, it turns out that it needs to be supplemented with additional powers.

First of all, any work cycle requires the presence of a device (machine) that is not the product of mere self-assembly or a similarly 'blind' physical force. It must be *constructed* so as to fit a given task and its construction requires the investment of work *provided by a pre-existing device*. At this point, we run into the first of many 'hen and egg' riddles that abound in living beings. In short, to use Kauffman's term, an autonomous agent must be 'prestated'. Unlike dissipative structures from the previous section, it does not emerge spontaneously, *ex nihilo*. It not only produces new agents but is itself the product of a parent autonomous agent. To put it simply, autonomous agents are always *born*, and with such an uninterrupted temporal chain of beings, evolution (history) enters the stage. Aristotle's definition is similar (see Chap. 2) but not identical, because evolution is not a concept we find in Aristotle's thought.

Reproduction, however, requires memory as well as a certain structural and informational background. In other words, to produce entities similar to itself, an autonomous agent must be a *replicator*. Note again that it does not come into existence by self-assembly, nor is it constructed by external agents. An autonomous

[4]As when a steam engine performs the energy transformation into work, and returns into initial position, ready to start a new cycle.

Box 8.2 The Concept of 'Agent' and 'Agency'

Morten Tønnessen (2015) completed an invaluable survey (questionnaire) concerning the definition of these terms. In an abstract of his report, he concludes that 'Although there is currently no consensus in the biosemiotic community on what constitutes a semiotic agent, [...] most respondents agree that core attributes of an agent include goal-directedness, self-governed activity, processing of semiosis and choice of action, with these features being vital for the functioning of the living system in question.' Such characterisation applies equally well to autonomous agents or living beings. Responses to the survey were, however, highly diverse and it is not easy to find their common denominator. As noted above, my suggestion in the survey was that the term 'agent' should be replaced simply by 'living being'. With respect to 'agency', I was more radical. In Tønnessen's words: 'Markoš [...] provocatively proposed to replace "agency" in the reviewed citations with "élan vital" – claiming that in this way "you get the same meaning". In Markoš' view this demonstrates that "agency" is a euphemism for what is meant by internal drive, subjectivity, interpretative disposition of the living, etc. By using this euphemism, he claims, "authors try to escape being labelled vitalists".' Yes.

agent is *born* and its development initiated based on and in accordance with memories and experiences inherited from its forebears. At the same time, though, it is confronted with and reacts to the current situation in its environment (by sensing, perceiving, and recording what is relevant). Its progeny is not identical: *descent with modification* (one of Darwin's principal concepts) enters the game. The individual autonomous agents have different characteristics, they become players in a game which Kauffman calls the *biosphere of autonomous agents*.[5] The agents by their simple presence influence and modify their environment, but by restructuring their internal parameters they also react to environmental factors (including other autonomous agents).

To consider the environment, however, we must add yet another factor to the concept of autonomous agents, namely *closure* towards the environment (see also Chap. 9). In short, the agent is equipped by *innerness*, which we should at this point take to mean that most of the agent's internal parameters (qualities) are maintained in states substantially different from those which are prevalent outside. The autonomous agent, however, and its environment are delimited – but not hermetically separated – by a boundary in the form of a *selective* barrier that allows *communication*

[5] At this level, another Darwin's concept, natural selection, enters the biosphere-forming game as well but for the moment being, let us limit our attention to the internal factors constituting autonomous agents and their biospheres.

[6] Additional concepts, such as information, codes, and structures – which are other basic attributes of life – are not discussed in connection with the concept (see Markoš, 2014).

(material, informational) between the autonomous agent and its environment. The image of autonomous agent is thus becoming more complex and below, we find further properties, which have been explicitly or implicitly attributed to it by Kauffman. From now on, we can state that Kauffmanian autonomous agent is a semiotic system commensurable, or even identical, with life.[6] Because life is the only system known to satisfy the requirements placed upon autonomous agents, from now on we can just speak about of living beings and about the actual biosphere as we experience it here on the Earth.

To work and reproduce, a living agent must be able to find and gather appropriate sources of energy and materials while avoiding noxious and dangerous elements and ignoring most of those that do not suit its way of being in the world. But the playground is dynamic: the situation changes according to the requirements of particular ontogenetic stages of the agent (e.g., the embryo, larva, or adult) and depending on external conditions (the seasons, presence of an enemy, etc.). Information that crosses the environment–agent boundary must be recognised, evaluated, transformed, amplified, and, above all, appropriately interpreted.

Other agents belonging to the same or different lineages, with their own requirements or worldviews, form a large part of the living agent's environment. Any life form must then embrace by its inputs not only banal information about physical parameters of its environment but above all knowledge of the pattern, behaviour, and intentions of the large and diverse community of autonomous agents it is born into. At the same time, the outputs of an autonomous agent are not only waste but importantly also signals and signs informing (eventually misinforming or even cheating, as in case of predation or mimicry) the others of its presence. I stress the word *signs*, not just physical appearances and/or hardwired signals. Such community of agents present here and now in the environment forms what Kauffman calls a *biosphere*.

The evolution of biosphere into the imminent future (or, to use Kauffman's term, the *adjacent possible*) is actively *negotiated*, that is, actively constructed from a myriad of possible alternatives, by the abovementioned community of agents. The biosphere is thus a self-constructing system, but one that is radically different from the dissipative systems discussed above. Its constituents are not passive elements moulded into a macroscopic state. Here, members of a biosphere to a large extent 'co-construct' (to use Kauffmanian terminology) the system from the inside (external forces and natural selection are also at work).

Kauffman continues his exposition by formulating the fourth (thermodynamic) law: 'Self-constructing systems tend to maximise their dimensionality, the number of types of events that can happen next' (p. xi), or to use a more elaborate wording:

> Emerging organization concerns the coming into existence in the evolving universe of entities carrying out the measurement of relevant rather than non-relevant properties of non-equilibrium systems. [...] As an average trend, biospheres and the universe, create novelty and diversity as fast as they can manage to do so without destroying the accumulated propagating organization which is the basis and nexus from which further novelty ids discovered and incorporated into propagating organization (Kauffman, 2000: 92–93).

A similar feeling is expressed by Joslyn (1999) when he speaks of 'semiotic agents', who are in most aspects similar to Kauffman's model. He stresses their 'autonomy and freedom of decision-making', thus presupposing properties listed above, and speaks of them being characterised by *deliberative* as opposed to merely *reactive* processes typical of the physical world. The innerness of a semiotic agent is manifested by its *dynamical opacity*, that is, by the fact that one cannot easily anticipate its next move or explain the past ones. Joslyn is also interested in the status of artificial intelligence, artificial life, and the 'simulation of natural systems such as organisms, humans, ecologies, economies, and societies'. At this point, let us leave the extremely interesting field of artificial intelligence and its semiotic status, and remain within the realm of the living.

Can such an autonomous system become the subject of (natural) science? Yes, says Kauffman, even if contemporary physics has no tools to study it as yet. Because autonomous agents are physical (embodied) systems, Kauffman concludes that physics needs to be broadened to accommodate them. My ambitions here are both more humble (I have no desire to disturb the circles of physics) and more ambitious, because I intend to investigate the interpretative powers of autonomous agents as a path towards meaning attribution.

This brings us to the concept of *Umwelt* (see also Box 8.3 below, also Chaps. 13 and 14). It goes without saying that individual *Umwelten* within a given lineage to some extent overlap, so they can reliably understand (and 'manipulate') each other's doings.[7] As a consequence, we can construe dynamic *Umwelten* encompassing entire species, biomes, and even the biosphere, where it would include all beings alive in it during a particular era (for more details, Markoš, 2002; Tønnessen, 2009; see Markoš & Švorcová, 2019).[8] All individuals then acquire their inner world (*innenwelt*) which corresponds to and interprets their being-in-the world. From this, it follows that all individuals are members of broader communities that can be viewed as a form of culture (Markoš, 2002, 2011, 2014, 2016; Markoš & Faltýnek, 2011).

From what has been said in previous two sections, we arrive at a scenario of two evolutionary drives. One, coming from the outside, is the classical Darwinian force based on descent with modification and natural selection. The other drive comes from the inside and its aim is to maximise the dimensionality of the system, as supposed by the Kauffmanian 'fourth law' (Kauffman, 2000).

---

[7] Uexküll was aware of a possibility of such 'zooming'. In his *Theoretische Biologie* (1928, 181), we read: 'Wenn wir die Funktionskreise aller Einzelwesen einer Art zusammen bauen könnten, so würden wir die gemeinsame Umwelt der ganzen Art erhalten, und diese würde entsprechend den Abweichungen der Einzelwesen größer und reicher sein als die Umwelt der einzelnen.' [If we could build up together the function-circles of all the individuals of one species, we should get the common surrounding-world of the whole species, and, in correspondence with the deviations of the individuals, this would be larger and fuller than that of any one of them.']

[8] What sets the human species apart is – as many philosophers and biologists point out – novelty introduced by acquisition of language by our species.

Box 8.3 *Umwelt*

Umwelt can be interpreted as a 'subjective, semiotic lifeworld' (Tønnessen, 2019: 403, also Tønnessen et al., 2016). According to John Deely, 'The concept of Umwelt [...] shows us how a given "environmental niche" is merely the physical part of a larger, objective, not purely physical whole which is, as it were, fully comprehensible only from the perspective of the particular life-form whose world it is' (Deely, 2010: 129–130; see also Deely, 2004, 2010). I agree but also note that it is unclear whether the concept refers to pre-stated ('hardwired') settings or requires the addition of an active interpretative *agency* (see also Chap. 13).

My co-workers and I (Markoš & Das, 2016; Švorcová et al., 2017; Švorcová & Kleisner, 2018) take *Umwelt* to denote a *field of possibilities* (chances, options) that encompasses all of the evolutionary and ontogenetic memory and experience of a particular lineage.[9] It is a repository, a record, a sediment of past manifestations of *agency* that may but need not become useful (retrieval of solutions if conditions return). Such *Umwelt* is also an arena for tinkering with evolutionarily novel phenomena (brought about by, e.g., climate change, epidemic, or invasive species).

An individual at a given time represents but a single point in such *Umwelt*. During its lifetime, it traverses a trajectory ('world line') across the *Umwelt* field. Its course across this field reflects its ontogeny, life history, and environmental inputs which made an impression on that individual as well as those which that individual actively *opted for*. Occasionally, a living entity can even broaden the field by extending its boundaries into new, hitherto unexplored areas beyond (which is how novel elements are accommodated) or invent new ways of living. What is important is that the abovementioned memories and experience (see Chap. 9) stored in the *Umwelt* field aid the interpretive effort of the individual, enable its optimal 'tuning' at a given point of its world line.

Memory and Experience, Signs and Meanings

In our further exploration of the area, let John Deely be our main guide (e.g., Deely 2004, 2009, 2010). The world is full of things that exist for themselves, regardless of being possibly sensed, perceived, or contemplated by its other denizens. This is the case of for instance stars, mountains, lightning, but especially all living beings. Things may not even exist anymore, like Napoleon, tyrannosaurus, or a continental glacier in Poland, or may occur in the future, like humans landing on the Mars.

[9] See also below how Koselleck (2020) understands history.

Finally, things can even exist merely as constructs of mind, e.g., the bottle imps or entropy. Things represent *subjects* of specific qualities.

Things can interact in two ways: intersubjectively or suprasubjectively. Intersubjective interaction is what can be described by physics, such as collisions of bodies, gravitation, refraction of light, ultrasound, or electric discharges in clouds. Suprasubjective interactions occur only in the presence of perceivers and, as far as we know, only living beings are capable of perceiving/interpreting. In such case, the interaction may acquire a new dimension, depending on the sensory and perceptual powers of the particular life form (such as its ability to hear ultrasounds, see in the ultraviolet range of light, or register the presence of a specific pheromone). Perception is an interpretive process. What is important is that it concerns only *some* properties of the thing perceived and that it combines the input with the added value of the memory and experience of the perceiver (Deely, 2004: 15). Such processing leads to the construction of an *object*, an image of a thing in the *Umwelt* of that particular perceiving creature (Deely, 2004: 19).

Deely, however, maintains that only human animals know that there are objects, while for non-human animals, only things are available (Deely does not consider other forms of life, like plants or bacteria). Only humans (*human animals*, as Deely would say) are aware of the existence of things that can be interpreted in many ways, and by confrontation of many such ways, they can arrive at better knowledge of that particular thing (and construct 'Umwelts' of objective reality and of narration; Švorcová et al., 2017). If so, one and the same thing can provide a plethora of different objects for different *human animals*, but only for them, and such object acquires different attributes in different contexts or at different stages of life.[10] It follows that all objects are *known*: the expression 'unknown object' is an oxymoron. Non-human animals, in contrast, do not have access to things themselves, to the realisation that an object is but one way of knowing a thing, one of many. Below, I argue that *all* forms of life, not only animals, are capable of perceiving and constructing objects (which is an interpretive process) and of attribution of meaning to their situation in the world (for further details, see Švorcová & Kleisner, 2018; Markoš & Švorcová, 2020).

The process of object construction can be illustrated by the triune relation scheme known from Peircean semiotics (e.g., Eco, 1997; Deely, 2010). A sign vehicle (Representamen, R) furnishes information concerning an Object (O) perceived (derived, as we saw above, from the thing) to the Interpretant (I). The Interpretant confronts incoming information with the contents of its memory and experience, thus creating a sign that will (*suprasubjectively*) better particularise the meaning of what the object is. As a result, the new assessment of O (via R) shifts the pattern of perception that is provided to I, and so on. The 'hypostasis' of R, O, and I, which

[10] For example, a horse (a thing) can elicit different objects in a young or old wolf, a botfly, a pathogen causing foot-and-mouth disease, a sparrow, in the minds of Montezuma, a butcher, a jockey, a horse breeder, etc.

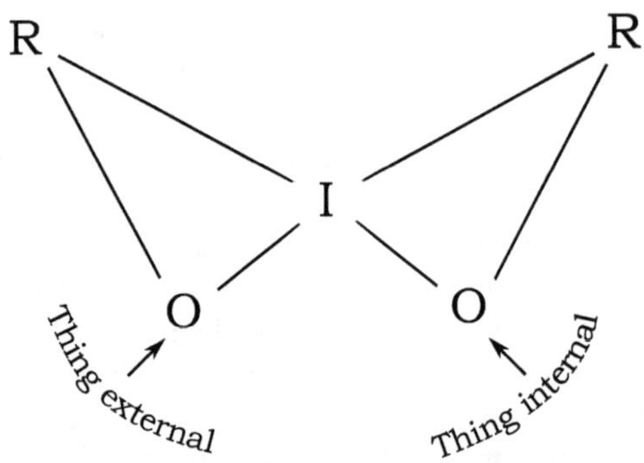

Fig. 8.1 Diagram of the semiotic process according to Deely (in our adaptation extrapolated to all life). On the left the world, on the right the interior of a living creature. A thing, either external or internal, is given in the form of an object (O) via a sign carrier (R) such that the interpreter (I) is able to perceive it. The interpreter integrates the particular inputs and puts them into interrelations, which enables him to render the object better. The two triadic relations are irreducible to their components and their superposition produces a sign that is thus 'suprasuprasubjective' (not bound to the subject's material)

cannot be studied separately,[11] leads via the sign to a never-ending semiotic spiral, unless of course the spiral ends up petrified in the form of a habit (see Chap. 9). In this way, the Object in the process of being confronted with both the external world and 'internal settings' of the Interpreter acquires ever clearer contours. At the same time, such internal settings are the subject of ceaseless alterations and amendments, and the same applies to the tuning of the sign vehicle (R). Moreover, the R, O, and I can change their places in the swirl of semiosis. Objects (and objective reality) are constructed, not pre-stated (Fig. 8.1).

The Peircean system was developed for humans as 'semiotic animals'. As we saw above, Deely broadens its scope to include also non-human animals (intending, I suspect, higher vertebrates), with a disclaimer that although animals recognise objects, they never learn about the existence of things. In contrast, I propose that the process of meaning construction/attribution outlined above describes how living beings – *any* living being and *nothing but* living being – can make sense of its position in the world. Life is a semiotic category, as we argue in Švorcová et al. (2017), and interpretation is thus a defining property of *all* life. The lesson to be taken is that all living beings *construct*, not mirror, their inner (objective) world, and the biosphere as a whole co-constructs its momentary state.

[11] Deely (2010: 385–6): 'That the *relationship itself* among or uniting the three terms, rather than the one of the three terms united (namely, the term representing the signified object to the power, the sign-vehicle or 'representamen'), is *one single relationship* – in short, a *triadic relationship,* irreducible to any dyadic (two-term) relationship or combination of dyadic relationships – and is what constitutes the being proper to signs was the further step that was not definitively constituted before the early seventeenth century *Tractatus de Signis* of John Poinsot, nor an idea visited again until the early twentieth century writings of Charles Peirce.'

Enter History

As an entry to this section, I suggest an analogy with a view of *human* history (attribution of one's meaning of/in the world) suggested by the historian Reinhart Koselleck (2020), who claims that a historical being, be it a human or a community, interprets its *present* situation in a world within a 'space of experience' (*Erfahrungsraum*) and a 'horizon of expectation' (*Erwartungshorizont*). The past and the future thus cooperate on advancing into the 'adjacent possible' (sensu Kauffman), and both are susceptible to new contexts, information, even hoaxes, etc. I will argue that this model can serve as a universal description of *here and now* of any lifeforms or their larger assemblages.

Our worldview, the worldview of all lifeforms, is grounded in history (evolution) that established the norms of the playground (*Umwelt*). The game results in this or that interpretation (understanding) the world and takes place under the auspices of life itself. All denizens of the biosphere share so many traits that it is universally accepted, at least in biology, that they all represent progenies of what one can call the Last Universal Common Ancestor (LUCA). There exist many speculations about the characteristics of LUCA. Let me here follow an outline given in Markoš and Švorcová (2019), where we speculate that LUCA in fact consisted of a highly promiscuous biosphere of (still undifferentiated) cells that gave rise of three extant lineages of cellular life: Bacteria, Archaea, and Eukarya. We assume that the scaffolding of protocols shared by those lineages was inherited from LUCA, whereby the sharing can take the form of not only as overt phenomena – just like different human languages can broadcast the same message with different meaning – but often also various miniscule details (the 'grammar', 'syntax', or 'vocabulary') of underlying structures and/or messages. All such processes make a substantial contribution to the collective memory (space of experience) across great assemblages of cells/organisms and even throughout the entire biosphere.

A major evolutionary breakthrough in this context came with the ability to save some experience into a reliable and stable medium that is not that experience. Such 'written (genetic) text' can be copied, disseminated to the progeny and across the biosphere without paying attention to its meaning, without the need to materialise what had been saved (see Chap. 9). The message can be easily 'read' and thus retrieved by any living being, and subsequently used as a program, a manual, or interpreted as a text. Note that reading a manual, and above all its interpretation, enables the interpreter/receiver and/or its community to take a variety of interpretative actions or even come up with new interpretations never tried before.

A language metaphor may be at this point illuminating: 'Language is something added [to the world]; it is a reduction of that world, and at the same time, it brings about its inflation', observed Eliška Fulínová in her lecture (Oct. 2020). *All* objects, not only words and propositions, are a sort of reduction but one that allows ordinary communication and orientation in the world. At the same time, though, they open new vistas, new interpretations, new constructions of the world/object that is being interpreted. This interpretative activity is endorsed by bodily functions and structures, which are likewise passed down from the shared beginnings: LUCA ensured that only *some* pathways canalise the flow of energy and matter – at the expense of many others. Such pathways make up as if the stem in all lineages, decorated by many variations on the common theme.

What is, however, especially important, are the shared means of communication via horizontal gene transfer, via communication at distance by chemical or physical signals and signs (hormones, pheromones, light, etc.), via membrane processes of fusion, division, or splitting away that involves cells or cell-derived structures, and finally, via cell-to-cell contacts which elicit responses in partners (differentiation, engulfing, growth, and the like). Such communication tools play a role not only in multicellular bodies but also in the bonds of cells from different organismal lineages. They constitute the basis of most kinds of symbiosis across the whole of the biosphere.

In Markoš and Švorcová (2019), we call the deep, stem homologies inherited from LUCA 'the Norm', whereby the entire evolution in subsequent differentiation into lineages takes the form of a ('cultural') interpretation of such norm within the given lineage, time, and environment. Again, a language analogy may help: a particular vocabulary and grammatic norms makes possible an infinity of utterances. Analogies to such play can be observed in different individuals, lineages, and species thriving in the biosphere. The presupposition of innerness of every living being is thus a necessary – and sufficient – condition of life.

Conclusion

The students of Kim Nasmyth, an outstanding molecular geneticist, were always after the presentation of their results prepared to answer the standard question posed by their boss: 'How will this help you to discover a "universal truth"?' (Tebb, 1998). I suggest a little amendment to the style of the question: 'How will this help you to discover *the* universal truth?' Formulation of universal truths that would hold for a given community is the common goal of science, linguistics, musicology, military, and many other branches of human understanding: it provides norms for such knowledge. Above such norms, however, there runs an endless game of narratives, dialogues, rituals, invention, structuration, etc. (for more on how similar fictional topics can be told across different times and nations, see, e.g., Auerbach, 1957). We suggest that all living beings with their traits as they appear here and now should be viewed (assessed) along the same lines. My aim was to stress that all forms of life can be viewed analogically to the abovementioned human activities.

References

Auerbach, E. (1957 [1946]). *Mimesis, the representation of reality in Western literature*. Doubleday

Deely, J. (2004). Semiotics and Jakob von Uexküll's concept of umwelt. *Sign Systems Studies, 32*(1–2), 11–34.

Deely, J. (2009). *Purely objective reality*. University of Scranton Press.

Deely, J. (2010). *Medieval philosophy redefined. The development of cenoscopic science, AD 354 to 1644 (from the birth of Augustine to the death of Poinsot)*. University of Scranton Press.

Eco, U. (1997). *Kant and the platypus. Essays on language and cognition*. Harcourt Brace & Co.

Joslyn, C. (1999). *Semiotic agent models for simulating socio-technical organizations*. At http://citeseerx.ist.psu.edu/viewdoc/download?doi=10.1.1.37.423&rep=rep1&type=pdf Accessed 25 Apr 2023

Kauffman, S. (2000). *Investigations*. Oxford University Press.

Koselleck, R. (2020 [1979]). *Vergangene Zukunft. Zur Semantik geschichtlicher Zeiten*, 11th ed. Suhrkamp. [English translation: (2004) *Futures past. On the semantics of historical time*. Columbia University Press.

Lopez, B. (1986). *Arctic dreams*. Vintage Books.

Markoš, A. (2002). *Readers of the book of life: Contextualizing developmental evolutionary biology*. Oxford University Press.

Markoš, A. (2011). Hermeneutics by the living. *Biosemiotics, 4*, 119–125.

Markoš, A. (2014). Biosphere as a semiosphere. Variations on Lotman. *Sign System Studies, 42*, 487–498.

Markoš, A. (2016). The birth and life of species-cultures. *Biosemiotics, 9*, 73–84.

Markoš, A., & Das, P. (2016). Levels or domains of life? *Biosemiotics, 9*, 319–330.

Markoš, A., & Faltýnek, D. (2011). Language metaphors of life. *Biosemiotics, 4*, 171–200.

Markoš, A., & Švorcová, J. (2019). *Epigenetic processes and the evolution of life*. Taylor and Francis.

Markoš, A., & Švorcová, J. (2020). Cutting down the Porphyrian tree: Objective reality as created by the innerness of living beings. In Ľ. Lacková, C. H. J. Rodriguez, & K. Kull (Eds.), *Gatherings in biosemiotics XX* (pp. 24–36). Tartu University Press.

Prigogine, I., & Stengers, I. (1985). *Order out of chaos*. Flamingo.

Seinfeld, J. H., & Pandis, S. N. (1998). *Atmospheric chemistry and physics: From air pollution to climate change*. Willey.

Švorcová, J., & Kleisner, K. (2018). Evolution by meaning attribution: Notes on biosemiotic interpretations of extended evolutionary synthesis. *Biosemiotics, 11*, 231–244.

Švorcová, J., Markoš, A., & Das, P. (2017). Origins of the cellular biosphere. In V. P. Sahi & F. Baluška (Eds.), *Concepts in cell biology – History and evolution* (pp. 271–290). Springer.

Tebb, G. (1998). Kim Nasmyth: The universal truth. *Current Biology, 8*, R257–R258.

Tønnessen, M. (2009). Umwelt transitions: Uexküll and environmental change. *Biosemiotics, 2*, 47–64.

Tønnessen, M. (2015). The biosemiotic glossary project: Agent, agency. *Biosemiotics, 8*, 125–143.

Tønnessen, M. (2019). What can be known about future Umwelten? *American Journal of Semiotics, 35*, 401–429.

Tønnessen, M., Magnus, R., & Brentari, C. (2016). The biosemiotic glossary project: Umwelt. *Biosemiotics, 9*, 129–149.

von Uexküll, J. (1928). *Theoretische Biologie* (2nd ed.). Verlag von Julius Springer.

Chapter 9
Plastic Ontogenesis: Memory, Closure, and Habitual Teleology in Development

Jana Švorcová

Abstract The aim of this chapter is to shed some light on the fascinating interplay between agency and memory in biological systems and to explore how these concepts can inform our understanding of organismal evolution. Memory phenomena can be observed on various levels of organismal organisation and evolution, from epigenetic marking on chromatin all the way to developmental regulatory networks. If we view such things as memory instead of mere heredity, we might be able to better explain the purposeful behaviour and biological teleology of the living, whereby any such explanation is necessarily linked to the concept of organismal closure.

Keywords Memory · Epigenetics · Closure · Kantian teleology · Habitual teleology · Plasticity

Introduction

In humans, agency (as defined by sociologists, e.g., Giddens, 1984) generally refers to the capacity of individuals to make free choices within a given (social) context and to affect their living environment, that is, to not just submit to actions of external forces. In the case of humans, agency is associated with intentionality and mental representations of what the subject wants to achieve. A general definition of *biological* agency was formulated by Stuart Kauffman (2000), who defined *autonomous agents* as agents who have the capacity to act on their own behalf. Autonomous agents act in a specific manner to preserve and enhance their existence, and

J. Švorcová (✉)
Department of Philosophy and History of Science, Faculty of Science, Charles University, Prague, Czechia
e-mail: jana.svorcova@natur.cuni.cz

J. Švorcová (ed.), *Organismal Agency*, Biosemiotics 28,
https://doi.org/10.1007/978-3-031-53626-7_9

Kaufmann viewed such actions as also goal-directed (i.e., such that agents have goals to achieve). According to Kauffman, an agent acts independently of any strict external control, driven rather by its internal principles and constant negotiations with other autonomous agents. Acting on one's own behalf involves finding a source of energy, minimising danger, learning about the world, gaining experience, learning new representations, but also deliberately manipulating the world.

In the context of defining the difference between the living and the non-living, one often encounters reasoning based on self-organisation (see Chaps. 2, 3, 4, and 5). As shown in this book, organisms/autonomous agents have the property of self-organisation (as well as self-maintenance and self-reproduction) when each part of the organism exists in a relationship of collective interdependence, the organism is an integrated whole, and its parts cannot be thought of without the whole and vice versa (see Chaps. 4 and 5). Moreover, an autonomous agent is also capable of performing at least one closed work cycle (see Chaps. 8 and 10). Sharov and Tønnessen (2021) include among autonomous agents humans, animals, plants, single-celled organisms, individual cells in multicellular organisms, families, colonies, populations, ecological consortia, human communities, businesses and nations, autonomous mechanisms, robots,[1] functional protein complexes in cells, viruses, but also various modified or engineered organisms. Nevertheless, our model of agency is slightly shifted: for us, agency starts on the level of an individual cell as a whole and is associated with the concepts of organisational closure, autonomy, and autopoiesis.[2] In other words, we do not consider for instance ribosomes to be agents *per se*.

The above mentioned Kauffmanian definition is mentioned also elsewhere in this book (e.g., Chaps. 8 and 10), but I need to repeat it because it is a point of departure for the phenomena of organismal agency I focus on in this chapter. Especially relevant to the present enquiry is the fact that autonomous agents can meaningfully react to their situation in the world and plan their future behaviour according to their previous individual and phylogenetic experience (Markoš, 2002). Such experiences become part of autonomous agents' internal organisation and memory, making them yet another principle which sets apart the realm of the living and the non-living. As Markoš emphasises (see Chap. 8), the existence of autonomous agents is historical, temporal, and evolutionary, which is also why biospheres composed of autonomous agents do not have a deterministic evolution but rather a history. Agents' manipulation within biospheres causes the universe to become more complex, to evolve with each passing moment. Negotiations between autonomous agents and biospheres (i.e., communities of autonomous agents), which for autonomous agents constitute the largest part of their world, are probably one of the most important aspects of interaction between autonomous agents and their environment (as apparent in, e.g., symbiotic relationships). It makes no good sense to study

[1] It is disputable whether robots and products of artificial intelligence are actually agents, given that their *telos* is always attributed only from the outside, while in real autonomous agents teleology manifests itself based on their internal organisation (see Chaps. 6 and 10).

[2] Naturally, these defining characteristics of agency are not exhaustive: they are rather just those which I work with in this chapter.

autonomous agents in isolation: we should always investigate them within a wider framework of their interactions with biospheres. Autonomous agents of the biosphere through their activity move towards further organised states and increase their diversity.

Other scholars also define agency as the capacity of a system to generate *end-directed behaviours* (cf. Sharov & Tønnessen, 2021). The aim of this chapter is to define such end-directed behaviour of biological systems through memory (stored experience), whereby memory is essentially associated with the idea of phenotypic plasticity and organisational closure, that is, with some – though not all – of the main defining attributes of agency.

Memory

Memory is the capacity of a system to retain and retrieve its own experiences. It is not the same as biological heredity (as defined by the 'classical' modern synthesis of evolutionary theory), which denotes a 'mere' transfer of information from one generation to the next. In biological systems, organic memory is essentially associated with learning about the environment and with forgetting. The notion of organic memory is not new in biology, but it has been defined in various ways. Building on previous work on this topic (Elsasser, 1987; Otis, 1994; Barbieri, 2003), I use this term to describe a form of memory that is not necessarily cerebral or neuronal; rather, it is fundamentally associated with living organisms from bacteria to humans. The concept of organic memory which appeared within the framework of the Lamarckian theory of heredity has been a subject of interest in biological discourse since the nineteenth century, as evidenced by the work of Hering (1870, 1897), Haeckel (1876), Semon (1904), Butler (1877, 1910), and many others (see Otis, 1994; Schacter, 2001; Švorcová, 2012). Although specific ideas about memory varied, it tended to be identified with the ability of organic matter to preserve past experiences, whereby memory localised in the brain is viewed as a specific manifestation of this more general potential. But ideas regarding the nature of these memory traces had been rather vague and after the formulation of Weismann's germ-plasm continuity theory, the concept of organic memory was abandoned. Still, the above-mentioned assumption I make here, namely that memory is somehow inherent in all living things and our brain memory is just one of many manifestations of this phenomenon, is consistent with the work of nineteenth-century neo-Lamarckians. It should be noted, though, that the original neo-Lamarckians had been accused of merely positing an analogy between memory and heredity, implying that their doctrine of organic memory actually brings nothing new (e.g., Johannsen, 1909; Rádl, 1930). Thanks to the current state of knowledge of epigenetics and evo-devo, we have, however, reached a better understanding of the processes which constitute cellular and supracellular memory – and can thus hopefully defend ourselves against this criticism.

Memory in Bacteria

Although bacteria do not have complex nervous systems like animals, they do have the ability (naturally in addition to the standard mechanisms of heredity) to store and pass on information between individuals and between generations through a process known as horizontal gene transfer (HGT). HGT is a common mechanism of genetic exchange among bacteria. It is widespread: it is estimated that up to 80% of bacterial genes have been involved in HGT at some point in their evolutionary history.

HGT can occur through three main mechanisms: transformation, conjugation, and transduction. Transformation involves the uptake of naked DNA from the environment, conjugation involves a transfer of DNA via direct cell-to-cell contact, and transduction involves the transfer of DNA via a virus (Boto, 2010). HGT is particularly common among bacteria which share the same environment or niche, because they are more likely to come into contact with each other and exchange genetic material. For example, some bacteria can develop resistance to antibiotics through HGT and once the bacteria acquire these resistance genes, they can pass them on to their offspring and even to other bacterial species, which helps them to quickly adapt to the presence of antibiotics (Michaelis & Grohmann, 2023).

The CRISPR-Cas system is a defence mechanism which allows bacteria to protect themselves from foreign genetic material, such as viruses or plasmids, by capturing and storing fragments of foreign DNA or RNA. These fragments are then used to recognise and destroy any future invasion of the same or similar foreign genetic material. In this sense, the CRISPR-Cas system allows bacteria to remember past infections and use this memory to defend themselves against future ones (Barrangou et al., 2007). Another example of memory in bacteria is associated with their ability to form biofilms, complex communities of bacteria which can resist antimicrobials and environmental stresses. Biofilm formation involves a coordinated response between bacteria, which is in turn based on their ability to acquire and store information about the environment and the presence of other bacteria (Flemming et al., 2016).

Most importantly, bacteria can pass information to their offspring through epigenetic inheritance, which can thus be thought of as a form of cellular memory. Bacteria can regulate gene expression and adapt to changes in their environment by chemically modifying their DNA. These modifications can be passed on to daughter cells during cell division, allowing them to inherit a 'memory' of previous environmental conditions and respond accordingly (Hu et al., 2018).

It should be noted that both HGT and CRISPR-Cas, as defined by contemporary biology, occur randomly but nonetheless exemplify bacterial memory, because DNA sequences carry information about the environment a cell had encountered. In these cases, the distinction between heredity and memory is not sharp. HGT facilitates the involvement of organisms in the biospheric pangenome by increasing the intersection of species and lineages' *Umwelten*. Bacteria are particularly proficient in gene sharing, having evolved specific mechanisms to facilitate regular gene flow through the HGT pathway. While many eukaryotes also engage in this 'gene market', their participation is less frequent. Animals, and possibly plants, stand out as being somewhat less connected to the regular turnover of pangenome genes in this respect. As we shall see below, epigenetic changes can likewise occur quite randomly, but they also take place in response to environmental signals. Bacterial systems are disposed to epigenetically modify certain loci in response to specific environmental conditions. These modifications then represent specific memory traces which emerged in reaction to the environment and can be viewed as analogical to learning and memory. Therefore, in accordance with Maturana and Varela (1980, Chap. 1), these representations emerge from an ongoing and mutual interaction between the organism and its environment.

Bacteria as Agent

From a biosemiotic point of view, even bacteria are agents: '*...all living organisms are semiotic agents capable of sensing and information-driven adaptive behavior. Thus, even the smallest creatures such as bacteria, the most abundant organisms on Earth, are agents*' (Sharov and Tønnessen (2021, 96). On the other hand, according to the protosemiosis model proposed by the same authors, bacteria lack the capability to perceive objects and, consequently, to engage in interpretation. They write: '*...bacteria do not 'know' that glucose exists outside. Their only observable is the binding of a chemoreceptor protein to glucose. The chemoreceptor system functions as a differentiator that distinguishes between the two possible states (i.e., glucose binding versus no binding). Such differentiation is not a representation because it does not provide any additional information about glucose as object, such as chemical composition.*' (Sharov and Tønnessen (2021, 100). Only the eukaryotes are capable of perceiving objects, that is, capable of eusemiosis. Sharov and Tønnessen draw this clear distinction between bacterial and eukaryotic life. I don't understand this dividing line. Defining objects for bacteria poses a challenge but in our model, agency and semiotic processes are thought of as starting already at the level of unicellular organisms, that is, the level of bacteria and unicellular eukaryotes (Chaps. 8 and 10). More importantly, living organisms perceive *signs* (not objects) that provide information about objects. So the binding of a molecule or a change in the receptor conformation can be seen as a sign that something like glucose is out there. Second, signalling in bacteria and eukaryotes has its differences, but it is not in principle that different. While membrane receptors in the bacteria and eukaryotes are in some respects different — bacterial receptors are typically integral membrane proteins, such as sensor kinases and chemoreceptors, while eukaryotic receptors can be found on the plasma or organelle membranes, such as G protein-coupled receptors, receptor tyrosine kinases, and ion channels — they also share some important features. Both bacterial and eukaryotic receptors rely on ligand binding to initiate the signal transduction. In both receptor types, ligand binding induces conformational changes that activate downstream regulatory processes. Moreover, both types of kinases act through phosphorylation and dephosphorylation. Despite structural and functional distinctions between bacterial and eukaryotic membranes and receptors, the fundamental concept of detecting external signals and transducing them into a cellular response is a norm shared across all forms of life (Markoš & Švorcová, 2019).

In addition, I believe, we can apply the concept of learning to non-neural organism, and learning does not have to be necessarily connected to consciousness and cognition (see further in the text). Moreover, there is yet another aspect where our concepts diverge from those presented by Sharov and Tønnessen (2021). In this book, agency is defined in association with autonomy, autopoiesis, teleology, and organisational closure. In the case of mitochondria or plastids, the functional organelles of eukaryotic cells, autonomy is limited. I would no longer call them *autonomous agents* since they had in a sense resigned on their agency and became secondary agents, too dependent on cellular processing as a whole. Similarly, I do not consider ribosomes in the cells to be autonomous agents: their autonomy is likewise limited, and they are allopoietic, that is, constructed and repaired by other cellular components (Sharov & Tønnessen, 2021). Nevertheless, these authors do consider ribosomes to be agents and we believe agency is a property of the whole cell – and this is where our model differs.

Epigenetic Memory

DNA, the molecule of heredity, is a necessary but not sufficient condition for the persistence of life in its various forms and living strategies. During the second half of the twentieth century, researchers discovered that the development of organisms does not depend solely on genetic information stored in DNA molecules. In eukaryotes, a substantial amount of parallel information is processed via various mechanisms such as DNA methylation, histone modification, RNA mediated processing, structural templating, prions, and the like. This information is inherited but is not part of the DNA script. Unlike mutations, which result from random changes in the DNA, this information is often induced by environmental conditions (although DNA methylation is believed to be also stochastic). A wide range of environmental factors, such as light, temperature, nutrition, pressure, gravity, traumatic stress, the presence of predators or symbionts, maternal care, or the presence of conspecifics, can significantly influence the phenotype (e.g., Agrawal et al., 1999; Nijhout, 2003; Waterland & Jirtle, 2003; Weaver et al., 2004; Cropley et al., 2006; Dias & Ressler, 2014). More importantly, some epigenetic changes can be passed onto future generations. Epigenetic inheritance can involve the inheritance of entire cellular structures, including organelles such as centrioles, mitochondria, plasma membranes, cilia, or microbial symbionts. In a more specific sense, though, epigenetic inheritance typically refers to the inheritance of chromatin structures and modifications, including DNA methylation and various histone modifications, as well as different types of RNA and exosomes.

As reviewed in Švorcová (2023), transgenerational epigenetic inheritance (TEI) has been demonstrated in plants *Arabidopsis thaliana*, *Taraxacum officinale*, *Linaria vulgaris* (in all three cases, inheritance of DNA methylation), yeast *Schizosaccharomyces pombe* (inheritance of histone modifications), *Drosophila melanogaster* (inheritance of the chromatin state mediated by a transcription factor), and nematode *Caenorhabditis elegan*s (inheritance of RNA molecules and histone modification). Mammals exhibit TEI as well, as evidenced in the differential methylations of the A^{vy} metastable epiallele in mice, which result in various coat colours in genetically identical individuals (Morgan et al., 1999). Heat conditioning in chicks led to transgenerational heat and immunological resilience accompanied by changed methylation patterns (Rosenberg et al., 2022). TEI has also been identified in cases of circumstances such as drug or alcohol abuse, exposure to toxins, or traumatic experiences including lack of maternal care, maternal separation, electroshocks, war, or famine (for a full review, see Švorcová, 2023). All such experiences can lead to changes on the level of epigenetic modifications, which consequently change the expression of particular genes (studies usually focus on genes associated with stress response, metabolism, or odour perception).

It should be noted that when we speak about experience with the external environment or environmental signals leading to epigenetic (sometimes hereditary) changes, this includes other organisms, which also represent a certain environment for us (and vice versa). For example, our microbiome influences the activity of histone deacetylases through the production of short-chain fatty acids, which in turn modulate gene transcription in our brain through the vagus nerve (Stilling et al., 2016). Inheritance is also possible through epididymosomes, which are small (30–150 nanometres in diameter) membrane-bound vesicles secreted by epithelia of the epididymis. As the ripening spermatozoa travel through the epididymis and mature, epididymosomes transfer to their surface proteins, lipids, and various RNAs. In consequence, the content and varieties of RNAs change dramatically, and studies suspect that such changes are due to these vesicles. The soma of the parent can thus inform the germ cell about previous experiences (Reilly et al., 2016; Sharma et al., 2018).

It should be noted, though, that TEI is as yet not generally accepted among biologists. Some scholars do not believe TEI to be significant and prioritise genetic inheritance over epigenetic inheritance. They argue that epigenetic inheritance is usually associated with the defence of sex cells against transposons, viruses, or transgenes, citing examples of secondary epimutations where DNA methylation of a gene is associated with the mutation of a neighbouring gene (Guéant et al., 2018). Opponents of the TEI claim that all epigenetic processing is gene-driven and the marks do not survive the reprogramming that takes place after fertilisation (which would make TEI impossible). Transgenerational epimutations are often considered merely noise and as such rarely heritable or adaptive (Mitchell, 2018). The pathways of signals from the brain to the sex cells are unknown, and evidence pertaining to any causal mechanisms of transmission is often missing. Scientists are therefore careful not to posit any causal links between environmental signals, molecular data, and transgenerational phenotypic traits, especially regarding the impact of complex events such as individual traumatic experience.

Nevertheless, based on our investigation of it we do consider TEI a significant biological phenomenon (Markoš & Švorcová, 2019; Švorcová, 2023) and believe its rejection is rather ideologically based (cf. Chap. 14 on scientific biases). TEI is potentially adaptive and has an evolutionary impact. Epigenetic modifications can result in changes in the gene expression, which affect the phenotype of an organism without changing the underlying DNA sequence. Such changes can lead to variations in traits that may be advantageous or disadvantageous to the organism in a particular environment. For example, epigenetic modifications can play a role in adaptive evolution by allowing organisms to respond quickly to environmental changes without having to wait for genetic mutations. Epigenetic changes can also generate phenotypic diversity, thus increasing the potential for evolutionary innovation and emergence of new traits in fluctuating environments (Lachmann & Jablonka, 1996; Uller et al., 2015). Over time, epigenetic modifications may become more stable and be passed on to subsequent generations, leading to epigenetic inheritance and even genetic assimilation (Waddington, 1952; Rutherford & Lindquist, 1998). Heritability of epigenetic modifications can thus affect the evolution of populations and species, but the extent to which epigenetic modifications contribute to long-term evolutionary changes and the mechanisms underlying the stability and heritability of epigenetic traits are subjects that require further research.

Developmental Memory

In developmental biology, the dynamics of the developing embryo is explained through developmental regulatory networks (DRNs), that is, complex interactions between genes and their products (transcription factors and regulatory RNAs), which control the expression of target genes and subsequently thus also the formation of cell types, tissues, and organs. Such control dynamics can change in response to various signals from the environment, such as hormones, growth factors, and nutrients, and these switches can also be mediated by epigenetic modifications. DRNs are conserved across different species, which points to a homological origin of regulatory genes, and similarities in interactions within the regulatory networks that control development have been found in a wide variety of organisms ranging from flies to humans. For example, the basic regulatory networks which control the development of *Drosophila melanogaster* are remarkably similar to those in vertebrates. This includes the genes controlling the formation of body segments (*Hox* genes), development of the nervous system, and the patterning of organs and tissues. Moreover, many of the key developmental processes, such as cell differentiation, cell migration, and tissue patterning, are fundamentally similar across different organisms. It seems therefore that although the DRNs are conserved across different species, these networks can lead in different organisms to different structures and organs. This is because while the underlying genetic and molecular mechanisms are similar, there is a great deal of variation in how these mechanisms are used and controlled (Davidson, 2006; Švorcová, 2012), that is, in how the same regulatory circuits are interpreted. One way this variation can arise is through changes in the timing, location, duration, or intensity of gene expression during development. Even small changes in the regulation of key genes or signalling pathways can have profound effects on the final outcome of development, leading to differences in organ size, shape, and function.

The underlying DRNs controlling limb development are also similar across different mammalian species, but changes in the regulation of these networks have led to the evolution of different limb types. For example, the *Sonic hedgehog* (*Shh*) gene is involved in limb development in mammals and helps control the formation of digits (fingers and toes). Changes in the regulation of *Shh* have been associated with the evolution of different numbers of digits in vertebrates (Saxena et al., 2017). The context in which each gene is deployed is thus crucial, and this context is controlled by the transcription module, i.e., a combination of promoter and enhancer elements as well as repressor elements that bind to specific transcription factors and other regulatory proteins. The activity of these elements is influenced by various factors, such as the availability of different transcription factors within the network module (see Chap. 11), the presence of epigenetic modifications, and by environmental influences. In addition, there are master regulatory genes, which control the expression of other genes within the DRN and can act in a context-specific manner, activating different downstream targets in different contexts. For example, the expression of a gene that controls heart development can be in one context affected by different combinations of regulatory elements (transcriptional factors and cis-elements) than in another context, for instance in the liver. This context specificity is due to interactions between different genes and regulatory elements within the DRN, but it is also influenced by environmental cues that can modulate gene expression (Davidson, 2006).

Another way such variation can arise is through the interaction of DRNs with the environment (Gilbert & Epel, 2015). Development is controlled not only by genes but also by interactions between genes and the external environment (again, including cohabitants of the biosphere). For example, the shape and structure of a bird's beak is influenced by the types of food available in its environment. Although the underlying genes and regulatory networks controlling beak development are similar across different bird species, differences in the diet and other environmental factors can lead to differences in beak shape and function (Abzhanov et al., 2004). The way that DRNs are used and controlled can thus change over evolutionary time as different species adapt to different ecological niches, which results in the evolution of different types of structures and organs although the underlying DRNs controlling these structures are similar.

The big mystery regarding the DRN is how the developing embryo can handle such complex and differentially interactive developmental networks based on the same genetic information. If the maternal input and DNA all it requires (as in the development of *Drosophila,* Hamm & Harrison, 2018), and then the development-based interactions between genes and their products run like a well-oiled machine?

I suggest that the regulatory processes of the abovementioned conserved genes show that organic memory cannot be fully described as relying only on information contained in the DNA molecules, because a very similar script leads to different morphological structures.

Organic memory in development is not merely a storage device in the form of a DNA molecule. Rather, it is an interactive network based on experience with the environment (and several other factors such as maternal input, genetic information, biophysical forces, etc). Such networks and changes they undergo are in principle *historical* and changes in the network's architecture are ultimately the determining factor in the evolution of the phenotype. In agreement with M. West-Eberhard (2003), I thus view genes as sometimes functioning as a secondary form of memory, as successors of form (as interpreted, e.g., by Newman & Bhat, 2009). Epigenetic modification can then provide some explanation of how such networks are stabilised, because they can switch on and off various genes and their products, and this pertains also to the zygote, which is equipped with the basic informational (DNA, various RNAs, chromatin modifications) and structural inheritance (membrane, mitochondria, endoplasmic reticulum, ribosomes) from both mother and father.

This view of DRNs as memory networks is consistent with a process Jesper Hoffmeyer calls *semiotic scaffolding*. It takes place when (1) *a pre-existing non-functional gene acquires a new meaning through integration into the functional (transcribed) part of the genome,* and (2) *the gene product* (e.g., a protein species) *would hit an unfilled gap in the semiotic needs of the cell or the embryo. In this way, a new gene becomes a scaffolding mechanism support a new kind of interaction imbuing some kind of semiotic advantage upon its bearer* (Hoffmeyer, 2011, 58). According to this concept, processes at every level of organismic function are regulated through semiotic interactions among the components, thus continuously adapting biochemical or physiological activity to changing circumstances. This intricate network of semiotic controls establishes a highly complex semiotic scaffolding for living systems. Its function is to ensure optimal performance of organisms by engaging in semiotic interactions with cue elements, which are typically present in dynamic situations. The process of ontogeny is thus supported by a web of internal sign processes which canalise the correct direction of the process. For example, at the cellular level, semiotic scaffolding plays a crucial role in ensuring a proper integration of the digital coding system (the genome) into a myriad of analogical coding systems operational across cell membranes and organelles (Hoffmeyer, 2008, 2011).

Developmental Plasticity

Developmental or phenotypic plasticity of organisms is what allows them to develop different phenotypes based on the same genotype in response to changes in the environment, be it the presence of a predator, food availability, social interactions, etc. Phenotypic plasticity can be continuous in nature and limited from above or from below, resulting in for instance different organ size, such as the horns of the male *Scarabaeus* beetle which change based on the quantity and quality of nutrition. These continuous changes (a continuous series of potential phenotypes determined by the environment) can also be non-adaptive and reversible, usually reflecting the effect of temperature or diet on body size (see Chap. 12).

A second type of phenotypic plasticity is polyphenism, which is discontinuous, that is, an either–or choice, discrete switching between phenotypes without any intermediate form. Polyphenism is assumed to be a permanent adaptation. In insects, it is called seasonal polyphenism, while in mammals or birds we speak rather of life cycles (Gilbert & Epel, 2015). For example, the locust *Schistocerca gregaria* produces two phenotypes depending on the number of individuals in a given area: at lower population densities, the locust is green and has only miniature wings, while at higher densities, the offspring are darkly pigmented and have wings and legs adapted for migration (Ernst et al., 2015). This polyphenism is mainly due to the degree of 'tactile stimulation' of the legs of the nymph. That leads to the production of serotonin in the locust brain, which eventually influences DNA methylation, gene transcription and development into a given phenotype (Fig. 9.1).

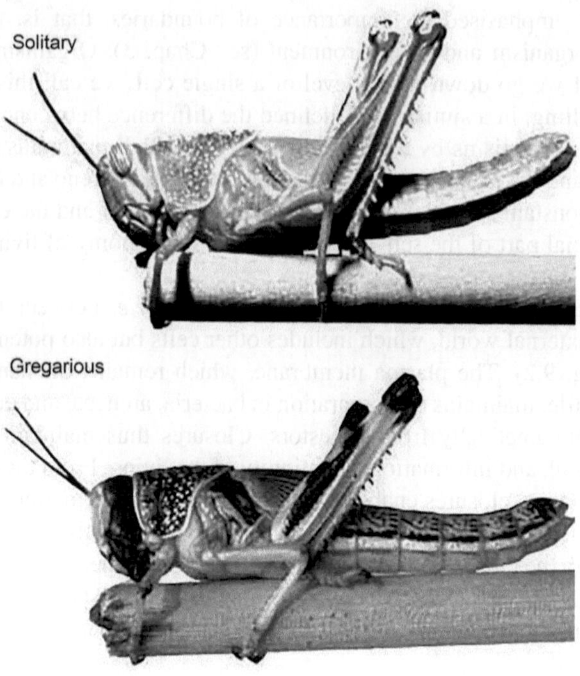

Fig. 9.1 Polyphenism of *Schistocerca gregaria*. Although the photographs of the locusts are in black and white, there is a noticeable difference between the cryptic, long-winged, and migratory gregarious phenotype (below) and the solitary, shorter-winged and usually green coloured phenotype (above). (Credit to Wikimedia commons)

All organisms on Earth are to some extent plastic (Pfennig, 2021) and plasticity can be observed on various levels of organisation, ranging from behaviour, physiology, or hormone production in specific organs all the way to gene transcription mediated by various chromatin modifications and RNA molecules. My interpretation of plasticity from an evolutionary point of view is processual: organismal forms are *inherently* plastic (taking inspiration from Newman et al. (2006), although they are not processualists). Being potentially exposed to constant change, they stabilise their existence through more and more fixed relationships between informational molecules, such as DNA and phenotypes – and memory is a crucial aspect of this stabilisation.

Organisational Closure

When we claim that organisms are capable of storing their previous experience with the environment, the concept of organisational closure is also highly relevant. A closure is more than just a demarcation between two environments; it encompasses all activities inherently connected to it, such as the activity of cellular receptors, gradient changes, microtubuli, and the like (see Švorcová & Markoš, 2023). Already Aristotle had emphasised the importance of boundaries, that is, the separation between an organism and its environment (see Chap. 3). Organisms define their boundaries. If we go down to the level of a single cell, we call this phenomenon *closure*. Schelling, in a similar way, defined the difference between inorganic entities and living organisms by a continuous activity which maintains the difference between self and the external world (see Chap. 5), while Moreno and Mossio (2015) noted that a constant setting of boundaries between oneself and the external conditions is a crucial part of the self-determination and autonomy of living beings (see Chap. 10).

The primary function of a closure is to act as a *barrier* between the inside of a cell and the external world, which includes other cells but also potentially harmful elements (Fig. 9.2). The plasma membrane, which remains unchanged since the beginning of life, maintains this separation in bacteria, archaea, and eukaryotes, and is inherited epigenetically from ancestors. Closures thus maintain the organisational, structural, and informational continuity of the living. From the perspective of catalytic processes, closures enable *selective* uptake, transformation, and release of substances, energy, and information, which can serve as signals for other organisms. Thus, although the closure is closed, it is also selectively open (for detailed analysis see Chap. 10). Closure is the basis of self-organisation towards order (Kauffman, 2000) and a prerequisite for autopoiesis, making it a *necessary condition for the emergence of life.*[3]

[3] The fundamental difference between the living and the non-living can be defined in different ways: for instance, for Schelling (see Chap. 5) or Jonas (2001), it is metabolism. We emphasise this qualitative difference in the form of the closure and the memory based on it (Švorcová & Markoš, 2023).

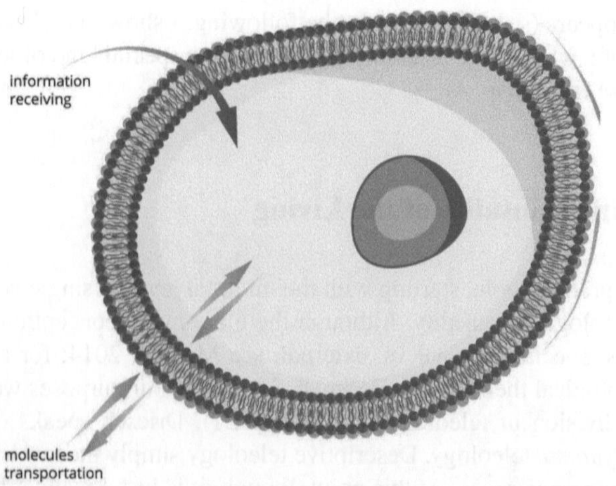

Fig. 9.2 Schematic depiction of a closure. Single cells have a membrane that functions as a topological closure between the inside (the cell) and the outside (the environment). This closure enables a constant *selective* uptake, transformation, and release of substances, energy, and information either through the membrane as such or via several types of receptors (not depicted)

The importance of closure and its association with agency has been explained and repeatedly stressed in biosemiotics by Jesper Hoffmeyer (1996, 2000; also Kull et al., 2011), who was also convinced that semiosis emerged only with the first cells (Hoffmeyer, 1996). Closure is in his view defined as a formation of processual asymmetry by membrane closure followed by selective semiotic interactions between the interior and the exterior across the membrane (or between a cell and non-cell or the self and non-self). Such asymmetry between an organism and its environment both separates and brings together. Organisms create models of their environment based on relevant information and meaning attribution ('difference that makes a difference' to the organism), but also change the environment. Closure is a necessary precondition for interpretation and semiosis. Cells interpret signs received from the outside as well as from the inside and put them into context with their memory structures and experiences (Švorcová & Markoš, 2023).

Such system is also self-referential, because it is both digital and analogous (see Chap. 1). Organism's relationship to operational closure is a process of identity formation (see Chap. 7). When it establishes an operational closure, an organism can be viewed as differentiated from other organisms, and this applies also to single cells, which are the first agents. Such autopoietic organisation is maintained through actions intrinsically linked to the closure. Closure also allows for inwardness. In fact, Portmann (2022) has argued that organisms are forms of being which integrate their experience into their mode of existence and process their experience independently of conscious processes. The inwardness of organisms, according to Portmann, constantly appears (see Chap. 14). In the following, I show that closure, as the first emergence of a whole, is a necessary condition for a special kind of teleology in the Kantian sense (see Chap. 4).

The Specific Causality of the Living

Only an integrated whole, starting with the minimal level of single cells, can create a special teleological causality. Although there are many conceptions of teleology (universal vs. special, internal vs. external, see Muraca, 2014; for a taxonomy of various teleological theories, see Perlman, 2010), for our purposes we shall use the Drieschian division of teleology (Driesch, 1921). Driesch speaks of *descriptive*, *static*, and *dynamic* teleology. Descriptive teleology simply means that we describe things as having a purpose. At this point, though, it is just a manner of description, as when we say that dolphins evolved fins so they could swim but do not mean to imply that this is a goal-oriented system per se. Static teleology is then just a special case of general laws, which describe the way the world functions, but their teleology is ultimately merely apparent. For instance, Monod (1971) in his conception of *teleonomy* admits that organisms are purposive in the sense of maintaining their viability and adapting to environmental conditions due to processes such as natural selection, that is, because natural selection and random genetic variation made them seem purposive. Ultimately, though, it is just an implementation of a programme, a kind of preformism, where for instance the development of an individual (ontogenesis) is determined not by its final cause or its execution but by a pre-programmed development. The behaviour of an organism seems purposive because in the course of natural selection the organism adapted to given conditions – but it is not a teleological system per se. Darwinism thus does not consider the teleological character of functions because natural selection would ultimately account for them through ordinary mechanistic causation (Hoffmeyer, 2011). The third kind of teleology, the dynamic one, is then according to Driesch (1921) the purposiveness of organisms and their autonomy that cannot be explained using solely the general laws of physics and chemistry. In this case, Driesch focuses on ontogenesis and uses an assumption derived from his study of lens regeneration in newts, when he realised that each cell in the embryo contains the potential of the whole. His concept of entelechy was actually an attempt to scientifically define a quality present in a developing organic body (Markoš, 2002). As shown in several places in this book (see Chaps. 2, 6, and 10), we do not consider teleological explanations in biology to be merely metaphorical, as is the case in what is above called descriptive teleology.

Some conceptions of teleology presuppose a mental representation of the goal, mental anticipation, or intentionality. But in 'lower' forms of organisms, such as plants or bacteria, we cannot work with mental states or conscious representations[4]

[4]When describing unicellular organisms, I use the term representation, which is often associated with mental representation. However, I deliberately avoid using terms and concepts such as mind or consciousness.

(though Chamowitz, 2012 talks about awareness in the case of plants). Nevertheless, already Aristotle noted that every living being has a purpose of life: its goal is to remain alive and realise its own form through ontogenesis (see Chaps. 2 and 3). Every organismal manifestation, activity, constitution, all that is adjusted so as to achieve this one goal. Even so, it would be wrong to interpret ontogenesis as having some sort of subjective motivation or intentions. The goal-oriented emergence of form is naturally existing, regardless of any intentionality (see the example with walking in Chaps. 2). As Denis Walsh puts it, purposive occurrences are robust across a range of different initial conditions and mechanisms, yet eventually they lead to the same goal. And this kind of causality, not linked to subjective action, is what Aristotle calls *telos* (Walsh, 2015).

Although to claim that bacteria have intentionality would be far-fetched, one can speak about representations on the level of cells. They are not mental; rather, they are representations of previous experience of the living environment (including other organisms)[5] as shown above using the example of epigenetic modifications. Also, it ought to be borne in mind that even mental representations are ultimately based on the cellular activity of neurons, which are not so different from the cells because they share the same cellular norm (Markoš & Švorcová, 2019, similar aspects are also emphasised by Cvrčková et al., 2009), that is, the same genes for intercellular communication. After all, conscious experiencing of the world is only a small portion of how the environment can be experienced. Thus, such epigenetic changes can be regarded as a form of anticipating the future environment (based on previous experience) an organism is born (or hatched) into. It is a specific and material embodiment of the end-directed tendencies (final causes) in organismal constitution as Aristotle argued (Deacon, 2013). Furthermore, in accordance with Maturana and Varela (1980), such representation is not solely a physical portrayal of an external environment; rather, it is an *embodied experience* that emerges through the ongoing reciprocal interactions between the organism and its surroundings.

Already Kant had emphasised the peculiarities of 'natural purposes' which our mind simply cannot fully comprehend (see Chap. 4; Toepfer, 2011; Muraca, 2014). Such natural purpose is based on interrelations of the parts that make up the whole and the whole that makes up the parts, because the organism is, according to Kant, both an organised and self-organising being (see Chap. 4). Our inability to grasp natural teleology is not due to the parts causing the whole (which is what mechanistic explanations of nature assume), nor is it due to the relationships between the parts themselves. Instead, we cannot truly grasp natural teleology because of the way in which the parts are caused by the whole. This kind of causation takes place

[5] This line of argumentation, i.e., how various, mostly symbiotic, organisms influence our evolution, is developed elsewhere (Markoš & Švorcová, 2019). For the sake of clarity, I reduce here these other organisms to the concept of 'environment', which they naturally represent for other organisms.

when the idea or concept of the whole precedes and forms the basis of the generation of the parts as their purpose. In other words, the whole is the cause of the parts because the parts are generated with the purpose of fulfilling the idea or concept of the whole (Förster, 2002; Muraca, 2014). This is what differentiates the realm of the living and non-living, and it encompasses the internal process of self-generation which is never guided from the outside.[6] Living organisms arise from within themselves, out of their whole, thus becoming the cause of themselves (see also Chap. 2), arising *out of their own closures.*

The premise of the whole as an agent is also important in the abovementioned argument to the effect that ribosomes – but also mitochondria or plastids – are not agents. Just as Brier (2015) argued that one cannot say that my brain has a specific experience but only that I, as a thinking and feeling subject, have an experience, so we could say that only the cell as a whole and autonomously existing subject with some form of rudimentary experience can experience the world. Similarly, Jaeger (Chap. 10) defines agency as irreducible, because it is a property of the organisation of an entire cell or organism.

Habitual Teleology

As Koutroufinis puts it (see Chap. 6), in order for an organism to maintain itself and sustain the flux of energy and matter from its surroundings, it must be able to interpret what parts of the physical surroundings are relevant to it. Similarly, Jaeger (in Chap. 10) emphasises that the behaviour of an organism is based on the internal organisation underlying its goals and needs. I would add that experience of the environment that informs the organism's goals and needs, in which it pursues them, is also necessarily part of this internal organisation. Organisms are plastic, and if a specific signal from the environment persists, such plasticity can be fixed by epigenetic marks across several generations, eventually leading to a fixation of developmental memory based on DRNs or to a selection for polyphenic traits. This activity may indeed be analogous to habitual evolution, and it can explain the purposeful end-directedness of ontogeny. The development or ontogeny can then be described and defined by tracing the memory footprints of epigenetic marking and changes in DRNs, which are based on experience and the historical continuity of the species. If we include memory in our explanatory framework of ontogenetic processes, we cannot explain away the goal-directedness of the living only through Monodian teleonomy (1971), i.e., by random changes in genetic information and selective

[6] For Kant, intrinsic purposiveness is not an objective principle but rather a regulative maxim of our reflective power of reason (see Chap. 4). He introduces a creator into his argument because it is only through the idea of an external force that we can grasp organismic purposiveness, that is, by analogy to our own human purposiveness. But then the *organism as subject* seems to disappear from his conception.

processes working on populations of genotypes. Such tending towards a goal is no longer a strange or vitalist explanation when it is based on stored memory traces.

The notion of evolution by habit was developed already by Lamarck, whose thoughts then inspired Charles Sanders Peirce. He considered habit an integral part of his evolutionary vision (for more detail, see Švorcová et al., 2023). Similarly, Sharov and Tønnessen (2021, 156) consider habit to be a crucial evolutionary process that ensures the goal-directness of living beings: '*The capacity to develop and preserve purposeful habits is a fundamental property of all self-reproducing agents. This creates a self-supporting and self-developing feedback loop between an agent and its purposes: The existence of an agent is supported by purposeful agential activity, which in turn is supported by the agents' existence. Thus, agents exist not only as they are, but also as being guided by the purposeful habits of their ancestors and their own. In other words, the existence of agents is purposeful, where purposes can be internal, as in organisms, and/or external, as in cell subagents and human artifacts.*' Nevertheless, a habit can often seem highly automatic, almost law-like. It is useful for living beings to delegate processes to automatisms but to view automatisms as the original state of affairs would be akin to the fallacy of misplaced concreteness (as defined by Whitehead, 1978), which describes a state where we confuse our concepts, schemes, or models with the way things really are (Whitehead aimed his argument primarily at the concepts of modern physics). We confuse the abstract with the immediate given. It is not just about knowledge: we also tend to see things where there are in fact stabilised processes.

It is preferable to view selection as working on both phenotypes and genotypes. Aside from the influence of random mutations in evolution, which are undeniable, I am thus inclined to think that organisms are inherently plastic and can respond to environmental changes by changing their phenotypes. These phenotypic changes can confer an adaptive advantage and lead to the emergence of new traits through phenotypic accommodation. The evolution of a new phenotype therefore does not necessarily require the development of new gene complexes. Instead, it can be achieved by modifications to the regulatory architecture of existing ones, leading to epigenetic changes or modular reorganisation of the phenotype (see Chap. 11). If the phenotype proves advantageous over multiple generations in a consistent environment, genetic changes can take place through various mechanisms, such as the Baldwin effect (when random mutations lead to phenotype fixation; Baldwin, 1896), Waddington assimilation (expression of previously hidden genetic variation; Waddington, 1952, 1956; Rutherford & Lindquist, 1998) or genetic accommodation (changes in gene frequency; West-Eberhard, 2003). Also, adaptation based on a tuned interplay between epigenetics and an organism's environment cannot be viewed as random: it is goal-oriented based on the function of memory. Indeed, organisms have been fine-tuned throughout their evolutionary history to regulate their genome in response to their environment, and they are likely to encounter similar signals repeatedly. Neither Aristotle (see Chap. 2) nor Schelling (see Chap. 5) considered explanations through random adaptation to be sufficient, and now we can finally put a finger on why they were right to think so.

Conclusion

In the following chapter, Johannes Jaeger will provide a detailed conceptual examination of organismal agency in the context of ontogeny and organisational closure. This analysis is closely tied to a concept of *organisational constraints*, that is, with the question of how various dynamic ontogenetic processes are limited or guided by the agency of an organism. The notion of *organisation as a basic way of imposing constraints* was also introduced by Terrence W. Deacon in his book *Incomplete Nature* (2013). In his quest to elucidate the emergence of mind from matter, Deacon notes that constraints are integral components of both biological and cognitive systems. In fact, '*a constraint is anything that restricts a system or a process from achieving its potential*' (Hoffmeyer, 2015, 63). Constraints in this sense impose limitations on all potential organisational states in a dynamic system and play a pivotal role in guiding such systems towards specific outcomes. As constraints consistently eliminate certain dynamical options, they contribute to the generation of reliable outcomes (Cassell, 2015). Deacon's emphasis on the causal impact of constraints enables him to articulate how general organisational patterns are manifested in nature. In absence of constraints, a system would be unable to achieve the specific and improbable states of organisation and self-maintenance that characterise living beings. This underscores the vital role of constraints in the emergence of teleodynamics. Moreover, by excluding particular processes, constraints inherently convey information by assigning significance to these specific processes. Constraints are also critical for the emergence of meaning, because interpretation can be thought of as an integration of externally available constraints that facilitates the organisation of work, subsequently contributes to the formation of additional constraints, and so on. The notion of constraints, which are also *experience-based*, thus covers a wide spectrum of memory-related phenomena discussed in this chapter. They range from genetic information constraining the cell's activity to epigenetic modifications limiting gene expression, cellular closure selectively interpreting external signs, and all the way to external environmental constraints imposing limitations on organisms. Even Peircean 'habit' can be then described as an expression of constraint (Hoffmeyer, 2015).

In this chapter, we have defined autonomous agents as agents with the capacity to act on their own behalf, driven by their own internal principles rather than steered by external control. Autonomous agents act in a goal-directed manner to preserve and enhance their existence, and they plan their future behaviour according to their previous experiences. In relation to this, we have also examined the concept of memory as the capacity of a system to retain and subsequently retrieve its experiences of the environment (including other forms of autonomous agents), starting with various forms of memory in bacteria and epigenetic memory in eukaryotes, and all the way to DRNs in ontogeny, which play a critical role in controlling development across different species. Organic memory in development is not just a storage device in the form of DNA. It is an interactive network based on experience of the environment and changes in the architecture of this network are the determining factor of the evolution of the phenotype.

This line of thought also involves the concept of phenotypic plasticity, meaning the ability of organisms to develop different phenotypes in response to changes in the environment via either continuous or discontinuous changes. All organisms exhibit some degree of plasticity, and their plastic existence is stabilised through fixed relationships between informational molecules, phenotypes, and the environment, with memory playing a crucial role in this process. To understand how organisms store their previous experience of the environment, we need to bring in the notion of organisational closure. Closure acts as a barrier between the inside of a cell and the external world, enables self-organisation towards order, and is a necessary precondition of interpretation and semiosis. This makes it a necessary condition for the emergence of life. Memory also helps us understand the teleology of the living: living organisms exhibit a special teleological causality that cannot be fully explained by the general laws of physics and chemistry. This causality is grounded in the interrelation of parts and the whole, where the whole in a sense precedes, grounds but also constrains the generation of the parts whose purpose is to fulfil this idea or concept of the whole: this makes living organisms both self-generating and self-causing. Such memory-based activity of organisms may, in our view, explain the purposeful end-directedness of ontogeny. It also presupposes selection working on phenotypes rather than genotypes and suggests that organisms can respond to environmental changes by changing their phenotype, which leads to the emergence of new traits through phenotypic accommodation. This adaptation is not random but rather goal-oriented based on the function of memory.

References

Abzhanov, A., Protas, M., Grant, B. R., Grant, P. R., & Tabin, C. J. (2004). Bmp4 and morphological variation of beaks in Darwin's finches. *Science, 305*(5689), 1462–1465.

Agrawal, A., Laforsch, C., & Tollrian, R. (1999). Transgenerational induction of defences in animals and plants. *Nature, 401*, 60–63.

Baldwin, J. M. (1896). A new factor in evolution. *The American Naturalist, 30*(354), 441–451.

Barbieri, M. (2003). *The organic codes. An introduction to semantic biology*. Cambridge University Press.

Barrangou, R., Fremaux, C., Deveau, H., Richards, M., Boyaval, P., Moineau, S., Romero, D. A., & Horvath, P. (2007). CRISPR provides acquired resistance against viruses in prokaryotes. *Science, 315*(5819), 1709–1712.

Boto, L. (2010). Horizontal gene transfer in evolution: Facts and challenges. *Proceedings of the Royal Society B: Biological Sciences, 277*(1683), 819–827.

Brier, S. (2015). Can biosemiotics be a "science" if its purpose is to be a bridge between the natural, social and human sciences? *Progress in Biophysics and Molecular Biology, 119*(3), 576–587.

Butler, S. (1877). *Life and habit*. A. C. Fifield.

Butler, S. (1910). *Unconscious memory*. A. C. Fifield.

Cassell, P. (2015). Incomplete Deacon: Why new research programs in the sciences and humanities should emerge from Terrence Deacon's incomplete nature. *Religion, Brain & Behavior, 5*(1), 60–65.

Chamowitz, D. (2012). *What plants know: A field guide to the senses*. Oneworld Publications.

Cropley, J. E., Suter, C. M., Beckman, K. B., & Martin, D. I. K. (2006). Germ-line epigenetic modification of the murine A^{vy} allele by nutritional supplementation. *PNAS, 103*(46), 17308–17312.

Cvrčková, F., Lipavská, H., & Žárský, V. (2009). Plant intelligence: Why, why not or where? *Plant Signaling & Behavior, 4*(5), 394–399.

Davidson, E. H. (2006). *The regulatory genome. Gene regulatory networks in development and evolution*. Academic.

Deacon, T. (2013). *Incomplete nature: How mind emerged from matter*. W. W. Norton & Company.

Dias, B. G., & Ressler, K. (2014). Parental olfactory experience influences behavior and neural structure in subsequent generations. *Nature Neuroscience, 17*, 89–96.

Driesch, H. (1921). *Philosophie des Organischen*. W. Engelmann.

Elsasser, W. M. (1987). *Reflections on a theory of organism*. The John Hopkins University Press.

Ernst, U. R., Van Hiel, M. B., Depuydt, G., Boerjan, B., De Loof, A., & Schoofs, L. (2015). Epigenetics and locust life phase transitions. *The Journal of Experimental Biology, 218*(1), 88–99.

Flemming, H. C., Wingender, J., & Szewzyk, U. (2016). Biofilms: An emergent form of bacterial life. *Nature Reviews Microbiology, 14*(9), 563–575.

Förster, E. (2002). Die Bedeutung von §§ 76, 77 der Kritik der Urteilskraft für die Entwicklung der nachkantischen Philosophie. *Zeitschrift für Philosophische Forschung, 56*(2), 169–190.

Giddens, A. (1984). *The constitution of society: Outline of the theory of structuration*. University of California Press.

Gilbert, S. F., & Epel, D. (2015). *Ecological developmental biology: The environmental regulation of development, health, and evolution*. Sinauer Associates, Inc.

Guéant, J.-L., Chéry, C., Oussalah, A., Nadaf, J., Coelho, D., Josse, T., Flayac, J., Robert, A., Koscinski, I., Gastin, I., et al. (2018). APRDX1 mutant allele causes a MMACHC secondary epimutation in cblC patients. *Nature Communications, 9*, 67.

Haeckel, E. (1876). *Die Perigenesis der Plastidule oder die Wellenzeugung der Lebenstheilchen*. Reimer.

Hamm, D. C., & Harrison, M. M. (2018). Regulatory principles governing the maternal-to-zygotic transition: Insights from *Drosophila melanogaster*. *Open Biology, 8*(12), 180183.

Hering, E. (1921 [1870]). Über das Gedächtnis als allgemeine Funktion der organisierten Materie. Vortrag gehalten in der feierlichen Sitzung der Kaiserlichen Akademie der Wissenschaften in Wien am XXX. Mai MDCCCLXX. Akademische Verlagsgesellschaft.

Hering, E. (1897). *On memory and the specific energies of the nervous system*. The Open Court Publishing Co.

Hoffmeyer, J. (1996). *Signs of meaning in the universe. The natural history of signification*. Indiana University Press.

Hoffmeyer, J. (2000). Code-duality and the epistemic cut. *Annals of the New York Academy of Sciences, 901*, 175–186.

Hoffmeyer, J. (2008). *Biosemiotics: An examination into the signs of life and the life of signs*. University of Scranton Press.

Hoffmeyer, J. (2011). Biology is immature biosemiotics. In C. Emmeche & K. Kull (Eds.), *Towards a semiotic biology: Life is the action of signs* (pp. 43–65). Imperial College Press.

Hoffmeyer, J. (2015). Constraints on matter are real agencies. *Religion, Brain & Behavior, 5*(1), 60–65.

Hu, L., Xiao, P., Jiang, Y., Dong, M., Chen, Z., Li, H., Hu, Z., Lei, A., & Wang, J. (2018). Transgenerational epigenetic inheritance under environmental stress by genome-wide DNA methylation profiling in cyanobacterium. *Frontiers in Microbiology, 9*, 1479.

Johannsen, W. (1909). *Elemente der exakten Erblichkeitslehre. [Elements of an exact theory of heredity]*. Gustav Fischer.

Jonas, H. (2001). *The phenomenon of life: Toward a philosophical biology*. Northwestern University Press.

Kauffman, S. (2000). *Investigations*. Oxford University Press.

Kull, K., Emmeche, C., & Hoffmeyer, J. (2011). Why biosemiotics? An introduction to our view on the biology of life itself. In C. Emmeche & K. Kull (Eds.), *Towards a semiotic biology: Life is the action of signs* (pp. 1–21). Imperial College Press.

Lachmann, M., & Jablonka, E. (1996). The inheritance of phenotypes: An adaptation to fluctuating environments. *Journal of Theoretical Biology, 181*(1), 1–9.

Markoš, A. (2002). *Readers of the book of life. Contextualizing developmental evolutionary biology*. Oxford University Press.

Markoš, A., & Švorcová, J. (2019). *Epigenetic processes and the evolution of life*. CRC Press/Taylor and Francis Group.

Maturana, H., & Varela, F. (1980). *Autopoiesis and cognition. The realization of the living*. Springer.

Michaelis, C., & Grohmann, E. (2023). Horizontal gene transfer of antibiotic resistance genes in biofilms. *Antibiotics, 12*, 328.

Mitchell, K. (2018). Grandma's trauma – A critical appraisal of the evidence for transgenerational epigenetic inheritance in humans. *Blog Entry*. Available online: http://www.wiringthebrain.com/2018/05/grandmas-trauma-critical-appraisal-of.html. Accessed on 11 Apr 2023.

Monod, J. (1971). *Chance and necessity: An essay on the metaphysics of life*. A. A. Knopf.

Moreno, A., & Mossio, M. (2015). *Biological autonomy: A philosophical and theoretical enquiry*. Springer.

Morgan, H. D., Sutherland, H., Martin, D. I., & Whitelaw, E. (1999). Epigenetic inheritance at the agouti locus in the mouse. *Nature Genetics, 23*, 314–318.

Muraca, B. (2014). Teleology and the life sciences: Between limit concept and ontological necessity. In S. A. Koutroufinis (Ed.), *Life and process. Towards a new biophilosophy*. De Gruyter.

Newman, S. A., & Bhat, R. (2009). Dynamical patterning modules: A "pattern language" for development and evolution of multicellular forms. *International Journal of Developmental Biology, 53*(5–6), 693–705.

Newman, S. A., Forgacs, G., & Müller, G. B. (2006). Before programs: The physical origination of multicellular forms. *The International Journal of Developmental Biology, 50*(2–3), 289–299.

Nijhout, H. F. (2003). Development and evolution of adaptive polyphenism. *Evolution and Development, 5*(1), 9–18.

Otis, L. (1994). *Organic memory: History and the body in the late nineteenth and early twentieth centuries*. University of Nebraska Press.

Perlman, M. (2010). The modern philosophical resurrection of teleology. In A. Rosenberg & R. Arp (Eds.), *Philosophy of biology. An anthology*. Wiley-Blackwell.

Pfennig, D. W. (2021). *Phenotypic plasticity & evolution: Causes, consequences, controversies*. CRC Press/Taylor and Francis Group.

Portmann, A. (2022 [1965]). New fronts of biological work. In F. Jaroš & J. Klouda (Eds.), *Adolf Portmann. A thinker of self-expressive life* (pp. 13–21). Springer.

Rádl, E. (1930). *The history of biological theories*. Oxford University Press.

Reilly, J. N., McLaughlin, E. A., Stanger, S. J., Anderson, A. L., Hutcheon, K., Church, K., Mihalas, B. P., Tyagi, S., Holt, J. E., Eamens, A. L., & Nixon, B. (2016). Characterisation of mouse epididymosomes reveals a complex profile of microRNAs and a potential mechanism for modification of the sperm epigenome. *Scientific Reports, 6*, 31794.

Rosenberg, T., Marco, A., Kisliouk, T., Haron, A., Shinder, D., Druyan, S., & Meiri, N. (2022). Embryonic heat conditioning in chicks induces transgenerational heat/immunological resilience via methylation on regulatory elements. *FASEB Journal, 36*, e22406.

Rutherford, S. L., & Lindquist, S. (1998). Hsp90 as a capacitor for morphological evolution. *Nature, 396*(6709), 336–342.

Saxena, A., Towers, M., & Cooper, K. L. (2017). The origins, scaling and loss oftetrapod digits. *Philosophical Transactions of the Royal Society B, 372*(1713), 20150482.

Schacter, D. L. (2001). *Forgotten ideas, neglected pioneers: Richard Semon and the story of memory*. Routledge/Taylor and Francis Group.

Semon, R. (1921 [1904]). *The mneme*. Allen and Unwin.

Sharma, U., Sun, F., Conine, C. C., Reichholf, B., Kukreja, S., Herzog, V. A., Ameres, S. L., & Rando, O. J. (2018). Small RNAs are trafficked from the epididymis to developing mammalian sperm. *Developmental Cell, 46*(4), 481–494.
Sharov, A. A., & Tønnessen, M. (2021). *Semiotic agency. Science beyond mechanism*. Springer.
Stilling, R. M., van de Wouw, M., Clarke, G., Stanton, C., Dinan, T. G., & Cryan, J. F. (2016). The neuropharmacology of butyrate: The bread and butter of the microbiota-gut-brain axis? *Neurochemistry International, 99*, 110–132.
Švorcová, J. (2012). The phylotypic stage as a boundary of modular memory: Non mechanistic perspective. *Theory in Biosciences, 131*(1), 31–42.
Švorcová, J. (2023). Transgenerational epigenetic inheritance of traumatic experience in mammals. *Genes, 14*, 120.
Švorcová, J., & Markoš, A. (2023). Closures as a precondition of life, agency, and semiosis. *Biosemiotics, 16*, 45–59.
Švorcová, J., Lacková, Ľ., & Fulínová, E. (2023). Evolution by habit: Peirce, Lamarck, and teleology in biology. *Theory in Biosciences, 142*, 411–422.
Toepfer, G. (2011). Zweckmäßigkeit. In G. Toepfer (Ed.), *Historisches Wörterbuch der Biologie. Geschichte und Theorie der biologischen Grundbegriffe* (pp. 786–834). J. B. Metzler.
Uller, T., English, S., & Pen, I. (2015). When is incomplete epigenetic resetting in germ cells favoured by natural selection? *Proceedings. Biological sciences, 282*(1811), 20150682.
Waddington, C. (1952). Genetic assimilation of an acquired character. *Evolution, 7*(2), 118–126.
Waddington, C. H. (1956). Genetic assimilation of the Bithorax phenotype. *Evolution, 10*(1), 1–13.
Walsh, D. M. (2015). *Organisms, agency, and evolution*. Cambridge University Press.
Waterland, R. A., & Jirtle, R. L. (2003). Transposable elements: Targets for early nutritional effects on epigenetic gene regulation. *Molecular and Cellular Biology, 23*(15), 5293–5300.
Weaver, I. C., Cervoni, N., Champagne, F. A., D'Alessio, A. C., Sharma, S., Seckl, J. R., Dymov, S., Szyf, M., & Meaney, M. J. (2004). Epi-genetic programming by maternal behavior. *Nature Neuroscience, 7*, 847–854.
West-Eberhard, M. J. (2003). *Developmental plasticity and evolution*. Oxford University Press.
Whitehead, A. N. (1978). *Process and reality. An Essay on Cosmology*. Free Press.

Chapter 10
Ontogenesis, Organisation, and Organismal Agency

Johannes Jaeger

Abstract At first sight, the empirical study of ontogenesis and the theoretical study of organismal agency seem to have little in common. In this chapter, I discuss why this initial impression is incorrect. First of all, ontogenesis and agency are indirectly connected at the level of the whole organism, because they are co-dependent on the peculiar organisation that characterises living systems. While ontogenesis is constrained by its own requirement to maintain living organisation in the form of organisational closure throughout the lifecycle, agency is grounded in the same phenomenon of organisational continuity. Secondly, cellular agency contributes more directly to various important processes of multicellular development in organisms with multiple levels of organisation. This leads to a view of ontogenesis that emphasises agency and variation in the underlying cellular dynamics and focuses on stability and reproducibility of ontogenetic processes as its main explanatory targets. I examine how these insights can help us bridge the explanatory gap between reductionist mechanistic empirical approaches and theoretical considerations regarding the organisation of an organism as a whole. I conclude that the two approaches are best used in a complementary manner. Only by placing ontogenetic mechanisms within the larger context of the evolving lifecycle can we arrive at an adequate understanding of their functionality and evolution.

Keywords Organismic agency · Ontogenesis · Biological organisation · Organisational continuity · Reproducer perspective · Organisational function · Viable variation · Cellular agent theory · Process

---

J. Jaeger (✉)
Department of Philosophy, University of Vienna, Vienna, Austria

Complexity Science Hub (CSH), Vienna, Austria

Ronin Institute, ronininstitute.org, U.S.A.
e-mail: yoginho@ronininstitute.org

J. Švorcová (ed.), *Organismal Agency*, Biosemiotics 28,
https://doi.org/10.1007/978-3-031-53626-7_10

Introduction

Why should an experimental biologist care about *organismal agency*? Some would argue that *every* biologist should care about agency, because that is what sets the domain of the living apart from the domain of the non-living (see, e.g., Moreno & Mossio, 2015; Walsh, 2015). It is not, however, my aim to make or defend this general claim here. I consider the agential perspective to be one of many potentially valid approaches in contemporary biology, and I am interested in the scope of its explanatory power. Does the agential perspective bring new questions, new approaches, and new insights to the table? To investigate this possibility, I will focus on a more specific question: is taking organismal agency seriously in some way relevant to the empirical practice of studying cellular and developmental regulatory processes and their role in evolution? And also *vice versa*: is the empirical study of cellular and developmental regulatory processes and their evolution relevant to a proper grasp of organismal agency? If there is a mutual interdependence, it is certainly not obvious or straightforward. These two areas of empirical versus theoretical inquiry seem unrelated; at this point in history, they exist and operate quite independently of each other.

A naturalistic conceptualisation of agency can be grounded in the peculiar way organisms are organised (as argued by Moreno & Mossio, 2015). Within this perspective, agency is a relational property: what life is depends not on what an organism is made of, but instead on how its component processes interact and sustain each other to create an organismic whole. In this sense, agency is irreducible: it is a property of the organisation of an *entire* cell or organism that cannot be localised to any particular subset of component processes within a living system.

In contrast, the perspective of modern cell biology and developmental biology is predominantly reductionist. Biological phenomena are usually explained in terms of their underlying molecular and genetic *mechanisms*. While the *function* of a mechanism may have to be considered against a wider background of a whole cell or developmental system and its evolution (see, e.g., Wright, 1973; Cummins, 1975; Boorse, 1976; Bigelow & Pargetter, 1987; Love, 2007; Mossio et al., 2009), the *activities* of a mechanism are to be explicated exclusively in terms of its inherent molecular and genetic components and their orchestrated interactions (e.g., Kauffman, 1971a; Wimsatt, 1974, 1976; Bechtel & Richardson, 1993; Glennan, 1996; Machamer et al., 2000; Bechtel & Abrahamsen, 2005; Craver, 2007; Bechtel & Abrahamsen, 2010; Bechtel, 2011; recently reviewed in Glennan & Illari, 2017; Craver, & Tabery, 2019, see Chap. 6). The tacit assumption behind this approach is that understanding of an increasing number of these mechanisms in sufficient detail should eventually add up to understanding the organism or the cell as a whole.

In this chapter, I critically examine this assumption and show that a strictly reductionist (bottom-up) approach is insufficient for a proper understanding of cellular and developmental processes and their contributions to organismic organisation and evolution. In turn, I also argue that a purely abstract (top-down) functional analysis of biological organisation makes it difficult to make a link with empirical work in cell and developmental biology. To maximise their effectiveness and usefulness, both approaches ought to complement and inform each other. I propose specific ways in which this can be achieved. On top of that, I illustrate how the empirical study of developmental processes in multicellular organisms would profit from a perspective that takes agency at the cellular level seriously. It is in these rather roundabout ways that the empirical study of regulatory mechanisms and the theory of the organism and its agency are interconnected.

To show how the two approaches affect each other—and to ultimately reconcile them—I must first reframe developmental and cellular regulatory activities within their wider organismic and evolutionary context. To this purpose, I adopt the account of biological organisation proposed by Moreno, Mossio, and colleagues (e.g., Moreno & Mossio, 2015), embedded in Griesemer's (2006) *reproducer perspective* on evolution (see Jaeger, 2023). In this view, a reproducer is the fundamental and minimal unit of evolution. It roughly corresponds to an organism's lifecycle, subdivided into processes of *reproduction* and *ontogenesis*, which are mutually dependent in the strong sense, meaning one cannot exist without the other. '*Reproduction*' is defined in a conventional way, but by '*ontogenesis*', I mean all aspects of the lifecycle – metabolic, physiological, developmental, and behavioural – involved in an organism's acquisition of the capacity to reproduce (Griesemer, 2006; Jaeger, 2023). This definition is much broader than the usual notion of development as 'embryogenesis' or 'morphogenesis'. It applies to lifecycle stages and behaviours beyond embryonic development and pertains to unicellular and multicellular organisms alike. For instance, it qualifies the growth and maturation of a protist between cell divisions as an ontogenetic process, because growth and maturation are required for its capacity to reproduce. Similarly, the courting behaviour of an adult organism can likewise be considered part of 'ontogenesis' in this sense. Equipped with this broad and basic definition of 'ontogenesis', we can now tackle two important questions, which lie outside the scope of any reductionist–mechanistic approach to biology.

First, we can ask how ontogenetic processes contribute to organismal agency, and how this is connected to acquisition of the capacity to reproduce. The key to answering this question lies in the realisation that both agency and completion of a lifecycle depend on a continuous self-maintenance of biological organisation (Saborido et al., 2011; Moreno & Mossio, 2015; Mossio & Pontarotti, 2020; DiFrisco & Mossio, 2020; Jaeger, 2023). We have already argued that this kind of biological organisation is irreducible. It is also highly non-aggregative: an organism considered as a whole behaves very differently from any sum of activities of its mechanistic processes. In fact, the component processes of a living system usually *require* the larger context of the organism to robustly behave the way they do. Considered from this top-down organismic perspective, the contributions of individual regulatory mechanisms to the overall organisation of a living system become a highly nontrivial and nonadditive problem that requires a close (re)examination. Put simply, we must ask what is missing from the reductionist–mechanistic picture that prevents us from understanding the regulatory and behavioural capabilities of the organism and its evolving lifecycle as a whole.

Second, we can investigate how (and to what extent) the organisation of the whole organism constrains and influences the structure and activity of particular ontogenetic mechanisms. This question is especially important in the context of regulatory evolution. For an organism to be evolvable, it must be able to generate variability within certain boundaries that keep its organisation functionally intact, thus ensuring that the organism remains alive and able to complete its lifecycle (Griesemer, 2006; Montévil et al., 2016a; Jaeger, 2023). The classical notion of *developmental constraints* to some extent captures such functional boundaries (Maynard Smith et al., 1985), but it does not relate them back to the organism's overall organisation. In other words, it does not distinguish between those functional constraints which are indispensable for the general maintenance of organisation and those which are more contingently required to sustain a particular cellular or developmental activity that could, in principle, be absent or vary without fatally disrupting the completion of the lifecycle. In what follows, I will explore whether it is possible to combine an organisational account of organisms with a novel empirical research programme to differentiate between these two alternatives. On top of that, I will touch upon the role of cellular agency during ontogenesis in multicellular organisms.

The ultimate aim of this chapter is to connect the theoretical and abstract high-level discussions about biological organisation and the basic units of evolution with empirical work on cellular and developmental dynamics. I hope to convince the reader that these two types of investigation not only can but *should* be integrated if we are to reach a deeper and more rigorous understanding of ontogenesis, agency, and their role in evolution. To achieve this, I start with a brief overview on biological organisation and organismic agency in the following section. Further, I discuss the fundamental importance of ontogenesis for the reproducer perspective on evolution, which also allows me to ask how ontogenesis contributes to organismic agency. In the following two sections, I examine how organismal agency constrains ontogenesis as seen from an evolutionary point of view and I argue that the view established in previous sections requires a fundamental rethinking of the way we study the role of cellular agents in ontogenesis. Finally, in the last section of the chapter, I present several important implications of the agential approach and contextualise it within the historical tradition of process structuralism.

Biological Organisation and Organismic Agency

What links ontogenesis and organismal agency is *biological organisation*. To understand this connection, we need a precise definition of 'organismal agency'. To take the agential perspective, in its most basic sense, is to acknowledge that organisms have the capacity to originate causal effects from within their boundaries or, to be more precise, from within their self-producing and self-maintaining (autopoietic) architectures (see, e.g., Varela et al., 1974; Maturana & Varela, 1980; Moreno & Etxeberria, 2005; Barandiaran et al., 2009; Moreno & Mossio, 2015; Jaeger, 2023).

But this is only part of it. A candle flame, for example, can also self-maintain and cause effects from within its internal organisation. That, however, does not mean it is alive or has agency in the sense of the term used here. To arrive at a more accurate definition, we need to further consider how organisms interact with their environment (Barandiaran et al., 2009). Some of the internally generated effects involved in an organism's interaction with its environment can be interpreted as *goal-oriented actions* (Walsh, 2015). It is the organism's repertoire of such actions that enables it to pursue ends by using opportunities and avoiding obstacles (that is, by leveraging *affordances*) in its perceived environment (Gibson, 1979; Walsh, 2015; Roli et al., 2022). That is what agency means (see Chaps. 6, 9, and 14). Examples of such goal-oriented behaviour range from a bacterium swimming up a glucose gradient, in search of nourishment to a deer dividing its time between grazing on an open meadow and hiding in the underbrush to avoid predators, all the way to the complex social interactions of human beings in pursuit of our intricate but often fleeting ambitions. To summarise in simple terms: organisms – from bacteria to all the way to humans – are capable of 'acting on their own behalf' in more or less sophisticated ways, each according to their own particular needs and abilities (Kauffman, 2000).

Agency requires a certain degree of *autonomy* from the environment (Moreno & Etxeberria, 2005; Moreno & Mossio, 2015), 'freedom from immediacy' (Gold & Shadlen, 2007). Unlike algorithmic 'agents' in computer science, true organismic agents are more than merely complicated input–output processing devices (Roli et al., 2022). Their responses are not necessarily automatic: agency, in the intuitive sense of the term, implies that at least some causes of an organism's behaviour are not directly determined by environmental inputs; rather, they are based on the internal organisation underlying its goals and needs. Moreover, and this is crucial here, the kind of agency we are talking about also implies that some of the causes of an organism's *ontogenesis* originate from within its own autopoietic organisation. There is a degree of freedom, a degree of *self-determination* (Mossio & Bich, 2017), not only in behaviour but also in ontogenesis. Self-determination is rooted in the historicity and the particular – self-referential and multilevel – architecture of biological organisation. The organism's current behavioural and ontogenetic state is contingent on and continuous with its own past internal states as well as (as we shall see) those of its ancestors (Montévil et al., 2016a, Chap. 9). This connection to the past is mediated by a form of iterative and dialectic self-constraint, a *dynamic presupposition*: at each moment of its existence, organisation of the organism establishes the conditions for its own continued existence (Bickhard, 2000; Moreno & Mossio, 2015; Mossio & Bich, 2017). This leads to an *organisational continuity* as a special form of causal continuity, which underlies the organism's ability not only to act but also to maintain its identity and reproduce itself (DiFrisco & Mossio, 2020; Mossio & Pontarotti, 2020; Jaeger, 2023).

Before we can better understand how organisational continuity is achieved and how it plays a role in ontogenetic and evolutionary processes, we should clarify what we mean by 'biological organisation'. Biological systems are organised in the sense that they realise a kind of causal closure (Piaget, 1967; Rosen, 1991; Moreno and Mossio, 2015; Mossio and Bich, 2017). *Organisational closure* is a property of the whole organism (Chaps. 8 and 9). It is complementary to the organism's thermodynamic openness and describes the fact that all functions of a living system which are required to maintain the conditions for its continued existence mutually depend on each other, supporting and maintaining one another through their collective interactions. Organisation is dynamic and multileveled, because these functions are actualised by a set of constraints imposed on the underlying open flow of matter and energy (see Chaps. 5, 7, 9, and 11). The exact structure of these constraints is not fixed either: it can, and indeed must, change over time. But such changes happen on a different timescale than the constrained physicochemical processes which underlie it. As an example, consider an enzyme whose concentration is constantly regulated by the cell, but at a much slower timescale than the metabolic reaction it catalyses. To be more specific, organisational closure implies that there is *a closure of constraints* within a living system (Montévil & Mossio, 2015; Mossio et al., 2016). The cell must be able to produce and regulate all the enzymes required to maintain its metabolic state which, in turn, must be sufficient for achieving enzyme production. In this circular and hierarchical way, organisms are *closed to efficient causation* (Rosen, 1991, cf. Chap. 2). This kind of closure allows organisms to achieve their autonomy, to maintain their own boundaries, and to reproduce themselves: it is *an autopoietic system* (Varela et al., 1974; Varela, 1979; Maturana & Varela, 1980).

The crucial question that concerns us here is how organisation is generated and how it persists over time. What does it mean to say that organisation provides the conditions for its own continued existence? The notion of organisational continuity is rooted in what Stuart Kauffman (2000) called *work-constraint cycles*: an organism, as a thermodynamically open system, releases energy in the form of physical work, which it uses to kinetically channel the underlying physicochemical processes in specific directions (see Chaps. 8 and 12). Let us illustrate this point with enzymes again: the presence of these catalysts, with each at a particular regulated level of concentration, determines which metabolic reactions (selected from an enormous range of possible ones) actually take place in the cell. In this sense, the cell kinetically 'lifts' the reactions that constitute its metabolism from an unimaginably vast configuration space of potential chemical reactions. The resulting set of actualised reactions must in turn be autocatalytic, meaning that that metabolism itself must be able to produce all the enzymes required for its continued existence (see above, and Kauffman, 1971b, 1986, 1993). From one moment to the next, the concentration of individual enzymes may vary but this autocatalytic property (*autocatalytic closure*) of the whole set of metabolic reactions must be maintained. If it is not, the cell dies. We could say that a living cell 'harvests' its flow of metabolic energy through iterative work-constraint cycles. Enzymes enable metabolism which generates more enzymes, and so on. Closure is maintained via this kind of dialectic dynamic both within the cell or the lifecycle (ontogenesis, in the broad sense defined in the Introduction) and from one generation to the next (heredity and reproduction; Saborido et al., 2011; DiFrisco & Mossio, 2020; Mossio and Pontarotti 2020). This establishes organisational continuity in entire lineages of dividing cells and evolving organisms.

Organisational continuity has two fundamental implications which reveal a subtle yet deep connection between organismal agency and ontogenesis. First, only a minuscule fraction of all imaginable sets of closed constraints are actually compatible with organisational continuity (DiFrisco & Mossio, 2020). At this point, we know very little about how such suitable sets of constraints are generated or selected or what they look like in terms of their physicochemical structure and dynamics. What is clear, though, is that selection of such viable sets is a process that iteratively and dialectically depends on all previous organised states in a living lineage. Organised states engender the propagation of further organised states through ontogenesis and reproduction (Jaeger, 2023). This is what we mean by autonomy through self-constraint (see above, and Mossio & Bich, 2017) or autonomy through historicity (Montévil et al., 2016a). Organisational continuity is a succession of organised states whose trajectory enjoys a degree of freedom from the organism's immediate environment. The internal causes of this trajectory remain opaque to an external observer: they are subject to closure and cannot be detected from outside the organism. On top of that, they are resistant to reductionist analysis because we cannot reduce or localise them to any component process within the organism. By taking a living system apart to focus on one or more of its component mechanisms we kill it, thus losing its overall organisation, the integrated whole which causes it to be alive in the first place.

All of this means that biological organisation goes beyond any simple reductionist–mechanistic analysis, where regulatory processes can be strictly delimited with a concrete beginning and outcome interconnected by a specific set of interacting components. Instead, biological organisation is a dynamic *holarchy* (Koestler, 1982), i.e., it is constituted from processes and constraints that are always both the producer and the produced, both a part and a whole, without any straightforward overall hierarchical polarity of 'up' or 'down' in their interactions. This holarchy has no particular beginning or end (naturally with the sole exception of the origin and eventual extinction of all life). Holarchic systems severely challenge our ability to grasp their potential. It has been argued that this is why it is impossible to formalise general rules for organismic behaviour and evolution (Roli et al., 2022). Although we may be able to predict and simulate most of an organism's dynamics and habitual behaviour using mechanistic approaches, we cannot capture the full range of its capabilities using an algorithmic model (Rosen, 1991; Roli et al., 2022). Organisms can behave, develop, and evolve in truly surprising ways. The causes of some of their behaviours cannot be predicted – they can only be retrospectively reconstructed. Ultimately, this is what we mean when we say that organisms have autonomy and agency.

The second implication is much simpler and easier to grasp but no less fundamental. Organisational continuity is what is necessary for the completion of a lifecycle. Constraint closure must be sustained not only throughout ontogenesis, but also during reproduction (Saborido et al., 2011; DiFrisco & Mossio, 2020; Mossio & Pontarotti, 2020). While continuous in this sense, reproduction and ontogenesis differ in two main features. First of all, constraints involved in reproductive processes affect not only the lifecycle of the parent organism, but also that of its offspring. They are part of an extended parent–offspring dyad that realises closure just as its individual component systems do (DiFrisco & Mossio, 2020). Secondly, during reproduction, the number of individualised organised systems changes. This takes place either through fission (e.g., cell division, or birth), or fusion (e.g., of gametes upon fertilisation), while during ontogenesis, this number remains constant (DiFrisco & Mossio, 2020). Completion of a lifecycle requires *both* ontogenesis and reproduction (see next section). And last but not least, the organisational perspective defines heredity as a conservation of functional constraints across generations (Saborido et al., 2011; Mossio & Pontarotti, 2020). To put it succinctly, an organism cannot survive, reproduce, or evolve without organisational continuity.

Taken together, what I have discussed so far outlines the following picture: Organisational closure must be maintained throughout ontogenesis. It is inherited by the offspring through processes involved in reproduction, which imposes strong restrictions on the underlying sets of constraints and physicochemical processes. At the same time, organisational continuity is what enables the organism to evolve, develop, and act. Because it is an irreducible property of the entire lifecycle, organisational continuity cannot be understood by purely reductionist–mechanistic approaches. In brief, organisational continuity is what connects ontogenesis with organismal agency and embeds them both in the larger process of evolution. To gain a proper understanding of these deep connections, we must take a high-level perspective on organismic evolution (Jaeger, 2023).

Ontogenesis from the Reproducer Perspective

Now is the time to contextualise what I have discussed so far in terms of its impact on evolution. I have demonstrated how organisational continuity underlies the completion of a lifecycle. This is a necessary condition for the evolvability of any organism, because a lifecycle constitutes the *reproducer*, which is the fundamental and minimal unit of evolution (see Introduction of this chapter, and Griesemer, 2006). To summarise: a reproducer consists of two tightly interwoven processes, namely ontogenesis and reproduction, which mutually depend on each other for their very existence.

But why are reproducers central to evolution? Let me briefly review here an argument which I presented in full detail elsewhere (Jaeger, 2023). The central point is that a suitable unit of evolution must be able to give rise to more of itself (Szathmáry & Maynard Smith, 1993). This can be explicitly formulated through a combination of basic principles: the *principle of multiplication* (unit **A** makes more **A**) and the *principle of heredity* (**A** makes **A**, not **B**). Traditionally, it has been held that the basic unit of evolution is some kind of *replicator*, defined as an entity (usually a specific molecule, such as DNA, or the genome as a subcellular organelle in a broader sense) capable of multiplying and transmitting its structure directly and (relatively) intact through a copying process. The problem is that no such entity (with a suitable copying process) exists outside its cellular context. Even prions and viruses need cellular hosts for replication. As explained in detail in Jaeger (2023), 'naked' replicators without an appropriate cellular environment quickly succumb to error catastrophes. They degenerate or fragment into a myriad of 'subspecies' within just a few iterations of the copying process. A suitable unit of evolution therefore also requires a suitable *principle of variability* (**A** sometimes makes **A'**, which may in fact be **B**). There is a lot packed into the 'sometimes' here: the amount and kind of variation produced must be quite specific. Not only do we need replication errors to occur above a certain base rate, but we also cannot have too many of them. Furthermore, they must produce a particular quality of phenotypic variability with a sufficient amount of potentially beneficial variants to be amenable to positive selection (Jaeger, 2023). On top of that, biological reproduction always includes material overlap between generations (Griesemer, 2006). The offspring necessarily arise from parts of their ancestors. But a simple copying process of replication does not involve such overlap: for instance, the bases to synthesise a new strand of DNA come from outside the existing structure of the replicator. Therefore, replicators cannot account for organisational continuity within a lineage (unless the organisation is completely programmed into the replicator, which is not the case in biological reproduction; see below).

To make a long story short: naked replicators cannot properly reproduce in any biologically accurate sense, and they cannot produce the kind of selectable heritable phenotypic variability needed for evolution unless they are embedded in the more complex dynamics of a complete cellular lifecycle with organisational continuity, that is, in a reproducer. And this lifecycle, by necessity, must include ontogenesis (*sensu lato*, as outlined above). Since the offspring arise from *parts* of their parents (a subset of component processes; **a** instead of **A**), there must be a process which enables them to replenish the missing components while maintaining organisational continuity in order to acquire the capacity to reproduce again (Griesemer, 2006). In the case of unicellular organisms, this means simply growth, multiplying organelles, and genome replication. In the case of more complex multicellular organisms, this includes the processes of embryogenesis, postembryonic development, metamorphosis, and so on. In both cases, behavioural patterns are also necessary, such as searching for nutrition or mates. In summary, a proper unit of evolution must, in addition to the principles of multiplication and variation mentioned above, include a *principle of ontogenesis* (**a** becomes **A**) and a broadened principle of heredity (organisation must be maintained across generations) (Griesemer, 2006; Jaeger, 2023).

One aspect of ontogenesis is of a special importance in this evolutionary context. Let me therefore emphasise again: ontogenetic processes are not only responsible for producing the mature form of a reproducer (**A**) but in doing so, they also produce the phenotypic variability which natural selection can act upon (**a** becomes **A'** or indeed **B,** Montévil et al., 2016a). Therefore, ontogenesis is central to an organism's evolvability in two complementary ways: first, it accomplishes the completion of the lifecycle, which is the basic prerequisite for evolvability (the general ability to evolve); second, it produces the kind of variability that can lead to adaptation (evolvability in the narrower sense of potential adaptability; see Wagner & Altenberg,

1996; Pigliucci, 2008). The first aspect constrains the second, and the second presupposes the first. The kind of variability that can be produced must be viable in the sense that it must occur within a bounded space of possibilities which allows for the maintenance of organisational continuity throughout the lifecycle. In what follows, I investigate both the contributions of ontogenesis to this maintenance and the constraints imposed on it by this fundamental condition for evolvability (see Chap. 11).

Organisational Function and Ontogenesis: A Top-Down Theoretical Analysis

In the above, I have introduced a conceptual framework which should enable us to assess the contribution of ontogenesis to organismal agency. Seen from this perspective, viable evolvable organisms must be reproducers. The completion of a reproducer lifecycle requires both ontogenesis and organisational continuity. Ontogenesis generates phenotypic variability that is viable as long as it enables the organism to acquire the capacity to reproduce. In turn, organisational continuity is necessary for acquiring that reproductive capacity, while also providing the basis for heredity, organismic autonomy, self-determination, and agency from the reproducer perspective. Our central question can thus be narrowed down to *how ontogenesis contributes to organisational continuity*.

A natural first step towards tackling this problem is to redefine what it means for an ontogenetic process to have a function from the perspective of the organism's overall organisation (Christensen & Bickhard, 2002; Mossio et al., 2009). Accordingly, *the organisational function of a component process is identified with its contribution to organisational closure and continuity in the context of the reproducer lifecycle*. This approach seems promising, but it is also fraught with a number of substantial conceptual challenges, especially if our aim is to inform the empirical study of ontogenesis by theoretical considerations about agency and organisation.

The main problem is that contributions of ontogenetic processes to organisational closure and continuity are generally difficult to pin down. Obvious exceptions to this are the core metabolic and physiological processes needed to maintain cellular milieu and integrity, as well as cellular and developmental processes directly involved in reproduction, such as genome replication and all aspects of mitotic division. In multicellular organisms, we can add gametogenesis and the associated generation and maintenance of suitable somatic niches (e.g., the development of primary sexual organs).

However, most ontogenetic processes fall outside these straightforward categories. They contribute only indirectly, contingently, and/or tangentially (if at all) to self-maintenance and reproduction. As an example, consider the processes driving flagellar motion in eukaryotic protists. First and foremost, they contribute to locomotion. Their role in self-maintenance and reproduction is rather indirect and contingent (they get the organism in position for feeding or photosynthesis). This latter contribution not only depends on the organism's intrinsic locomotor ability, but crucially also on external factors outside its boundaries and beyond its control: did it manage to actually increase food or light intake through movement? The answer radically hinges on the context and to some degree on chance. Given these uncertainties, we ought to ask ourselves what is more useful and tractable for empirical research: to identify the function of flagellar processes in terms of their contribution to locomotion (function in a traditional systemic sense; Cummins, 1975) or in terms of their much more tenuous and indirect contribution to organisational continuity (Mossio et al., 2009)? How can we distinguish the latter from the former? Can we do so at all? Clearly, the two largely overlap. Under such circumstances, how are we to isolate, determine, and quantify contributions to self-maintenance and reproduction? Under most natural conditions, the task seems ill-defined.

Let us consider another example: in all organisms, growth and reproduction obviously rely on the intake of nutrients or other forms of energy from the environment. All metabolic, physiological, and behavioural processes involved in feeding or photo/chemosynthesis therefore fall under the gamut of ontogenesis broadly defined as 'acquiring the capacity to reproduce'. On the other hand, not all the energy invested in these processes actually contributes to self-maintenance and reproduction. How large a part of it does so once again heavily depends on contingent environmental conditions. First of all, different metabolic pathways need to be activated under different nutrient or light conditions. An organism living in a variable environment therefore requires many more pathways than those constituting its core metabolism. Other secondary metabolic pathways can be used to synthesise a range of toxins and repellents directed against potential predators, or attractants and signals used for finding a mate or for essential mutualistic or symbiotic interactions. Whether such secondary pathways help the organism acquire the capacity to reproduce depends on the presence, or absence, of the corresponding energy sources, mates, mutualists, symbionts, or predators.

In both of the cases presented above, the main problem is that the contribution of a given component process can be rigorously established only in retrospect, when we know whether the organism actually managed to maintain its organisation and reproduce in a particular situation. At the present moment, not even the organism itself can be certain that its energy is well invested, that is, that the processes it engages in actually contribute to creating the conditions of its continued existence. It needs to hedge its bets, thus necessarily wasting energy on tasks that will turn out to be futile. Living is risky and predictions are hard, especially when they are about the future. The property of 'contributing to organisational closure' is afflicted by predicative vagrancy (Rescher, 2009). It undoubtedly exists (since organisational continuity is maintained throughout the reproducer lifecycle), but we cannot identify it unless we reconstruct it in retrospect. This seems to imply that the attribution of organisational function will be hard or even impossible to operationalise in the context of specific empirical research projects.

But this is not the only difficulty. Further complications arise from the dynamic physical structure of the complex regulatory processes that contribute to biological organisation. Such processes often exhibit high amounts of redundancy and distributed robustness (Wagner, 2005a, b, 2011). For instance, many nutrients can be metabolised through parallel pathways with overlapping activities or routed through alternative non-equivalent metabolic channels (e.g., glycolysis vs. the pentose phosphate shunt). This enables the cell to compensate for a perturbation or interruption of specific pathways. Gene regulatory networks exhibit a similar degree of redundancy and distributed robustness. The fact that a regulatory or metabolic process can be bypassed complicates not only the interpretation of genetic or environmental perturbations but also the assessment of its contribution to organisational closure and continuity. If the organism remains viable after a genetic factor or a metabolic pathway is knocked out, can we really still say that it contributed to self-maintenance? After all, organisational continuity was maintained despite the removal of the factor. In fact, it seems reasonable to assume that some degree of redundancy and distributed robustness is a fundamental prerequisite of maintaining organisational continuity in the first place. Since organisational continuity requires continuous structural changes in the underlying physicochemical networks (DiFrisco, 2014; DiFrisco & Mossio, 2020), specific component processes must be replaceable, at least under certain circumstances.

As if all of these complications were not enough, our third and last example highlights the additional challenges posed by multicellularity. Is the agency of a multicellular organism the consequence of the agency of its cells or is there an additional level of organisation? The answer to this question depends on the kind of organism we are looking at: different multicellular organisms, such as plants, fungi, and animals, show radically different degrees of organisational integration (Arnellos et al., 2014; Arnellos & Moreno, 2015). Only in the case of animals can a convincing argument be made that agency depends on higher-level (or second-order) organisation involving the nervous system, which mediates the autonomous movements required for the kind of interactions with the environment we call animal behaviour (Arnellos & Moreno, 2015; see also Varela et al., 1974; Varela, 1979; Maturana & Varela, 1980; for further discussion of organisation in cognitive systems). These movements are still connected to basic self-maintenance and reproduction (they are necessary for, e.g., nutrition or finding a mate) and nerve tissue obviously depends on the basic metabolic machinery of its cells. Nevertheless, actions based on motility have a degree of independence from their underlying metabolic, physiological, and developmental processes, and they usually take place at a much faster timescale (Arnellos & Moreno, 2015). They are certainly more independent than movements in plants and fungi, which are usually based on modified growth processes. Thus, at least in metazoa, we must distinguish between contributions to first- and second-order organisation, which are not necessarily aligned (see following section on cellular agency).

All of this taken together means that it will be extremely difficult to assign specific organisational functions to those regulatory and metabolic processes involved in ontogenesis which are not part of the self-maintaining and reproductive core of metabolism and cellular dynamics. On the one hand, their contributions to organisational closure and continuity radically depend on context and chance. On the other hand, indirect functional contributions are deeply entangled in a highly redundant, complex, and everchanging mesh of regulatory interactions and environmental inputs. While biological organisation can be convincingly formulated in precise and explicit terms at an abstract functional level, it has proven difficult to map it down to the level of physicochemical regulatory processes and their transient and dynamic interactions that implement the closure of constraints.

Nevertheless, efforts aimed at such a top-down mapping are currently underway at the level of single cells, where it can be hoped that we could connect abstract relational models of organisation to particular metabolic and cellular processes (Hofmeyr, 2017, 2021). This work has already yielded some interesting insights into the relation of mechanism (defined in the Introduction as the orchestrated local interaction of physical parts) to whole-cell organisation (see also the section on organisational constraints on ontogenesis). For instance, it requires the systemic property of 'cellular milieu' as a key variable of the model, highlighting the limitations of a reductionist approach, which tries to explain biological organisation solely in terms of molecular components (Hofmeyr, 2021). Moreover, it allows us to separate those aspects of cellular organisation that can be captured by algorithmic simulations and are *sensu lato* mechanistic from the co-emergent dialectical self-maintaining cycle at the heart of cellular dynamics, whose mechanistic/algorithmic descriptions necessarily remain incomplete (Hofmeyr, 2021). Both of these insights hint at a fundamental explanatory gap between mechanistic component processes and whole-cell dynamics, which calls for investigation. Nevertheless, it remains doubtful that we can close this gap in an exclusively top-down manner, by assigning organisational functions to component processes of an organism. Our examples show that usefulness of this approach is limited to investigation of the core processes directly involved in metabolism and cell division. The predicative vagrancy of organisational function outlined above makes it difficult to envision how its application could be extended to cellular processes which only indirectly contribute to self-maintenance and reproduction but are nevertheless necessary for survival in an unpredictable environment – not to mention any further complications of its application to the multicellular context of embryogenesis and post-embryonic development.

Organisational Constraints on Ontogenesis: A Bottom-Up Empirical Approach

While the practical prospects of top-down functional mapping currently remain uncertain, tackling the problem pragmatically, from the bottom up, may be more feasible. It could be done by empirically probing how organismic agency – via the requirement of organisational continuity – *constrains* the dynamic processes underlying ontogenetic processes and their evolution. The thesis I put forth here is simple: to properly understand these processes, we must not only understand the specific mechanisms that realise them in a given individual or species but also the constraints placed on their dynamic structure by the general condition that a lifecycle must be complete – with organisational continuity maintained during its entire duration – for an agential reproducer to be evolvable (to be a proper unit of evolution; see above). I argue that a comparative approach to ontogenesis can offer an empirical entry point to a deeper understanding of this relationship between ontogenesis and organisational continuity.

Constraints imposed by the overall organisation of the organism on ontogenesis are different from classic developmental constraints (see above, also cf. Maynard Smith et al., 1985). As noted above, classic developmental constraints are defined with regard to the proximate outcome of an ontogenetic process (e.g., wing development in insects is constrained by the need to generate functional wings), which itself need not be necessary for completion of the lifecycle (wingless mutant insects can survive and thrive, at least under laboratory conditions). Furthermore, it may be more productive to think about such constraints from a positive angle, to ask what kind of *variational properties* of ontogenetic processes (Wagner & Altenberg, 1996; Salazar-Ciudad, 2006) are enabled within the boundaries set by organisational continuity. We shall see that at least part of this variability is empirically accessible, unlike the elusive functional contribution of ontogenetic processes to organisational continuity described in the previous section.

Contemporary research in empirical cell and developmental biology employs one dominant mode of explanation: ontogenetic regulatory processes are explicated *in mechanistic terms* (see Introduction). Let me explain in a bit more detail what I mean by 'mechanism'. Often, the term is mistakenly interpreted as a requirement for explanation at the molecular level (especially in the form of a molecular genetic mechanism). What I have in mind here, though, when speaking about mechanism, is 'an explanatory mode in which we describe what are the parts [of a system], how they behave intrinsically, and how those intrinsic behaviors of parts are coupled to each other to produce the behavior of the whole' (von Dassow & Munro, 1999: 309). Note that mechanistic explanation is not necessarily reductionistic: the relevant parts of a mechanism belong to multiple levels of organisation. On top of that, I have argued elsewhere that mechanisms must provide an explanation in processual terms (DiFrisco & Jaeger, 2019): 'A mechanism is a structure performing a function in virtue of its component parts, component operations, and their organisation. The orchestrated functioning of the mechanism, manifested in patterns of change over time in properties of its parts and operations, is responsible for one or more phenomena' (Bechtel & Abrahamsen, 2010: 323). Simply put, the aim of the mechanistic approach is to explain the activities of ontogenetic processes in terms of relevant components and how they are structured. This presupposes that activities of cellular and developmental systems must be determined by their regulatory structure, relatively independently of the specific context in which those activities take place.

Despite its tremendous empirical success, the mechanistic approach has a number of limitations. We have already illustrated (in the previous section) its inability to tackle truly irreducible organism-level properties such as organisational continuity. Another problem is that the activity of an ontogenetic process is generally not determined, but merely relatively vaguely constrained, by its regulatory structure, i.e., its components and their interactions. There are many different regulatory structures capable of generating the same dynamic behaviour, and in turn, one particular regulatory structure can behave in very different ways depending on its organismic and environmental context (see Jaeger et al., 2012; DiFrisco & Jaeger, 2019; Jaeger & Monk, 2021a, b, and references therein). In other words, the structure–dynamics mapping of most ontogenetic mechanisms is highly degenerate in both directions, which makes them difficult to define as specific sets of components and interactions. Secondly, mechanistic decomposition into components and interactions always presupposes a specific activity of the process that is to be decomposed (Kauffman, 1971a; Wagner & Laubichler, 2000). To give a rather obvious example, we can decompose an early embryo of the fruit fly in radically different ways depending on whether we are interested in patterning along its major (anteroposterior) or minor (dorsoventral) axis. Given these distinct patterning activities, we end up with different components and different interactions driving the process. Yet, these mechanisms are not entirely separable: their activities overlap and influence each other in subtle ways, and both processes form part of the same primary *morphogenetic field*, the same developmental system. Simultaneous patterning along both axes is required for organisational continuity. In general, identification of mechanisms requires a predefinition of their function. It should be clear then that there is no 'objective' or global way of mechanistically decomposing ontogenetic processes without prior functional analysis. Though far from arbitrary, delimitation of mechanisms (and even the levels of abstraction or organisation we assign them to), fundamentally depends on what we are looking for in the first place (see also Wimsatt, 2007).

None of these caveats, however, affect the fact that we have powerful and principled empirical methods to perform a functional decomposition of ontogenetic systems into *dynamical modules*, which are distinguishable elementary activities which combine to yield the overall dynamics of the system (Jaeger & Monk, 2021a, b; see Chap. 11). Genetic approaches allow us to disrupt these activities in a targeted and specific manner (often through a combination of particular interventions), while data-driven dynamical systems modelling helps us reveal how specific components contribute causally through their nonlinear interactions to generate the observed systems-level behaviour (DiFrisco & Jaeger, 2020). Despite the daunting complexity and redundancy of developmental systems, it is therefore increasingly possible to disentangle the complex causal interactions that ultimately contribute to organisational continuity (at least under laboratory conditions). In particular, if an alteration or elimination of a specific activity has a lethal or sterilising effect, it is required for the completion of the lifecycle and thus for organisational continuity. If it does not, then the effect of our manipulation is within the viable range of variation in at least one particular (laboratory) environment. If we target a whole range of cellular or developmental activities under different conditions, we can begin to systematically identify constraints imposed by organisational continuity and start mapping the permissible range within which ontogenetic processes may vary without disrupting the completion of the lifecycle.

So far, this is not very different from classical genetic analysis. However, there is an additional aspect of ontogenesis that must be considered: when we identify a dynamic mechanism for an ontogenetic process in a specific organism under specific conditions, we cannot tell which of the components and interactions we have identified are essential for its activity and which are present merely due to evolutionary/historical contingencies. If our aim is to map variational properties of a process within the constraints imposed by organisational continuity, we need to understand which aspects of the mechanism are free to vary – and which ones are not. This can be achieved by a comparative approach to ontogenesis, based on the notion of *the homology of process* (Gilbert & Bolker, 2001; DiFrisco & Jaeger, 2021). This approach uses a set of overlapping criteria, such as the homology of parts and phenotypic outcomes, topological position (compared to other ontogenetic processes), dynamical properties and their complexity, as well as transitional forms across evolutionary lineages, to homologise the activity of ontogenetic processes. To put it simply, it allows us to determine whether a given ontogenetic process *is the same activity* in different evolutionary lineages.

Once we have established the homology of ontogenetic processes, we can compare and contrast their underlying mechanisms. I have outlined elsewhere what such a research programme could look like (Jaeger & Crombach, 2012; Crombach & Jaeger, 2021). It would have to use a combination of functional decomposition and model-based recomposition to reconstruct dynamic mechanisms in related evolutionary lineages. On the one hand, this would allow us to come up with a sequence of mechanistic changes which explain the evolution of the mechanism (so-called *cross-lineage explanations*, DiFrisco & Jaeger, 2021). On the other hand (and more to the point of my current argument), this would also help us identify those aspects of a mechanism that can vary while preserving homology of the process. If applied to an ontogenetic process whose knocking out or substantial alteration has a lethal or sterilising effect, this kind of comparative analysis enables us to map the kind of variation in the underlying mechanism that is activity-preserving and thus viable (at least for environments we can reproduce in a laboratory). As a proof of principle, my research group performed such an analysis for the evolution of the gap gene regulatory network in different species of flies, which revealed conserved and variable dynamical modules that govern distinct modes of gene expression (patterning activities) in the embryo (see, e.g., Crombach et al., 2016; Verd et al., 2019; reviewed in Jaeger, 2018).

Although still laborious and technically challenging, such analyses that combine genetic approaches with data-driven modelling are now eminently feasible, even outside the species already established as classic laboratory models (Crombach & Jaeger, 2021). Ideally, they should be complemented by complementary quantitative genetic approaches to statistically map the contribution of genetic causes to phenotypic variation (see Nunes et al., 2013, for review). As we perform more and more of these comparative case studies in an expanding range of evolutionary lineages and developmental contexts, they will reveal the rough contours of the kind of viable variability that is amenable to natural selection. The better the sampling, the more refined the resulting map. This is why to be successful, such a research programme would have to be implemented as a broadly coordinated and concerted effort.

Despite the considerable time and energy required, it would surely be worth our while. The potential payoffs are considerable. Mapping of viable ontogenetic variability provides a mechanistic and empirical foundation for our theoretically derived principle of ontogenesis and its fundamental role in evolution (see the section on the reproducer perspective; cf. also Wagner et al., 2000; DiFrisco et al., 2020). Such a research programme enables us to gain mechanistic insights into both macroevolutionary patterns (DiFrisco & Wagner, 2022) and the evolutionary origins of novel traits (see, e.g., Wagner, 2011; DiFrisco et al., 2022). It reveals unexpected connections between robustness and evolutionary innovations (Wagner, 2011) and provides a deeper understanding of the modularity of ontogenetic systems (Jaeger & Monk, 2021a, b). Ultimately, it may help us discover regularities or even generalisable principles underlying the nonisotropic distribution of selectable variability, which would help us understand how this constrained variation contributes to differential evolvability in different evolutionary lineages (Wagner & Altenberg, 1996). But most importantly, it should show how the requirement of organisational continuity, the necessity to complete the lifecycle, shapes the distribution of ontogenetic variability. The reproducer cycle (ontogenesis and reproduction) must remain undisrupted for any evolution to take place at all. This is how the top-down constraints imposed by biological organisation shape the space of possible ontogenetic mechanisms, and thereby also the course of evolution.

Cellular Agency and the Biological Default State

Apart from contextualising and embedding ontogenetic processes within organismic organisation and evolution, an agential perspective can be of help even more directly when it comes to the empirical study of development in multicellular organisms. The traditional reductionist approach to developmental biology not only focuses on mechanisms as the dominant mode of explanation (which need not be reductionist, cf. the previous section), but is also centred around the metaphor of a *genetic* or *developmental programme* (see Peluffo, 2015, for the historical origin of the metaphor; and Goodwin, 1985; Nijhout, 1990; Webster & Goodwin, 1996; Lewontin, 2000; Oyama, 2000; Fox Keller, 2002; Noble, 2008; for exemplary criticisms of the concept). Such a programme is thought to consist of a set of intrinsic (mostly genomic) instructions, executed according to some regulatory logic (epigenetic or encoded in the genome), that determine cell fate decisions in the presence of particular intercellular or environmental signals which are regarded as instructive triggers for particular branches of the program. On this view, variation in the activity of ontogenetic processes is seen as noise: an aberration from the correct execution of the program. Cells are treated as inert and passive vehicles for their genomes, which carry the instructions determining their behaviour, and the default cellular behaviour in the absence of any forcing causes is quiescence.

Rather than relying on reductive programmes, the analysis presented above shifts our focus to biological organisation as imposing top-down constraints on intrinsically active cellular behaviour (see esp. the previous section). This is where agency and ontogenesis truly meet. Nevertheless, it is important to emphasise again that multicellular organisms possess several distinguishable levels of biological organisation (cf. section on organisational function and ontogenesis). At the lower level, these constraints enable cellular agency and, at the same time, ensure the completion of mitotic division cycles (and hence tissue growth by cell proliferation) by providing each cell in a tissue its own organisational continuity. At the higher level, the intrinsic agency of individual cells is constrained to ensure organisational continuity at the level of the multicellular system. This generally happens by reducing or channelling the intrinsic activity of individual cells. We have moved from a picture of cells needing active instruction to a view of cells needing continuous restraint.

This kind of agential perspective has been carefully elaborated in a series of papers that put forth a conceptual framework for the study of ontogenesis which we may call *cellular agent theory* (Soto et al., 2016a, b; Mossio et al., 2016; Montévil et al., 2016a). It postulates *a biological default state*, which – in the absence of outside interference – consists of cells proliferating and actively moving around (Soto et al., 2016b). This default state is rooted in the view of cells as active biological agents which are subject to organisational closure (*ibid.*). This is what the authors call the *principle of organisation* (Mossio et al., 2016). It requires an internal closure of constraints (the genome providing an important set of constraints, Montévil & Mossio, 2015), which must be maintained throughout the mitotic cycle to ensure organisational continuity in the developmental lineages of dividing cells (DiFrisco & Mossio, 2020). In addition, there is a higher level of constraints which act on the behaviour of these actively dividing cells in their multicellular tissue-level context. These constraints may be physical, for instance such that they inhibit cellular movement. An extreme example of this occurs in plant and fungal tissues where rigid cell walls inhibit cell motility almost completely. Alternatively, constraints may be chemical: signals or hormones, for example, that evoke altered cellular behaviour upon interpretation by the cell which deviates from the default state (e.g., quiescence, unequal divisions, cell shape changes, or cell death by apoptosis). Together, these higher-level constraints ensure organisational continuity and the completion of the lifecycle at the higher level of the multicellular organism (Mossio et al., 2016). Their breakdown can result in uncontrolled cell proliferation, and end in lethal disruptions of the lifecycle, as observed with many forms of malignant carcinogenesis (Sonnenschein & Soto, 2016).

There is a major difference between how the traditional view of ontogenesis and cellular agent theory treat variability. Instead of considering it a nuisance (noise, a deviation from ideal system behaviour) the latter treats it as a foundational principle: the *principle of variation* (Montévil et al., 2016a; related to, but not to be confused with the evolutionary principle of variation introduced above in the section on the reproducer's perspective). From Darwin's descent with modification to the behaviour of individual cells in growing tissues which divide into similar but slightly distinct daughter cells, variation is an ever-present feature of all living agents

(Montévil et al., 2016a). In fact, far from being mere noise, it often plays a constructive role in ontogenesis. This should come as no surprise to someone who adopts the cellular agent view. Variation is there to be harvested by the maturing organism, while stability (in quiescence or proliferative growth) is a deviation from the default state that is in need of explanation. Organisation is what provides stability (Mossio et al., 2016). In this way, variation and organisation can be seen as opposing but complementary principles, with living organisms caught forever in a delicate dance between the two.

While an analysis of organisational constraints on ontogenesis (see the two preceding sections) requires a concerted comparative research programme to map out viable variability, cellular agent theory emphasises the empirical and computational study of complex cellular behaviour and its role in specific ontogenetic processes. This fits well with the reemergence of a modern-day *mechanobiology*, whose main aim is an integrated understanding of tissue-level phenomena and gene-regulatory dynamics (see, e.g., Heisenberg & Bellaïche, 2013; Ayad et al., 2019). Still, most of these recent mechanobiological approaches abstract away from cells as agents by representing tissues as simple lattices or interconnected springs (cf. Heisenberg & Bellaïche, 2013). In contrast, cellular agent theory focuses specifically on the agential roles of cells in ontogenesis. It emphasises spontaneous variation and context-dependent autonomous interactions at the cellular level, leading to nonidentical iterations of ontogenetic processes at the multicellular level, which nevertheless manage to produce robust outcomes. Now, the stress is on explaining how such robustness arises, and how it is able to not only tolerate but harvest the intrinsic variability in the agential cellular behaviour underneath it. Classical approaches to the study of development, based on reductionist–mechanistic empirical methods and deterministic dynamical systems modelling, will completely miss these fundamental aspects of ontogenesis.

To wrap up our discussion, let me illustrate the application of cellular agent theory using several practical examples. An empirical example is provided by the interpretation of filopodial extrusions in tip cells of growing blood vessels as an act of active (agential) perception (Zakirov et al., 2021): through their exploratory movements, the filopodia enrich local concentration of the relevant tissue signal, enhancing uptake, which thus creates a positive feedback loop that accelerates the process of angiogenesis. In another example, this time a study using evidence-based computational modelling (Montévil et al., 2016b; Montévil & Soto, 2023), the authors investigated the tissue-level dynamics of mammary organogenesis. They were able to robustly reproduce patterning as observed in the wildtype and hormonally induced tissue malformations. Another modelling study (of glycemia regulation, Bich et al., 2020) investigated the role of higher-level constraint closure on homeostatic regulation. The focus of this study was on how such higher-level constraints set the parameters – and thus restrict the behaviour – of the lower-level feedback-driven regulatory system.

While it is true that such empirical and modelling studies are still preliminary, together they strongly suggest that insights into the behaviour of cellular agents are an important foundation for understanding the dynamics of many developmental processes in multicellular organisms with multiple levels of biological organisation. In such systems, organism-level agency constrains ontogenesis, while cellular level agency enables it. It will be interesting to investigate this enabling role in studies that combine empirical and modelling approaches based on cellular agent theory.

Conclusions

In this chapter, we have examined the connection between the theoretical study of organismic agency, or rather, organisational continuity, and the empirical study of cellular and developmental processes – broadly circumscribed by understanding ontogenesis as 'acquiring the capacity to reproduce'. This connection is twofold. First, there is an indirect link (a *co-dependence*, to be precise) between organismic agency and ontogenesis at the level of the whole organism. Ontogenesis contributes to organisational continuity and is, in turn, constrained by it. Organisational continuity, maintained via ontogenesis and reproduction, is not only a fundamental prerequisite for the completion of the lifecycle (and hence basic evolvability), but also provides the foundation of organismic agency in biological organisation. Second, there is a more direct connection between agency and ontogenesis in organisms organised above the level of individual cells. In multicellular development, for example, agential behaviour of individual cells can play an important constructive role during ontogenesis. At the same time, though, it is restricted by top-down constraints that ensure organisational continuity at the level of lifecycle of the whole organism. This leads to a new conceptual framework for the study of development, which is based on three concepts: biological default state, organisation (implying cellular agency), and variation (Soto et al., 2016a).

Based on my argument, I can now revisit the two questions I set out to answer in the Introduction: on the one hand, is organismal agency relevant to the empirical practice of studying cellular and developmental regulatory processes and their role in evolution? The answer I would give here is clearly 'yes', both in terms of the role of cellular agency in multicellular development and, more indirectly, in terms of the constraints imposed on the space of possible ontogenetic mechanisms by organisational continuity and the completion of the lifecycle. Both of these aspects deserve more attention from experimental biologists and modellers of cellular and developmental processes.

On the other hand, is the empirical study of cellular and developmental regulatory processes and their evolution relevant for a proper grasp of organismal agency? The answer here must be an even more emphatic 'yes'. It is far from obvious how the abstract sets of relations and constraints that define organisational closure can be mapped onto the highly dynamic physicochemical processes and interactions that constitute an actual living system. Although attempts at such a mapping are being undertaken (Hofmeyr, 2017, 2021), it is still early days and a general assignment of organisational function beyond the core processes of self-maintenance and reproduction may prove difficult to implement rigorously. As a pragmatic alternative, we have suggested a comparative mechanistic research programme (see also DiFrisco et al., 2020; DiFrisco & Jaeger, 2021) to study ontogenetic processes across evolutionary lineages through experimental interventions and data-driven mathematical modelling (Jaeger & Crombach, 2012; Crombach & Jaeger, 2021). Its aim is to begin mapping out the constraints that delimit and define viable variation. Although this will not give us direct access to measuring or predicting the specific contributions of particular processes to organisational continuity, it should at least show us what kind of variability is permissible within its constraints. As a result, we should arrive at a better understanding of how organisational constraints may affect the dynamic structure of ontogenetic mechanisms and influence the dynamics and direction of evolutionary processes.

What I have *not* attempted here is to reframe ontogenesis in organisational terms (cf. Bich & Skillings, 2023). Instead, I see my analysis as firmly located within the Darwinian tradition that considers evolution at the level of the organism and views the reproducer (relying on both ontogenesis and reproduction) as the basic unit of evolution (Griesemer, 2006; Jaeger, 2023). This evolutionary perspective has recently been criticised as excessively focused on the adult (reproductive) stage of the lifecycle, with ontogenesis becoming teleologically oriented towards acquiring the capacity to reproduce (Bich & Skillings, 2023). I do not think this criticism applies to the argument I have presented here. In fact, I concur with Bich and Skillings who note that ontogenesis is constrained in important ways by organisational continuity (DiFrisco & Mossio, 2020). Seen from the reproducer view, reproduction and ontogenesis are processes which are distinguishable, yet intricately intertwined. The main criterion that sets them apart is whether the number of organised systems remains constant (ontogenesis) or not (reproduction; DiFrisco & Mossio, 2020). Ontogenesis does not 'work towards' the adult reproductive stage, whatever that may mean. Instead, it simply describes those phases of organisational continuity which occur between reproductive events. While Bich and Skillings (2023) attempt to provide a narrower definition of multicellular development, I am not interested in such distinctions here. In fact, due to the complications described in the section dedicated to organisational function and ontogenesis, I believe this to be a futile task. The set of constraints that contribute to the reproductive goals of the organism at any given moment simply cannot be precisely defined – so we should not even try. Instead, I see organisational continuity occurring in the presence of constant changes in the set of physicochemical processes and constraints that constitute organisational closure. As outlined above, the path towards a description of these changes leads less through theoretical redefinitions of what ontogenesis means and more through pragmatic empirical approaches to measure viable variability during ontogenesis.

From an historical perspective, the approach of viewing ontogenesis as constrained by higher levels of organisation builds on the research programme of *process structuralism* in developmental and evolutionary biology (see, e.g., Goodwin, 1982a, b, 1985; Webster & Goodwin, 1982, 1996; Goodwin et al., 1993). Both frameworks focus their efforts on the discovery of general structural constraints on ontogenetic processes (or morphogenetic fields; cf. section on organisational constraints). Furthermore, both can be seen as attempts to properly position the role of genetic causes within a much more complex systems-level context. Finally, although each approach does this in its own distinct way, both research programmes aim at distinguishing functionally constrained features of ontogenesis from those that are historically contingent. But this is where the similarities end. In fact, there are several fundamental differences. While structuralism aims at abstracting away the variational aspects of biological processes, I emphasise variation as a foundational principle. Similarly, structuralists tend to downplay the role of history and historical contingency in evolution and development, while my approach puts historicity centre stage in the form of dynamic presupposition and organisational continuity in developmental and evolutionary lineages (see Bickhard, 2000; DiFrisco & Mossio, 2020). If process structuralism can be seen as the antithesis of the mechanistic approach to ontogenesis, then my current proposal is their synthesis. It is the central point of my argument that we need *both* the bottom-up mechanistic experimental practice *and* the top-down agential perspective grounded in biological organisation to properly understand ontogenesis and its role in evolution.

Acknowledgments I thank my late mentor and supervisor, Brain Goodwin, whose spirit can strongly be felt in this manuscript. Matteo Mossio, Denis Walsh, and James Griesemer provided moral and intellectual support on various occasions and came up with key ideas that I use in my argument, which is based on numerous extensive discussions with my philosophical collaborator James DiFrisco, who provided detailed feedback on the manuscript, even though the views presented here and the responsibility for potential errors remain exclusively mine.

References

Arnellos, A., & Moreno, A. (2015). Multicellular agency: An organizational view. *Biology & Philosophy, 30*(3), 333–357.

Arnellos, A., Moreno, A., & Ruiz-Mirazo, K. (2014). Organizational requirements for multicellular autonomy: Insights from a comparative case study. *Biology & Philosophy, 29*(6), 851–884.

Ayad, N. M. E., Kaushik, S., & Weaver, V. M. (2019). Tissue mechanics, an important regulator of development and disease. *Philosophical Transactions of the Royal Society B: Biological Sciences, 374*(1779), 20180215.

Barandiaran, X. E., Di Paolo, E., & Rohde, M. (2009). Defining agency: Individuality, normativity, asymmetry, and spatio-temporality in action. *Adaptive Behavior, 17*(5), 367–386.

Bechtel, W. (2011). Mechanism and biological explanation. *Philosophy of Science, 78*(4), 533–557.

Bechtel, W., & Abrahamsen, A. (2005). Explanation: A mechanist alternative. *Studies in History and Philosophy of Science Part C: Studies in History and Philosophy of Biological and Biomedical Sciences, 36*(2), 421–441.

Bechtel, W., & Abrahamsen, A. (2010). Dynamic mechanistic explanation: Computational modeling of circadian rhythms as an exemplar for cognitive science. *Studies in History and Philosophy of Science Part A, 41*(3), 321–333.

Bechtel, W., & Richardson, R. C. (1993). *Discovering complexity: Decomposition and localization as strategies in scientific research* (Original ed.). Princeton University Press.

Bich, L., & Skillings, D. (2023). There are no intermediate stages: An organizational view of development. In M. Mossio (Ed.), *Organization in Biology*. Springer.

Bich, L., Mossio, M., & Soto, A. M. (2020). Glycemia regulation: From feedback loops to organizational closure. *Frontiers in Physiology, 11*, 69.
Bickhard, M. H. (2000). Autonomy, function, and representation. *Communication and Cognition – Artificial Intelligence, 17*, 111–131.
Bigelow, J., & Pargetter, R. (1987). Functions. *Journal of Philosophy, 84*(4), 181–196.
Boorse, C. (1976). Wright on functions. *The Philosophical Review, 85*(1), 70–86.
Christensen, W. D., & Bickhard, M. H. (2002). The process dynamics of normative function. *The Monist, 85*(1), 3–28.
Craver, C. F. (2007). *Explaining the brain: Mechanisms and the mosaic unity of neuroscience*. Clarendon Press.
Craver, C., & Tabery, J. (2019). Mechanisms in science. In E. N. Zalta (Ed.), *The Stanford encyclopedia of philosophy* (Summer 2019). Metaphysics Research Lab, Stanford University. https://plato.stanford.edu/archives/sum2019/entries/science-mechanisms
Crombach, A., & Jaeger, J. (2021). Life's attractors continued: Progress in understanding developmental systems through reverse engineering and in silico evolution. In A. Crombach (Ed.), *Evolutionary systems biology* (pp. 59–88). Springer.
Crombach, A., Wotton, K. R., Jiménez-Guri, E., & Jaeger, J. (2016). Gap gene regulatory dynamics evolve along a genotype network. *Molecular Biology and Evolution, 33*(5), 1293–1307.
Cummins, R. (1975). Functional analysis. *Journal of Philosophy, 72*(20), 741–765.
DiFrisco, J. (2014). Hylomorphism and the metabolic closure conception of life. *Acta Biotheoretica, 62*, 499–525.
DiFrisco, J., & Jaeger, J. (2019). Beyond networks: Mechanism and process in evo-devo. *Biology & Philosophy, 34*, 54.
DiFrisco, J., & Jaeger, J. (2020). Genetic causation in complex regulatory systems: An integrative dynamic perspective. *BioEssays, 42*(6), 1900226.
DiFrisco, J., & Jaeger, J. (2021). Homology of process: Developmental dynamics in comparative biology. *Interface Focus, 11*(3), 20210007.
DiFrisco, J., & Mossio, M. (2020). Diachronic identity in complex life cycles: An organizational perspective. In A. S. Meincke & J. Dupré (Eds.), *Biological identity: Perspectives from metaphysics and the philosophy of biology*. Routledge.
DiFrisco, J., & Wagner, G. P. (2022). Body plan identity: A mechanistic model. *Evolutionary Biology, 49*, 123–141.
DiFrisco, J., Love, A. C., & Wagner, G. P. (2020). Character identity mechanisms: A conceptual model for comparative-mechanistic biology. *Biology & Philosophy, 35*, 44.
DiFrisco, J., Wagner, G. P., & Love, A. C. (2022). Reframing research on evolutionary novelty and co-option: Character identity mechanisms versus deep homology. *Seminars in Cell & Developmental Biology, 145*, 3–12.
Fox Keller, E. (2002). *Making sense of life: Explaining biological development with models, metaphors, and machines*. Harvard University Press.
Gibson, J. J. (1979). *The ecological approach to visual perception*. Houghton Mifflin.
Gilbert, S. F., & Bolker, J. A. (2001). Homologies of process and modular elements of embryonic construction. *Journal of Experimental Zoology Part B: Molecular and Developmental Evolution, 291*(1), 1–12.
Glennan, S. S. (1996). Mechanisms and the nature of causation. *Erkenntnis, 44*, 49–71.
Glennan, S., & Illari, P. (Eds.). (2017). *The Routledge handbook of mechanisms and mechanical philosophy*. Routledge.
Gold, J. I., & Shadlen, M. N. (2007). The neural basis of decision making. *Annual Review of Neuroscience, 30*, 535–574.
Goodwin, B. C. (1982a). Biology without Darwinian spectacles. *The Biologist, 29*, 108–112.
Goodwin, B. C. (1982b). Development and evolution. *Journal of Theoretical Biology, 97*(1), 43–55.
Goodwin, B. C. (1985). What are the causes of morphogenesis? *BioEssays, 3*(1), 32–36.
Goodwin, B. C., Kauffman, S., & Murray, J. D. (1993). Is morphogenesis an intrinsically robust process? *Journal of Theoretical Biology, 163*(1), 135–144.

Griesemer, J. (2006). Genetics from an evolutionary process perspective. In E. M. Neumann & C. Rehmann-Sutter (Eds.), *Genes in Development* (pp. 199–237). Duke University Press.

Heisenberg, C.-P., & Bellaïche, Y. (2013). Forces in tissue morphogenesis and patterning. *Cell, 153*(5), 948–962.

Hofmeyr, J.-H. S. (2017). Basic Biological Anticipation. In R. Poli (Ed.), *Handbook of anticipation* (pp. 1–15). Springer.

Hofmeyr, J.-H. S. (2021). A biochemically-realisable relational model of the self-manufacturing cell. *Biosystems, 207*, 104463.

Jaeger, J. (2018). Shift happens: The developmental and evolutionary dynamics of the gap gene system. *Current Opinion in Systems Biology, 11*, 65–73.

Jaeger, J. (2023). The fourth perspective: Evolution and organismal agency. In M. Mossio (Ed.), *Organization in Biology*. Springer.

Jaeger, J., & Crombach, A. (2012). Life's attractors. In O. S. Soyer (Ed.), *Evolutionary systems biology* (pp. 93–119). Springer.

Jaeger, J., & Monk, N. (2021a). Dynamical modularity of the genotype-phenotype map. In A. Crombach (Ed.), *Evolutionary systems biology: Advances, questions, and opportunities* (pp. 245–280). Springer.

Jaeger, J., & Monk, N. (2021b). Dynamical modules in metabolism, cell and developmental biology. *Interface Focus, 11*, 20210011.

Jaeger, J., Irons, D., & Monk, N. (2012). The inheritance of process: A dynamical systems approach: The inheritance of process. *Journal of Experimental Zoology Part B: Molecular and Developmental Evolution, 318*(8), 591–612.

Kauffman, S. A. (1971a). Articulation of parts explanation in biology and the rational search for them. In R. C. Buck & R. S. Cohen (Eds.), *PSA 1970* (pp. 257–272). D. Reidel Publishing Company.

Kauffman, S. A. (1971b). Cellular homeostasis, Epigenesis and replication in randomly aggregated macromolecular systems. *Journal of Cybernetics, 1*(1), 71–96.

Kauffman, S. A. (1986). Autocatalytic sets of proteins. *Journal of Theoretical Biology, 119*(1), 1–24.

Kauffman, S. A. (1993). *The origins of order: Self-organization and selection in evolution* (1st ed.). Oxford University Press.

Kauffman, S. A. (2000). *Investigations*. Oxford University Press.

Koestler, A. (1982). *The ghost in the machine*. Hutchinson.

Lewontin, R. (2000). *The triple helix: Gene, organism, and environment*. Harvard University Press.

Love, A. C. (2007). Functional homology and homology of function: Biological concepts and philosophical consequences. *Biology & Philosophy, 22*, 691–708.

Machamer, P., Darden, L., & Craver, C. F. (2000). Thinking about mechanisms. *Philosophy of Science, 67*(1), 1–25.

Maturana, H. R., & Varela, F. J. (1980). *Autopoiesis and cognition: The realization of the living*. Springer.

Maynard Smith, J., Burian, R., Kauffman, S., Alberch, P., Campbell, J., Goodwin, B., Lande, R., Raup, D., & Wolpert, L. (1985). Developmental constraints and evolution. *The Quarterly Review of Biology, 60*(3), 265–287.

Montévil, M., & Mossio, M. (2015). Biological organisation as closure of constraints. *Journal of Theoretical Biology, 372*, 179–191.

Montévil, M., & Soto, A. M. (2023). Modeling organogenesis from biological first principles. In M. Mossio (Ed.), *Organization in Biology*. Springer.

Montévil, M., Mossio, M., Pocheville, A., & Longo, G. (2016a). Theoretical principles for biology: Variation. *Progress in Biophysics and Molecular Biology, 122*(1), 36–50.

Montévil, M., Speroni, L., Sonnenschein, C., & Soto, A. M. (2016b). Modeling mammary organogenesis from biological first principles: Cells and their physical constraints. *Progress in Biophysics and Molecular Biology, 122*(1), 58–69.

Moreno, A., & Etxeberria, A. (2005). Agency in natural and artificial systems. *Artificial Life, 11*(1–2), 161–175.

Moreno, A., & Mossio, M. (2015). *Biological autonomy*. Springer.
Mossio, M., & Bich, L. (2017). What makes biological organisation teleological? *Synthese, 194*, 1089–1114.
Mossio, M., & Pontarotti, G. (2020). Conserving functions across generations: Heredity in light of biological organization. *The British Journal for the Philosophy of Science, 73*(1), axz031.
Mossio, M., Saborido, C., & Moreno, A. (2009). An organizational account of biological functions. *The British Journal for the Philosophy of Science, 60*(4), 813–841.
Mossio, M., Montévil, M., & Longo, G. (2016). Theoretical principles for biology: Organization. *Progress in Biophysics and Molecular Biology, 122*(1), 24–35.
Nijhout, H. F. (1990). Metaphors and the role of genes in development. *BioEssays, 12*(9), 441–446.
Noble, D. (2008). *The music of life: Biology beyond genes*. Oxford University Press.
Nunes, M. D. S., Arif, S., Schlötterer, C., & McGregor, A. P. (2013). A perspective on Micro-Evo-Devo: Progress and potential. *Genetics, 195*(3), 625–634.
Oyama, S. (2000). *The ontogeny of information: Developmental systems and evolution* (Revised ed.). Duke University Press.
Peluffo, A. E. (2015). The "genetic program": Behind the genesis of an influential metaphor. *Genetics, 200*(3), 685–696.
Piaget, J. (1967). *Biologie et connaissance*. Idées/Gallimard.
Pigliucci, M. (2008). Is evolvability evolvable? *Nature Reviews Genetics, 9*, 75–82.
Rescher, N. (2009). *Unknowability: An inquiry into the limits of knowledge*. Lexington Books.
Roli, A., Jaeger, J., & Kauffman, S. A. (2022). How organisms come to know the world: Fundamental limits on artificial general intelligence. *Frontiers in Ecology and Evolution, 9*, 806283.
Rosen, R. (1991). *Life itself: A comprehensive inquiry into the nature, origin, and fabrication of life*. Columbia University Press.
Saborido, C., Mossio, M., & Moreno, A. (2011). Biological organization and cross-generation functions. *The British Journal for the Philosophy of Science, 62*(3), 583–606.
Salazar-Ciudad, I. (2006). Developmental constraints vs. variational properties: How pattern formation can help to understand evolution and development. *Journal of Experimental Zoology Part B: Molecular and Developmental Evolution, 306B*(2), 107–125.
Sonnenschein, C., & Soto, A. M. (2016). Carcinogenesis explained within the context of a theory of organisms. *Progress in Biophysics and Molecular Biology, 122*(1), 70–76.
Soto, A. M., Longo, G., Miquel, P.-A., Montevil, M., Mossio, M., Perret, N., Pocheville, A., & Sonnenschein, C. (2016a). Toward a theory of organisms: Three founding principles in search of a useful integration. *Progress in Biophysics and Molecular Biology, 122*(1), 77–82.
Soto, A. M., Longo, G., Montévil, M., & Sonnenschein, C. (2016b). The biological default state of cell proliferation with variation and motility, a fundamental principle for a theory of organisms. *Progress in Biophysics and Molecular Biology, 122*(1), 16–23.
Szathmáry, E., & Maynard Smith, J. (1993). The origin of genetic systems. *Abstracta Botanica, 17*(1–2), 197–206.
Varela, F. J. (1979). *Principles of biological autonomy*. North Holland.
Varela, F. G., Maturana, H. R., & Uribe, R. (1974). Autopoiesis: The organization of living systems, its characterization and a model. *Biosystems, 5*(4), 187–196.
Verd, B., Monk, N. A., & Jaeger, J. (2019). Modularity, criticality, and evolvability of a developmental gene regulatory network. *eLife, 8*, e42832.
von Dassow, G., & Munro, E. (1999). Modularity in animal development and evolution: Elements of a conceptual framework for EvoDevo. *The Journal of Experimental Zoology, 285*(4), 307–325.
Wagner, A. (2005a). Distributed robustness versus redundancy as causes of mutational robustness. *BioEssays, 27*(2), 176–188.
Wagner, A. (2005b). *Robustness and Evolvability in living systems*. Princeton University Press.
Wagner, A. (2011). *The origins of evolutionary innovations: A theory of transformative change in living systems*. Oxford University Press.

Wagner, G. P., & Altenberg, L. (1996). Complex adaptations and the evolution of Evolvability. *Evolution, 50*(3), 967–976.
Wagner, G. P., & Laubichler, M. D. (2000). Character identification in evolutionary biology: The role of the organism. *Theory in Biosciences, 119*, 20–40.
Wagner, G. P., Chiu, C.-H., & Laubichler, M. (2000). Developmental evolution as a mechanistic science: The inference from developmental mechanisms to evolutionary processes. *American Zoologist, 40*(5), 819–831.
Walsh, D. (2015). *Organisms, agency, and evolution*. Cambridge University Press.
Webster, G., & Goodwin, B. C. (1982). The origin of species: A structuralist approach. *Journal of Social and Biological Structures, 5*(1), 15–47.
Webster, G., & Goodwin, B. C. (1996). *Form and transformation: Generative and relational principles in biology*. Cambridge University Press.
Wimsatt, W. C. (1974). Complexity and organization. In K. F. Schaffner & R. S. Cohen (Eds.), *PSA 1972* (pp. 67–86). D. Reidel Publishing Company.
Wimsatt, W. C. (1976). Reductive explanation: A functional account. In R. S. Cohen, C. A. Hooker, A. C. Michalos, & J. W. Van Evra (Eds.), *PSA 1974* (pp. 671–710). D. Reidel Publishing Company.
Wimsatt, W. C. (2007). *Re-engineering philosophy for limited beings: Piecewise approximations to reality*. Harvard University Press.
Wright, L. (1973). Functions. *Philosophical Review, 82*(2), 139–168.
Zakirov, B., Charalambous, G., Thuret, R., Aspalter, I. M., Van-Vuuren, K., Mead, T., Harrington, K., Regan, E. R., Herbert, S. P., & Bentley, K. (2021). Active perception during angiogenesis: Filopodia speed up Notch selection of tip cells in silico and in vivo. *Philosophical Transactions of the Royal Society of London B, 376*(1821), 20190753.

Chapter 11
Biological Modularity and the Origins of Agency

Jan Toman

Abstract From cells and multicellular bodies to offices, empires, or computer programs, complex entities around us usually consist of numerous largely independent modules. This is especially true of entities that could be called agents, organisms included. In this chapter, we summarise what modularity is and why we encounter it in living nature so often. A search for the origins and subsequent development of a modular arrangement in systems that undergo evolution will lead us to a realisation that it can potentiate transitions to even higher levels of modularity. Modularity and its changes are therefore a key factor in the emergence of new agents at different levels of organisation.

Keywords Biological evolution · Modularity · Agency · Frozen evolution theory · Macroevolutionary potential · Stability-based sorting · Integration · Parcellation · Robustness · Evolvability · Hierarchical complexity

Introduction: Modular Systems

Agents can be defined as an organised physical structure that acts on its own behalf (see Chaps. 8 and 9). In a broad sense, an agent can be a tissue-forming cell, a dog sniffing in the garden, or even a trading company trying to launch a new product. On the other hand, proteins – regardless of how important the cellular functions they perform, stones, or pages of this book under your fingers are not agents. Agents can evolve at various levels, but we can identify the most basal one. It is a cell, the

J. Toman (✉)
Department of Philosophy and History of Science, Faculty of Science,
Prague, Czech Republic
e-mail: tomanj@natur.cuni.cz

J. Švorcová (ed.), *Organismal Agency*, Biosemiotics 28,
https://doi.org/10.1007/978-3-031-53626-7_11

smallest unit acting on its own behalf (Kauffman, 2000).[1] Obviously, agents are *not* simple systems that could be described by just a handful of parameters. On the contrary, it is clear that organisms and their parts, communities, human societies, institutions, and other entities capable of acting in their environment as coherent units with a certain goal are, alongside galaxies, computers, or languages, typical *complex systems*.

Such considerations place us squarely on the thin ice of systems theory, an interdisciplinary field ranging from theoretical physics to economics. But there is no need to panic. The key features of complex systems were summed up in an excellent way even for non-mathematicians by Herbert A. Simon (1962), an American economist, cognitive scientist, and Nobel laureate in economics. For the moment, let us leave aside the fact that complexity of a system is always a relative property: a recording of Beethoven's Ninth Symphony has a greater complexity than a blank sheet of paper, but it cannot hold a candle to a living cell. Likewise, let us not delve into the debate over whether complexity can be universally defined.[2]

Let us thus limit ourselves to stating that complex systems, especially agents, to a significant degree share several nontrivial properties: they are generally large and have an internal structure consisting of numerous subunits that can mutually interact in various nontrivial ways. In such an arrangement, the whole becomes more than the sum of the parts (see Chap. 9). This holds even on a practical level: due to the large number of various interactions between the different parts, it is difficult to predict the behaviour of such entities over a longer period of time. In this context, Simon (1962) speaks about hierarchically organised systems. According to him, complex systems often consist of interconnected subsystems, components, which can be divided in ever more parts all the way down to the lowest level that is still of interest to us. For a sociologist, this level can be individuals within a population; for a biologist, it may be the cells within a body or the macromolecules of which the cells are composed; while for a physicist, it may be the level of atoms or even subatomic particles. But such an arrangement does not automatically indicate a hierarchy of control. It may also exemplify heterarchy or another arrangement, which is why we speak of *modularity* rather than a hierarchical arrangement today.

The importance of modularity in terms of agents and their traits cannot be overstated. Its functional and systematic significance was after all emphasised already by Aristotle in the fourth century BCE (see Chap. 3). As we will see below in this chapter, practically all agents (in our case organisms) have a modular structure and often combine into higher units, which can again be described as modular. These

[1] This can obviously be to a certain extent disputed. What about viruses? At least at the stage of virioplasm in a living cell, they also act in their own interest. Hypothetical pre-cellular organisms might also meet this requirement. On the other hand, these borderline examples of agency should not be mixed together with cases such as prions that change the conformation of other proteins to match their own or exclusive memes that replace their alternative variants in cultural evolution. These entities do not act – they merely replicate themselves.

[2] The reader can find some general outputs from a long discussion on this subject in, e.g., McShea (1991), Lloyd (2001), or Mitchell (2009).

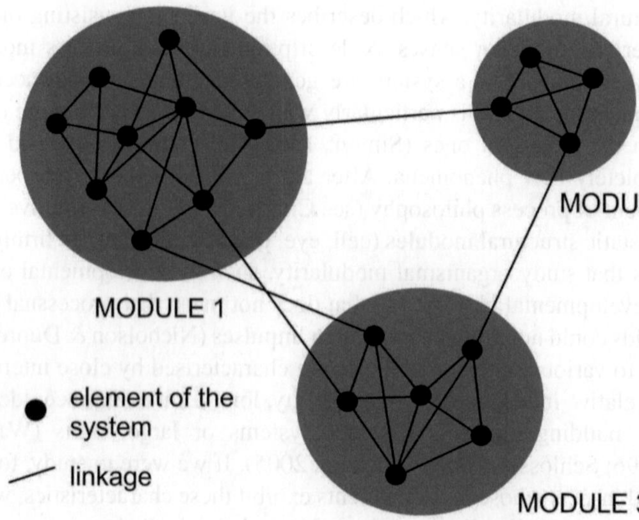

Fig. 11.1 Arrangement of the modular system

agent-based multilevel modular systems are much more complex than complex systems of entities that are not agents, and they display a tendency to further complexification. Modularity is therefore a motif that permeates all entities identifiable as agents and can tell us a lot about their origin, changes, and further evolution (Schlosser, 2002).

A module can be defined as an element of the arrangement of a complex system, such as a biomechanical unit or an organ, which exhibits far more intense connections (interactions) within itself than with other modules (Needham, 1933; Wagner & Altenberg, 1996, see Fig. 11.1). As a result, it maintains a considerable degree of internal integration and external autonomy: it preserves its structural or functional identity (Schlosser, 2002). These properties are evident in various examples, such as tissue-forming cells or individual ants within an anthill. From the process perspective, modules are relatively independent developmentally or functionally integrated subprocesses of a more general process, for example a physiological pathway or individual development (Schlosser, 2002). Examples of such modules include biosynthetic pathways or fixed action patterns of behaviours.

At the first glance, the second, processual, definition may appear different from the first, but a closer look shows that the two definitions are equivalent. After all, each structure had to be created by some ontogenetic process, be it the work of a mason, the activity of an artist, or the individual development of an organism. Therefore, from a certain perspective, we can see the static, physically existing modules as mere metastable stops within a dynamic process. What matters is that we must be able to define them in relation to another process, a reference process (Schlosser, 2002). The more general nature of process modules could make us consider whether we should not completely abandon the potentially confusing language of static structural modularity, which describes the world as consisting of objects currently experienced by our senses. A description that uses process modules to explain how the elements of a system are generated might be more accurate. It applies to all phenomena and is particularly well suited to describing the dynamic as opposed to the structural ones (Simon, 1962). Furthermore, it could help us uncover completely new phenomena. After all, similar arguments are behind the whole movement of process philosophy (see Chaps. 6 and 7). The intuitive vocabulary based on static structural modules (cell, eye, pelvic girdle, etc.) is firmly rooted in many fields that study organismal modularity, such as developmental or evolutionary and developmental biology, but that does not mean that processual insights from these fields could not bring new research impulses (Nicholson & Dupré, 2018).

According to various scholars, modules are characterised by close internal connections and relative independence, repeatability, long-term endurance, identity, or reusability as building elements in other systems or larger units (Wagner & Altenberg, 1996; Schlosser, 2002; Callebaut, 2005). If we were to study, for example, articulated insects whose body segments exhibit these characteristics, we could probably compile a similar list ourselves. Nevertheless, all these characteristics result from one universal property, which is the *quasi-independence* of each module on other modules within the whole. It is due to the fact that the individual modules engage in weak but not negligible interactions with each other that complex modular systems have such interesting properties.

Thanks to their quasi-independence, modules can play a specific role within a higher whole, such as an organism, whereby this role is context-sensitive only to a limited degree. Complex modular systems composed of quasi-independent modules are therefore *nearly decomposable* (Simon & Ando, 1961; Simon, 1962), which means that some of their components have a strong tendency to react in conjunction (modules), while other components do so to a lesser degree (components of different modules). The manner in which modules, as relatively independent units, are interconnected by their inputs and outputs with each other and with their environment then facilitates their coordinated operation within systems and association into higher units (Simon, 1962; Schlosser, 2002). Without this quasi-independence of individual parts, the emergence of complex entities would be significantly limited. An excellent example of such a system is the body of multicellular animal, which consists of individual organ systems.

A system can be described as modular to the extent to which each of its components operates preferentially based on its own, internally determined rules (Callebaut, 2005). At first glance, the spatial, structural modularity is more noticeable. An example of such an arrangement is a model built from a Lego kit. In the same context, though, we can also talk about functional modularity, which can be deduced from the intensity and nature of mutual interactions. An office is not an office because its employees are located in one building but because they share information and make decisions based on mutual collaboration in an organised way. In practice, an office can be just as effective even when its employees are located hundreds of kilometres apart and communicate via the internet. On the other hand, we can convert structural modularity to a functional one without much difficulty.

Physical proximity is, after all, only one function of modules. Particularly strong interactions over long distances (regardless of whether mediated by signalling molecules, nerve fibres, or telephone wires) are the characteristic of agents, that is, systems that originated in biological or cultural evolution.

Modular systems may consist of any entities, regardless of whether they are agents or not. There is, however, something special about agent-based modular systems. According to Simon (1962), we can distinguish between flat hierarchies (e.g., linear polymers that consist of repeated molecular units) and strongly multilevel hierarchies in terms of the number of their potentially interesting levels. Agents are characterised by the fact that they make up and tend to be members of particularly strong hierarchies. One example for all: Within an individual, we can identify organ systems and body segments. Both consist of individual organs, within which we can distinguish tissues formed by individual cells. But modularity does not end here. Within a cell, we can distinguish a number of compartments and organelles that consist of macromolecules. The lower levels are typically less relevant to biologists (in contrast to, for instance, chemists or physicists) although we can identify modular arrangements even there. The same pattern is found also at levels higher than an individual. Individual members of a species form (sub)populations, various species gather into mutually integrated communities, and, at the highest level, this leads to the emergence of the terrestrial biosphere. Agents, and especially organisms, are thus exemplary modular units, and the fact that they form particularly strong hierarchies, together with the specifics mentioned above, means that they form a unique, qualitatively different type of complex systems. In this publication, we are interested in organismal agency. In what follows, we shall therefore focus on living systems: organisms, their parts, or, conversely, the higher units they form. Analogously, though, one could also consider similar factors (with the necessary specifications) in relation to entities that emerge in the course of cultural evolution, such as companies, institutions, or empires (see, e.g., Toman & Flegr, 2017, 2018b).

From the perspective of entities that originated in a biological (or cultural) evolution, modularity has a close relationship with evolvability (Toman & Flegr, 2018b). Not all systems undergo evolution with the same ease. While terrestrial organisms have been proving this ability for almost four billion years, other complex systems, such as conventional computer programs, are incapable of evolution (see, for example Wagner & Altenberg, 1996; also note that they are not agents). If for no other reason than because even a small change to the precisely programmed code can damage the operation of the entire program, it would be foolish to expect new types of software to 'evolve' in our notebooks. Rather interestingly, though, even virtual organisms simulated in computers to study evolution reach their limits once they undergo a simple selection process. Unlike real organisms, they do not generate significant evolutionary novelties without targeted interventions (see, e.g., de Vladar et al., 2017).

Various systems therefore differ not only in how they evolve over time in response to changes in internal and external conditions, but also in how effectively they undergo evolution. In line with that, evolvability can be defined as, for instance, 'the ability of random variations to sometimes produce improvement' (Wagner &

Altenberg, 1996). Still, it would be a mistake to understand evolvability simply as the ability to be subjected to natural selection, because even systems in which it is extremely difficult to generate significant evolutionary novelties have this property. Evolvability thus seems to have at least two levels: one is the ability to be selected on the basis of simple gradual changes (determined by their occurrence, for example through mutations), while the other is the ability to change the system's internal organisation based on previous experience, which thus channels further possible changes and affects the potential for radically new innovations (Wagner & Altenberg, 1996; Pigliucci, 2008).[3]

At this second level, evolvability demonstrates itself as an exemplary biosemiotic process (Toman & Flegr, 2018b). The abovementioned internal organisation, often referred to as the *genotype-phenotype map*, ensures that phenotypic responses to selective pressures are informed by genetic, epigenetic, structural, and other organic memories (Markoš, 2002; Sharov, 2014; Watson et al., 2014; Markoš & Švorcová, 2019). Possible reactions to environmental pressures and mutations range from complete buffering against them, through 'hardwired' – or habitualized – solutions and polyphenic-like multifurcations, all the way to full plasticity. It is yet to be determined which of these states is ancestral but it should be noted that several influential authors align with the biosemiotic perspective, suggesting that it is typically characterised by 'plastic' adaptability to interpretation (Newman et al., 2006). Furthermore, evolvability is subject to its own evolution (Hansen, 2006; Pigliucci, 2008), a concept surprisingly analogous to learning (see, e.g., Watson & Szathmary, 2016), which implies that the evolutionary process may deviate from the traditional image of an 'opportunistic tinkerer' and instead experience a form of higher-level meta-evolution. The way particular lineages evolve then can be viewed as their way of interpreting current conditions (or alterations in these conditions) by drawing from past experiences, all with the goal of creating the most effective adaptations.

An important property of systems which influences their evolvability is robustness, i.e., resistance to both external and internal disturbances and disruptions. As emphasised already by Simon (1962), modularity is one of the most important ways in which systems can optimise their robustness. While in systems that do not show significant modularity, one error can easily jeopardise the functioning of the whole, in modular systems, the effect of an error is usually limited to one module. A modular arrangement thus increases the system's endurance – in our case the viability of organisms – while allowing for improvement of particular, already present, and largely independent subsystems without collapsing the whole (Wagner & Altenberg, 1996; Simon, 2005). Modular organisation also enables redundancy, i.e., the mutual backup of important functions, and improves the potential of complex regulation. Finally, modularity increases evolvability by allowing the system to proceed step by step, that is, to modularly construct higher units by connecting lower, already tested, functional ones (Simon, 1962). It is extremely unlikely that an anthill would form directly from a community of unicellular flagellates. The possibility that several

[3] For a different view of evolvability, see, e.g., Brown (2014).

protists combine into a multicellular organism and a few hundred million years later a number of multicellular individuals form a colony of eusocial insects is far more conceivable. According to Simon (1962), a modular arrangement radically reduces the number of steps required to adapt or complexify the whole. In fact, it is a fundamental requirement for problem-solving through trial and error, which, by the way, also encompasses natural selection. It is thus well possible that modularity is so frequent simply because 'among possible complex forms, hierarchies are the ones that have the time to evolve' (Simon, 1962).

Certain authors go even further, arguing that modularity may be absolutely necessary for at least more complex organisms to evolve, function, and successfully adapt. If all genes, cells, developmental and regulatory processes, and other modules within organisms were equally interconnected in their arrangement, control, or development, then the whole would succumb to a *combinatorial explosion* driven by far-reaching pleiotropic effects of even the slightest changes. Modularity reduces the number and nature of interactions between the elements of a system, paradoxically thus facilitating adaptive evolution at a given level of organisation (Wagner & Altenberg, 1996; Callebaut, 2005). In evolutionary developmental biology, evolutionary interconnectedness of elements that form a module compared to the elements that do not belong to a module is called *dissociated coevolution* (Schlosser, 2002; Schlosser & Wagner, 2004). It enables context-independent loss or duplication of modules, deployment of modules in new places, in new contexts, or with new functions, as well as heterochronic changes in the timing of developmental processes, or coordinated changes in the size, shape, and activity of body elements (see Chap. 9). Modules themselves remain variable only to a limited extent but by multiple copying, different combination or regulation, deployment in another place or at a different time, and similar processes, organisms can achieve highly diverse solutions. Modularity thus optimises evolvability and the resulting evolution of complex modular agents – in this case organisms – may resemble playing with a Lego kit.

The Origin and End of Organismal Modularity

Interconnected modules at multiple organismal levels mediate a network of genotype–phenotype relationships that allows organisms to successfully complete individual development, to function in their environment, and to undergo adaptive evolution over time. The terms *genotype–phenotype map* (Abrech, 1991) or *genetic architecture* of organisms (Hansen, 2006) have been adopted for this network of relationships that stands at the very foundations of evolvability. This map or architecture has many levels and ranges from genome organisation and regulation of individual genes, through direct interactions of their products, eventually regulation and interconnections of entire developmental, morphological, functional, and evolutionary modules, all the way to feedback loops between the external environment or individual experience and lower levels of development, structure, and functioning of organisms.

We saw above that a modular arrangement, or quasi-independent genetic representation of functionally distinct traits that allows for a dissociated coevolution, increases the evolvability of a system. On the other hand, modularity does not have a positive effect on evolvability automatically. Modularity at a genetic and developmental level must also correspond to the functional and evolutionary modularity at the level of the phenotype (Espinosa-Soto, 2014). The genotype–phenotype map thus achieves optimal evolvability when it has few links between traits serving different functions and many links between traits that form individual functional complexes (Wagner & Altenberg, 1996). For example, if a change in a single gene or developmental pathway affects both the forelimbs and the hind limbs, they are difficult to differentiate significantly, and the organism consequently exhibits limited evolvability. It was the interruption of this bond (and the related emergence of far less interdependent evolutionary modules of the forelimbs and the hind limbs) that probably enabled the evolution of wings in bats (Wang et al., 2010) or bipedality in humans (Young et al., 2010). The key remaining question is how the evolutionary modules within the genotype–phenotype map originate and change (Wagner et al., 2005; Hansen, 2006). This issue was discussed in detail by, e.g., Wagner et al. (2005) but a good introduction to the subject is found also in an older article by Wagner and Altenberg (1996).

In the simplest case, we deal with traits which are naturally modular: they are coded by a small number of genes with low pleiotropy, i.e., genes that directly affect only a given trait and do not require special explanation. Frequently cited examples of such features are, for example, the ABO blood system or hair colour. For most traits, though, the situation is more complicated. According to some authors, modularity can be the primary property of how organisms are built, for instance due to the physical and chemical processes associated with their origin or because self-sustaining systems can work well only if they are modular. Still, the study of genotype–phenotype maps shows that the modular structure of organisms does change over time, which indicates that it is probably an evolved property, regardless of whether this arrangement results from an advantage in individual selection, the sorting of clades, or a simple statistical necessity in the sense of Simon (1962). On the other hand, it does not seem likely that the evolution of modularity would lead to its uniform increase due to the effects of natural selection. It is thus most likely that modularity, like evolvability, develops by a systematic association of genetic architecture with traits under selection and subsequent sorting of evolutionary lineages based on their evolvability (Wagner & Altenberg, 1996; Espinosa-Soto, 2014).

From a general point of view, the specific processes by which the modular organisation of genotype–phenotype maps originate can always be classified as either parcellation or integration (Wagner & Altenberg, 1996). Parcellation refers to the 'pruning' of bonds between the elements which belong to different complexes within the originally integrated whole. Integration, on the other hand, can be defined as connection of the various originally independent elements into complexes that fulfil a common role (Callebaut, 2005) or as the strengthening of interactions within modules (Force et al., 2004). Both of these processes participated, for example, in the evolution of multicellular animals. Although they started by integration, i.e., by interlinking of originally looser colonies of choanoflagellates, parcellation (specialisation of repeated originally identical parts – cells, modules, body segments, castes, etc.) played a major role in their further evolution. The specific models and conditions under which the increase in modularity can be explained were summarised for example by Toman (2020: 141–152).

Moreover, a form of gradual integration probably also constantly takes place in the course of evolution of evolvability at all levels of complex organisms, even at the cost of reducing the modular character of the whole. Even Williston's age-old law, which describes the trend of differentiation and specialisation (i.e., subdivision) of serially repeated body parts in evolution, emphasises that their differentiation is immediately followed by integration and reduction (Williston, 1914). Specifically, this phenomenon has been documented in the case of integration of different types of skeletons of multicellular animals from their first occurrence in the Cambrian towards the present (Thomas, 2005). Integration of the skeletal elements of the skull of fishlike vertebrates (Schaefer & Lauder, 1996) or the integration which accompanies specialisation of arthropod appendages (Adamowicz & Purvis, 2006) are at least as impressive. From a global perspective, this phenomenon can be described as consolidation of a given constitutive level accompanied by differentiation and specialisation of elements at the level immediately lower and a significant simplification – sometimes even called 'machinification' – of modular elements at the lowest levels (McShea, 2015).

Although it is not the conventional interpretation, this gradual integration can also be described as the establishment and increasing agency of units at a given level, for example of multicellular organisms which consist of originally individual protists or of communities of eusocial species which are made up of previously solitary individuals of hymenopterans. This growing internal interconnectedness is driven by the fact that a more integrated body structure or function can provide its bearers with immediate benefits in terms of biological fitness and, paradoxically, also certain benefits resulting from higher evolvability. First and foremost, it is about increasing the efficiency of bodies and the robustness of their development and functioning. Every viable individual must inherently possess a degree of robustness, a trait that curtails its susceptibility to fluctuations in external conditions and variations in its genome (see, e.g., Wagner, 2005; Masel & Siegal, 2009). By mitigating the detrimental side effects of mutated alleles on the phenotype, robustness can also increase evolvability (Hansen, 2006). Last but not least, robustness increases the potential for accumulation of cryptic variation, that is, mutations which are temporarily (during stabilising selection) silent but under certain circumstances (such as directional selection or stress) manifest themselves, thus allowing evolutionary lineages to overcome valleys in the adaptive landscapes (see, e.g., Wagner, 2005; Masel & Siegal, 2009). On the other hand, absolute robustness, that is, a state in which the same phenotype would arise based on any alleles and under all environmental conditions, would halt the functioning of natural selection. The relationship between robustness and evolvability is therefore anything but straightforward, whereby mean values of robustness are generally best for evolvability.

In addition to these essentially selective explanations, another and far more general process, namely stability-based sorting, also takes place in the process of gradual integration. This phenomenon, which has been so far almost completely neglected in evolutionary biology (see Toman & Flegr, 2017), affects all material and immaterial entities. Crucially, though, it leads to the predominance of such entities – in our case alleles, genes, their modules, whole organismal traits, and evolutionary lineages – in the system (in our case biological evolution), which have a greater persistence and a lower probability of changing into something else. In the case of gradual integration, it is a matter of sorting out those connections and elements which are less likely or unlikely to be affected by changes, especially by the effect of directional selection. In the long run, this factor has the upper hand even against the effects of natural selection.

What ultimately prevails are not the traits which provide their bearers with the greatest benefits in terms of biological fitness or those organisms and evolutionary lineages that have the largest populations, greatest biological fitness, or the highest frequency of speciation. Instead, what comes up on top are evolutionary lineages that managed to survive while carrying properties that cannot further change (while increasing the persistence of their carriers, otherwise these carriers would become extinct) (Shcherbakov, 2012). The once successful trilobites or American passenger pigeons (*Ectopistes migratorius*) are an eloquent proof of the relentless nature of stability-based sorting. At the other side of this spectrum, we encounter organisms such as tardigrades (water bears) which have been around with no significant morphological changes for over 500 million years and can be expected to outlive most multicellular organisms in the future due to their low resource requirements and exceptional resilience (Ward & Brownlee, 2002). Paradoxically, therefore, the outcome of all these integration processes is that while evolvability is optimised (reducing the risk of new mutations being significantly deleterious or incompatible with life, increasing the likelihood of them being adaptive, and reducing the number of mutations needed to create an adaptive phenotype), the effectively unchangeable elements of the genotype-phenotype map and, consequently, also the evolutionary constraints, accumulate. The original modularity thus gradually vanishes and the potential to create fundamental evolutionary innovations decreases.

We have named this phenomenon ‘macroevolutionary freezing’ or ‘reduction of the macroevolutionary potential’ (Toman & Flegr, 2018b). But a similar conclusion was reached by several other researchers, e.g., Riedl (1977, 1978) or Wimsatt (Schank & Wimsatt, 1986; Wimsatt & Schank, 2004; Wimsatt, 2013) and it finds support in various findings from palaeontology, ecology, or developmental and molecular biology, such as decreasing intraspecific and interspecific disparity, retreat of older evolutionary lineages to peripheral environments, or slowing pace of evolution (Toman & Flegr, 2018a, b). Some researchers therefore even argue that natural selection is only a side effect of systems’ tendency to evolve towards higher (static) stability. In their view, the ultimate attractor of anagenesis is evolutionary stasis associated with the highest possible speed and accuracy of reproduction

(Shcherbakov, 2012). On the other hand, the explosive diversification of organisms since their origin almost four billion years ago contradicts this conclusion. It therefore seems that organisms are, or at least have been, able to maintain the potential to create evolutionary innovations, which is reflected in the emergence of often sophisticated adaptations, including new levels at which evolution can occur. In other words, during their evolution, organisms have probably discovered ways to preserve or restore their macroevolutionary potential.

Transitions to a Higher Level of Hierarchical Complexity

Reduction of macroevolutionary potential and its causes and consequences are the central points of the theory of frozen evolution (Flegr, 2008, 2013; Toman & Flegr, 2018a, b). According to this concept, lineages that retained their ability to create fundamental evolutionary innovations are, from a macroevolutionary perspective, at a great advantage because they die out less often when conditions change and are more often subject to adaptive radiations, i.e., rapid diversifications into significantly different daughter species. For the same reasons, such species are more likely to colonise completely new biomes – such as land due to terrestrialisation or air due to the emergence of active flight. It seems, therefore, that in addition to reducing macroevolutionary potential, there is an increasing pressure to preserve the evolutionary lineages that are less susceptible to macroevolutionary freezing or at least capable of reversing its effects over time.

There are, however, two things that stand in the way of an effective elimination of evolutionary lineages with a lower remaining macroevolutionary potential. We saw that integration associated with macroevolutionary freezing can bring organisms significant short-term benefits. Even worse, macroevolutionary freezing usually proceeds in small steps which affect the immediate success of the lineage only slightly or not at all. The increasing number of macroevolutionary frozen elements is therefore in the short term usually neutral with respect to selection and resembles an accumulation of slightly deleterious mutations by the principle of Muller's ratchet in finite populations of asexual species (Toman & Flegr, 2018b). This 'macroevolutionary ratchet' is likely to be effectively slowed down or even fully stopped by bacteria and archaea. Because each of their asexual individuals *de facto* establishes its own evolutionary lineage and because these organisms usually form huge populations, selection can still capture even those elements which reduce the macroevolutionary potential only very slightly. Eukaryotes, and especially their complex multicellular representatives, on the other hand, form much smaller populations where individual selection is the dominant factor. The competition or contest between entire evolutionary lineages, in this case species selection based on the highest remaining macroevolutionary potential, is the less dominant factor here. Accordingly, the trend of decreasing of macroevolutionary potential appears to be characteristic of eukaryotes and especially their multicellular representatives with a complex genetic architecture (Toman & Flegr, 2018a, b).

Nevertheless, even sexual eukaryotic lineages can slow down or even temporarily reverse macroevolutionary freezing (Toman & Flegr, 2018a, b; Toman, 2020). The first possibility is to create a new combination of largely frozen traits that had not yet undergone stability-based sorting. As noted above, this is greatly facilitated by the modular arrangement of organisms. In addition, some seemingly irreversibly frozen elements of the genotype–phenotype map can sometimes 'thaw' and start responding to directional selection again. There are several ways in which this can happen. For example, as a result of certain later adaptations, functional relationships between genes or higher-level modules may change and the internal pressures that held the module in a frozen state for most of the lineage's existence may be released (see e.g., Wimsatt & Schank, 2004; Budd, 2006) – this is actually a special case of changing genetic architecture.

An event of a similar order occurred independently several times in the evolution of multicellular animals, each time involving a change in germline cell determination. An ancestral and to date taxonomically most widespread way of determining which tissues give rise to germ cells is induction by a cascade of genetic signals at a relatively late stage of development (seen today for instance in salamanders, sturgeon, paddlefish, crocodiles, lizards, or turtles). Some evolutionary lineages have, however, abandoned this method and determine the cells that give rise to gametes much earlier in the embryonic development determinatively, i.e., through specific substances synthesised into the cytoplasm (seen today for example in frogs, teleost fishes, snakes, or birds). It is interesting that this transition almost always led to an increased pace of evolution and greater species, morphological, and functional richness of groups that adopted a deterministic designation of the germinal line, even in comparison with their sister lineages. It seems therefore that this change led to the release of some of the accumulated evolutionary constraints, in particular the seemingly irreversibly frozen module determining the early development of the posterior of the body, which led to a significant diversification of the resulting clades (Crother et al., 2016).

Significant thawing can also be associated with a radical simplification of individual development in a given group. Typical examples are crustaceans from the group Rhizocephala (Glenner & Hebsgaard, 2006), cnidarians belonging to Myxozoa (Canning et al., 2004), or transmissible mammalian cancers (Murchison, 2008). These radically simplified organisms may become the basis of new, initially macroevolutionary very plastic but gradually irreversibly freezing clades. Less significant changes of this type are heterochronic changes in development, i.e., shifts in the timing or the order of development of different parts of organisms. These, too, may be associated with the release of modules for new purposes and an overall relaxation of the regulation of developmental processes (Budd, 2006). Such processes may have played a role for example in the early evolution of chordates (Hu et al., 2017). Overall, however, the processes mentioned above are of a limited importance – they can increase macroevolutionary potential only temporarily and only in a small group of probably not highly frozen traits, they require specific ecological conditions, or remain limited to taxonomically narrow groups of parasites (Toman & Flegr, 2018a).

A more effective way of restoring macroevolutionary potential therefore seems to be a transition to higher levels of hierarchical complexity. We have dealt extensively with this topic elsewhere, where we have also summarised the extensive literature related to this subject (Toman & Flegr, 2018a). It is certain that as a result of these events the original macroevolutionary frozen units become quasi-independent, limited-variable modules within a higher-level parcellated whole. An emerging agent of this higher level (e.g., a multicellular organism or a colony of eusocial insects, if we move to an even higher level) can regulate, multiply, or combine these modules largely independently, deploy them elsewhere, during other stages of development, or in other contexts, to benefit from the restored macroevolutionary potential and from all the other advantages a modular arrangement provides. But macroevolutionary potential starts decreasing also at the level of the higher modular unit. It is due to the same processes as at the lower level: consolidation of agency on a new level through the selective advantage of more integrated units, increase in certain aspects of evolvability, and accumulation of macroevolutionary frozen, further effectively unchangeable elements of genetic architecture by stability-based sorting. The whole process of transition to a higher level of hierarchical complexity may therefore be eventually repeated, leading to the multilevel nature of the genotype–phenotype map (Toman & Flegr, 2018a, b).[4] This leads to a conclusion that integration and parcellation tend to alternate at subsequent hierarchical levels (Wagner & Altenberg, 1996; Kirschner & Gerhart, 1998; Eble, 2005). We have described an instance of this phenomenon above on the example of evolution of multicellular organisms, but it seems this could be a much broader, perhaps even general, rule.

Regarding the specific possibilities of transition to a higher hierarchical level, the first *de facto* corresponds to the establishment of modules within the genotype–phenotype map. It is important to note that a significant integration associated with the reduction of macroevolutionary potential at one level can lead to a parcellated modular arrangement at a higher level consisting of multiple copied integrated units from the original level. A clear example of such an event is the integration of natural groups of genes with a closely related influence on the phenotype into evolutionary, morphological, developmental, and other kinds of modules, and their possible multiplication and integration into a higher domain. Externally, this process can manifest itself for instance in the formation of body segments or serially repeated appendages (Wagner, 1989a, b). An extreme case of such transformation of genomic architecture is the duplication of a whole genome. These events occur relatively commonly in some evolutionary lineages, such as plants (Ren et al., 2018).

In certain evolutionary lineages, there even emerged completely new ways of inheritance, i.e., the storage, transmission, and, most importantly, (re)interpretation

[4] Note that modular rearrangements, transfers to higher levels of hierarchical complexity, etc., do not result from direct selective pressures of the environment. These macroevolutionary processes are the consequence of the everlasting sorting of evolutionary lineages for optimal evolvability. The resulting trend thus could be described, in line with McShea (1996), as 'driven at the large scale'.

of information. According to some authors (Szathmáry & Maynard Smith, 1995; Maynard Smith & Szathmáry, 2010; Szathmary, 2015), these transitions include the formation of chromosome, prokaryotic and later eukaryotic cells, embryonic development of multicellular organisms, the comparatively complicated development and functioning of complex eukaryotic colonies (such as termite mounds), or human language and culture. This phenomenon could also encompass the emergence of other interpretative processes, such as gene editing, epigenetic regulation of the expression of different genes and gene circuits, or individual learning and complex patterns of behaviour (Markoš, 2002; Jablonka & Lamb, 2006; Kurismaa, 2018). Finally, a number of taxon-specific adaptations are of a similar character, such as post-transcriptional modification of mRNA in the cephalopods or macronucleus assembly in ciliates (see, e.g., Markoš & Švorcová, 2019: 46–47).

Another type of transition through which evolutionary entities can restore their macroevolutionary potential is the integration of several originally individual entities of the original level into an agent at a higher level of hierarchical complexity. There are basically two ways in which organisms can achieve this result: either closely related individuals or unrelated organisms from even very distant lineages of evolution can fuse. Queller (1997, 2000) calls these two types of transitions 'fraternal' and 'egalitarian' transitions in individuality, respectively. The first kind is exemplified for instance by the formation of colonies of unicellular prokaryotes and eukaryotes or the formation of multicellular organisms and their colonies (cf. Chap. 14). The latter kind of integration is exemplified by various types of symbiosis – in fact, symbiogenetic events are typical egalitarian transitions in individuality (Švorcová et al., 2018). The result is, as in previous cases, a modular organism of a higher level with a renewed macroevolutionary potential. Even in this case, however, processes that lead to reduction of the macroevolutionary potential will begin to take effect, and thus the transition may eventually be repeated.

The specific traits which facilitate or limit the evolution of organisms at a higher level may vary. Nedelcu and Michod (2004) have clearly illustrated the whole issue on numerous independent transitions of the Volvocales algal group to multicellularity. Among Volvocales, we find unicellular representatives (e.g., *Chlamydomonas*), loosely colonial species where each cell retains the ability to reproduce and establish its own colony (e.g., *Eudorina*), but also organisms that evolved complicated colonial forms with germ and somatic lineage division, specialised cells, and intercellular matter, organisms which thus undergo rudimentary germinal development and can be viewed as emerging agents (e.g., *Volvox carteri*). The limitations characteristic of this group – the impossibility of simultaneous movement and division of cells (known as the flagellation constraint), together with division into multiple cells (multiple fission) – seem to have put a strong pressure on the formation of coloniality in this group. For this reason, they specialise certain cells for movement and other cells for reproduction, enabling them to attain significant size and produce offspring while in motion.

Still, the Volvocales did not leave their original habitat of temporary water bodies and did not establish a major successful clade or reach the levels of complexity and phenotypic diversity typical of multicellular plants. This is because when forming a higher level of hierarchical complexity, they rebuilt their genetic architecture in a simple way that included a number of strong evolutionary constraints. Unlike multicellular plants (as well as fungi and animals), they did not significantly dissociate the genotype–phenotype map of lower-level units and rearrange it at a higher level in a way that would effectively co-opt lower-level properties (e.g., cell growth, division, and specialisation) while maintaining many degrees of freedom. *Volvox carteri* seems to have achieved the division of the somatic and germ cell lines, and thus the establishment of an agent of the new level, via a series of simple restrictions on lower-level processes including growth, division, and cell specialisation. The result is a genotype–phenotype map with significantly limited robustness and flexibility, which can provide the basis for only a limited number of phenotypes due to a strictly deterministic development involving only a few developmental processes.

It is clear that at different levels of hierarchical complexity, various specific features can act as strong evolutionary constraints. In general, though, macroevolutionary potential will always depend on the extent to which the higher-level organism managed to dissociate the basic properties of lower-level units, such as growth, reproduction, or differentiation, and on what basis it established a genotype–phenotype map. At least equally important are the new regulatory processes through which the organism integrates lower-level modules at a higher level. Higher-level agents with a complex regulation of their development and functioning therefore show the greatest macroevolutionary potential. Last but not least, we should keep in mind that it is practically impossible *a priori* to identify all evolutionary constraints or, conversely, the traits that increase evolvability. Both largely depend on the context of other traits of the organisms in question and their environment. Certain features of organisms of a higher hierarchical level may thus at first sight seem to present significant evolutionary constraints but in the context of other properties or a certain environment, they actually increase the evolvability of a given lineage. A typical example is the abovementioned deterministic type of germ cell determination. At first glance, this seems to significantly limit the evolution of germ cells, especially in comparison with induction by developmental processes at a later stage of development (as was, in fact, the first expert opinion, see Buss, 1988). However, empirical observations show that the locking of this process and its separation from other developmental mechanisms actually led to an increased pace of evolution, as well as greater diversity and disparity of groups with deterministic germline determination (Crother et al., 2016).

A fundamental aspect of transitions to higher levels of hierarchical complexity is that they also transfer agency – and in the process there emerges a new, higher-level agent. The key moment in this process is believed to be the time when an entity of a higher hierarchical level begins to reproduce as an individual (Michod, 2007). On the other hand, agency is not an absolute but rather a relative property whose depth is subject to evolution (see, e.g., Queller & Strassmann, 2009). A consolidation of agency through gradual integration at a higher level seems less problematic for organisms that originated by internal modularisation or fraternally. Their parts share either all or a vast majority of genes right from the beginning and thus face a lower (although non-zero, remember, e.g., cancer) risk of intraindividual conflict resulting from the reproductive selfishness of originally separate lower-level modules. On the contrary: their origin facilitates their possible cooperation and integration within the higher-level whole.

The establishment and consolidation of agency is a fundamental problem for entities that originated by the egalitarian path. The partners are not related and a gain of one partner in such a relationship can, and often does, take place at the expense of the other(s), which leads to conflict. Entities of this kind thus often for a long time consist of just loosely connected consortia. Such interactions in the grey zone between mutualism and parasitism, which depending on the conditions swing to one side or the other, are characteristic of for example the coexistence of corals and their symbiotic algae (Wooldridge, 2010). But empirical observations show that such obstacles are not insurmountable. For example, our own eukaryotic cells were formed by the association of an archaeal organism with the α-proteobacterial ancestor of the mitochondria. The key in such cases seems to be a fair distribution of reproduction costs, respectively the moment when the partners start producing joint reproductive propagules and thus join together their reproductive success (Michod & Nedelcu, 2003).

The demise of agency is an equally interesting process. At the immediately lower level, as part of reconstruction of the genotype–phenotype map, the units diversify, but also create new functional bonds and integrate. As macroevolutionary frozen units of lower level, these modules are variable only to a limited extent. They cannot easily evolve new features. Still, if a higher-level organism includes many originally identical modules that mutually back up their functions – as is common mainly in the fraternal transitions or internal modularisation – some of these modules may lose certain functions while finetuning the rest. There is a lot of pressure, especially on increasing the economy and robustness of their operation. At the lowest levels, habitualization, specialisation and simplification reach extreme values. There is a maximal reduction in complexity and evolvability, complete macroevolutionary freezing, and streamlining to automated units, i.e., machinification. Agency disappears completely, or almost completely, for instance at the level of elements of genetic material, while semi-autonomous organelles are far along on this path. Trends in increasing hierarchical complexity by differentiation of modules at the immediately lower level and machinification at the lowest level are so closely related in biological evolution that McShea (2002, 2015) calls them jointly the 'evolutionary syndrome'.

Conclusion

While modularity is found all over the natural world, it is much more prominent in systems consisting of agents. Agents (e.g., organisms) form particularly strong hierarchies, they do it fast (compared to non-agents), and their systems often exhibit a trend towards higher complexity. We could say that while modularity can exist without agency, there is no agency without modularity. First and foremost, the emergence of non-modular agents (or any complex systems) is statistically extremely improbable (Simon, 1962). Secondly, systems most likely need to be modular in order to be self-sustainable because it reduces their sensitivity to errors. This makes modularity an absolute necessity for all living systems and, consequently, also agents.

The unusual prominence of modularity in biological systems is probably due to the interaction between sorting for a higher static stability on the one hand and sufficient macroevolutionary potential on the other. In longer time scales, the evolution of especially sexual eukaryotic lineages is characterised by integration, a process accompanied by consolidation of agency but also by reduction of the likelihood of emergence of fundamental evolutionary innovations. It brings significant short-term benefits to organisms and evolutionary lineages while acting as a conservative factor that slows down their diversification. This inevitably gives an advantage to entities that have been able to restore their macroevolutionary potential, whereby the most common way of achieving this is parcellation, that is, transition to a higher hierarchical level of organisation combined with multiple copying or combination of lower-level units.

Nevertheless, integration is inevitable even on the new level. The opposing effects of parcellation and integration can therefore lead to a cycle of emergence of new agents at ever-higher levels and consequently also the formation of strong hierarchies which are so typical for systems consisting of agents. Paradoxically, the contradictory influences of integration and parcellation may underlie the trend toward greater hierarchical complexity and serve as one of the main mechanisms for the emergence of new agents (Toman & Flegr, 2018a).

References

Abrech, P. (1991). From genes to phenotype: Dynamical systems and evolvability. *Genetica, 84*(1), 5–11.

Adamowicz, S., & Purvis, A. (2006). From more to fewer? Testing an allegedly pervasive trend in the evolution of morphological structure. *Evolution, 60*(7), 1402–1417.

Brown, R. (2014). What evolvability really is. *The British Journal for the Philosophy of Science, 65*(3), 549–572.

Budd, G. (2006). On the origin and evolution of major morphological characters. *Biological Reviews, 81*(4), 609–628.

Buss, L. (1988). Diversification and germ line determination. *Paleobiology, 14*(4), 313–321.

Callebaut, W. (2005). The ubiquity of modularity. In W. Callebaut & D. Rasskin-Gutman (Eds.), *Modularity: Understanding the development and evolution of natural complex systems* (pp. 3–28). MIT Press.

Canning, E., Okamura, B., Baker, J., Muller, R., & Rollinson, D. (2004). Biodiversity and evolution of the myxozoa. *Advances in Parasitology, 56*, 43–131.

Crother, B., White, M., & Johnson, A. (2016). Diversification and germ-line determination revisited: Linking developmental mechanism with species richness. *Frontiers in Ecology and Evolution, 4*, 26.

de Vladar, H., Santos, M., & Szathmáry, E. (2017). Grand views of evolution. *Trends in Ecology & Evolution, 32*(5), 324–334.

Eble, G. (2005). Morphological modularity and macroevolution: Conceptual and empirical aspects. In W. Callebaut & D. Rasskin-Gutman (Eds.), *Modularity: Understanding the development and evolution of natural complex systems* (pp. 221–238). MIT Press.
Espinosa-Soto, C. (2014). Evolution of modularity. In M. Benítez, O. Miramontes, & A. Valiente-Banuet (Eds.), *Frontiers in ecology, evolution and complexity* (pp. 205–213). CopIt-arXives.
Flegr, J. (2008). *Frozen evolution: Or, that's not the way it is, mr. Darwin – Farewell to selfish gene*. Charles University.
Flegr, J. (2013). Microevolutionary, macroevolutionary, ecological and taxonomical implications of punctuational theories of adaptive evolution. *Biology Direct, 8*, 1.
Force, A., Cresko, W., & Pickett, F. (2004). Informational accretion, gene duplication, and the mechanisms of genetic module parcellation. In G. Schlosser & G. Wagner (Eds.), *Modularity in development and evolution* (pp. 315–337). The University of Chicago Press.
Glenner, H., & Hebsgaard, M. (2006). Phylogeny and evolution of life history strategies of the parasitic barnacles (Crustacea, Cirripedia, Rhizocephala). *Molecular Phylogenetics and Evolution, 41*(3), 528–538.
Hansen, T. (2006). The evolution of genetic architecture. *Annual Review of Ecology, Evolution, and Systematics, 37*, 123–157.
Hu, H., Uesaka, M., Guo, S., Shimai, K., Lu, T., Li, F., Fujimoto, S., Ishikawa, M., Liu, S., Sasagawa, Y., Zhang, G., Kuratani, S., Yu, J., Kusakabe, T., Khaitovich, P., Irie, N., Consortium, E., & Consortium, E. (2017). Constrained vertebrate evolution by pleiotropic genes. *Nature Ecology & Evolution, 1*(11), 1722–1730.
Jablonka, E., & Lamb, M. (2006). The evolution of information in the major transitions. *Journal of Theoretical Biology, 239*(2), 236–246.
Kauffman, S. (2000). *Investigations*. Oxford University Press.
Kirschner, M., & Gerhart, J. (1998). Evolvability. *Proceedings of the National Academy of Sciences of the United States of America, 95*(15), 8420–8427.
Kurismaa, A. (2018). Assimiliating an associative trait: From eco-physiology to epigenetics. *Biosemiotics, 11*, 199–229.
Lloyd, S. (2001). Measures of complexity: A nonexhaustive list. IEEE Control Systems. *Magazine, 21*(4), 7–8.
Markoš, A. (2002). *Readers of the book of life: Contextualizing developmental evolutionary biology*. Oxford University Press.
Markoš, A., & Švorcová, J. (2019). *Epigenetic processes and the evolution of life*. CRC Press, Taylor & Francis Group.
Masel, J., & Siegal, M. (2009). Robustness: Mechanisms and consequences. *Trends in Genetics, 25*(9), 395–403.
Maynard Smith, J., & Szathmáry, E. (2010). *The major transitions in evolution*. Oxford University Press Inc.
McShea, D. (1991). Complexity and evolution: What everybody knows. *Biology and Philosophy, 6*, 303–324.
McShea, D. (1996). Perspective metazoan complexity and evolution: is there a trend? *Evolution, 50*(2), 477–492.
McShea, D. (2002). A complexity drain on cells in the evolution of multicellularity. *Evolution, 56*(3), 441–452.
McShea, D. (2015). Three trends in the history of life: An evolutionary syndrome. *Evolutionary Biology, 43*(4), 1–12.
Michod, R. (2007). Evolution of individuality during the transition from unicellular to multicellular life. *Proceedings of the National Academy of Sciences of the United States of America, 104*(1), 8613–8618.
Michod, R., & Nedelcu, A. (2003). Cooperation and conflict in the origins of multicellularity and the eukaryotic cell. In A. Moy & E. Font (Eds.), *Evolution: From molecules to ecosystems* (pp. 195–208). Oxford University Press.
Mitchell, M. (2009). *Complexity: A guided tour*. Oxford University Press.

Murchison, E. (2008). Clonally transmissible cancers in dogs and Tasmanian devils. *Oncogene, 27*, 19–30.
Nedelcu, A., & Michod, R. (2004). Evolvability, modularity, and individuality during the transition to multicellularity in volvocalean green algae. In G. Schlosser & G. Wagner (Eds.), *Modularity in development and evolution* (pp. 466–489). The University of Chicago Press.
Needham, J. (1933). On the dissociability of the fundamental processes in ontogenesis. *Biological Reviews, 8*(2), 180–223.
Newman, S. A., Forgacs, G., & Muller, G. B. (2006). Before programs: The physical origination of multicellular forms. *The International Journal of Developmental Biology, 50*(2–3), 289–299.
Nicholson, D. J., & Dupré, J. (2018). *Everything flows: Towards a processual philosophy of biology*. Oxford University Press.
Pigliucci, M. (2008). Opinion – Is evolvability evolvable? *Nature Reviews Genetics, 9*, 75–82.
Queller, D. (1997). Cooperators since life began. *The Quarterly Review of Biology, 72*(2), 184–188.
Queller, D. (2000). Relatedness and the fraternal major transitions. *Philosophical Transactions of the Royal Society of London Series B-Biological Sciences, 355*(1403), 1647–1655.
Queller, D., & Strassmann, J. (2009). Beyond society: The evolution of organismality. *Philosophical Transactions of the Royal Society B-Biological Sciences, 364*(1533), 3143–3155.
Ren, R., Wang, H., Guo, C., Zhang, N., Zeng, L., Chen, Y., Ma, H., & Qi, J. (2018). Widespread whole genome duplications contribute to genome complexity and species diversity in angiosperms. *Molecular Plant, 11*(3), 414–428.
Riedl, R. (1977). A systems-analytical approach to macro-evolutionary phenomena. *Quarterly Review of Biology, 52*(4), 351–370.
Riedl, R. (1978). *Order in living organisms: A systems analysis of evolution*. Wiley.
Schaefer, S., & Lauder, G. (1996). Testing historical hypotheses of morphological change: Biomechanical decoupling in loricarioid catfishes. *Evolution, 50*(4), 1661–1675.
Schank, J., & Wimsatt, W. (1986). Generative entrenchment and evolution. *PSA: Proceedings of the Biennial Meeting of the Philosophy of Science Association, Number Two: Symposia and Invited Papers (1986), 1986*(2), 33–60.
Schlosser, G. (2002). Modularity and the units of evolution. *Theory in Biosciences, 121*(1), 1–80.
Schlosser, G., & Wagner, G. (2004). Introduction: The modularity concept in developmental and evolutionary biology. In G. Schlosser & G. Wagner (Eds.), *Modularity in development and evolution* (pp. 1–11). The University of Chicago Press.
Sharov, A. A. (2014). Evolutionary constraints or opportunities? *Biosystems, 123*, 9–18.
Shcherbakov, V. (2012). Stasis is an inevitable consequence of every successful evolution. *Biosemiotics, 5*, 227–245.
Simon, H. (1962). The architecture of complexity. *Proceedings of the American Philosophical Society, 106*(6), 467–482.
Simon, H. (2005). Foreword: The structure of complexity in an evolving world: The role of near decomposability. In W. Callebaut & D. Rasskin-Gutman (Eds.), *Modularity: Understanding the development and evolution of natural complex systems* (pp. ix–xiii). MIT Press.
Simon, H., & Ando, A. (1961). Aggregation of variables in dynamic systems. *Econometrica, 29*(2), 111–138.
Švorcová, J., Markoš, A., & Das, P. (2018). Origins of the cellular biosphere. In V. Sahi & F. Baluška (Eds.), *Concepts in cell biology – History and evolution* (pp. 271–290). Springer.
Szathmary, E. (2015). Toward major evolutionary transitions theory 2.0. *Proceedings of the National Academy of Sciences of the United States of America, 112*(33), 10104–10111.
Szathmáry, E., & Maynard Smith, J. (1995). The major evolutionary transitions. *Nature, 374*, 227–232.
Thomas, R. (2005). Hierarchical integration of modular structures in the evolution of animal skeletons. In W. Callebaut & D. Rasskin-Gutman (Eds.), *Modularity: Understanding the development and evolution of natural complex systems* (pp. 239–258). MIT Press.
Toman, J. (2020). *Evoluce³: Evoluční trendy, evolvabilita a teorie zamrzlé evoluce*. Academia.

Toman, J., & Flegr, J. (2017). Stability-based sorting: The forgotten process behind (not only) biological evolution. *Journal of Theoretical Biology, 435*, 29–41.
Toman, J., & Flegr, J. (2018a). A virtue made of necessity: Is the increasing hierarchical complexity of sexual clades an inevitable outcome of their declining (macro)evolutionary potential? *Evolutionary Biology, 45*, 374–394.
Toman, J., & Flegr, J. (2018b). Macroevolutionary freezing and the Janusian nature of Evolvability: Is the evolution (of profound biological novelty) going to end? *Biosemiotics, 11*, 263–285.
Wagner, G. (1989a). The biological homology concept. *Annual Review of Ecology and Systematics, 20*, 51–69.
Wagner, G. (1989b). The origin of morphological characters and the biological basis of homology. *Evolution, 43*, 1157–1171.
Wagner, A. (2005). *Robustness and evolvability in living systems*. University Press Princeton.
Wagner, G., & Altenberg, L. (1996). Perspective: Complex adaptations and the evolution of evolvability. *Evolution, 50*, 967–976.
Wagner, G., Mezey, J., & Calabretta, R. (2005). Natural selection and the origin of modules. In W. Callebaut & D. Rasskin-Gutman (Eds.), *Modularity: Understanding the development and evolution of natural complex systems* (pp. 33–49). MIT Press.
Wang, Z., Dong, D., Ru, B., Young, R., Han, N., Guo, T., & Zhang, S. (2010). Digital gene expression tag profiling of bat digits provides robust candidates contributing to wing formation. *BMC Genomics, 11*, 619–631.
Ward, P., & Brownlee, D. (2002). *The life and death of planet earth*. Henry Holt and Company.
Watson, R., & Szathmary, E. (2016). How can evolution learn? *Trends in Ecology & Evolution, 31*(2), 147–157.
Watson, R., Wagner, G., Pavlicev, M., Weinreich, D., & Mills, R. (2014). The evolution of phenotypic correlations and "developmental memory". *Evolution, 68*(4), 1124–1138.
Williston, S. (1914). *Water reptiles of the past and present*. University of Chicago Press.
Wimsatt, W. (2013). The role of generative entrenchment and robustness in the evolution of complexity. In C. Lineweaver, P. Davies, & M. Ruse (Eds.), *Complexity and the arrow of time* (pp. 308–331). Cambridge University Press.
Wimsatt, W., & Schank, J. (2004). Generative entrenchment, modularity, and evolvability: When genic selection meets the whole organism. In G. Schlosser & G. Wagner (Eds.), *Modularity in development and evolution* (pp. 359–394). The University of Chicago Press.
Wooldridge, S. (2010). Is the coral-algae symbiosis really 'mutually beneficial' for the partners? *BioEssays, 32*(7), 615–625.
Young, N., Wagner, G., & Hallgrimsson, B. (2010). Development and the evolvability of human limbs. *Proceedings of the National Academy of Sciences of the United States of America, 107*(87), 3400–3405.

Chapter 12
Agential Patterns in Development and Evolution: Towards an Anti-entropic Approach to the Divergence of Altricial and Precocial Mammals

Andres Kurismaa

Abstract The problems of organismal agency and phenotypic plasticity present significant interest for modern developmental and evolutionary biology, as well as comparative and biosemiotic approaches. At the same time, the question of how behavioural and other proximal phenotypic factors may potentially drive developmental and evolutionary variation remains open, particularly in relation to life-history and allometric theory (where the issue of agency has been seldom explored). To fill these gaps, this chapter revisits the anti-entropic (negentropic) approach to ontogeny developed by I.A. Arshavsky (1903–1996) and his school. Within this tradition, diverse developmental processes and organisms were studied in order to understand how biological agency and phenotypic plasticity can be interrelated and together affect the formation and integration of biological traits – including the remarkable physiological and morphological features distinguishing primarily altricial (immaturely born) from primarily precocial (maturely born) eutherians. This chapter revisits these concepts and results in the light of current research. In particular, that Arshavsky's non-equilibrium approach may still help to frame new questions and integrative concepts is shown with respect to certain unresolved problems of metabolic scaling, including the apparent biophysical paradox of increased work capacity and bioenergetic reserves in larger/more precocial organisms. Finally, we shall also consider a novel anti-entropic hypothesis regarding possible agential origins of the prolonged gestation and reduced offspring numbers in such animals, which may have a variety of implications for current work in biology and biosemiotics.

Keywords Agency · Developmental plasticity · Metabolic scaling · Life-history · Biological non-equilibrium · Ilya A. Arshavsky

---

A. Kurismaa (✉)
School of Natural Sciences and Health, Tallinn University, Tallinn, Estonia

J. Švorcová (ed.), *Organismal Agency*, Biosemiotics 28,
https://doi.org/10.1007/978-3-031-53626-7_12

Introduction

Most received theories of biological organisation and evolution remain firmly rooted in mechanistic and object-oriented approaches to scientific explanation, going back to the paradigms of Descartes and Newton (Kauffman, 2000, 2012; Agutter & Wheatley, 2008; Walsh, 2015). While conspicuously successful in conceptualising the non-living world, the physicalistic foundations of these paradigms have left little room for integrative approaches to biology. This includes the question of the possible agency of living organisms, which is here taken to mean their capacity to (co-) shape their own behaviour and development, and possibly, even evolutionary patterns. To analyse such aspects, new biological and philosophical concepts are therefore necessary.

In the present chapter, we consider this problem with a particular focus on comparative biological theory, and specifically, the allometric and life history aspects of bioenergetic processes in mammals. By these means, we hope to show how agential processes may embody a fundamental, if rarely considered driver of ontogenetic and evolutionary variation. This may have wide implications for current work in diverse areas, including semiotic biology (Sharov & Tønnessen, 2021; Pattee & Rączaszek-Leonardi, 2012; Markoš & Švorcová, 2019).

How to ground an account of living systems based on the concept of biological agency? It is important to note that while considerable efforts have been made to articulate the concept of organismal agency in terms of its minimal biological and semiotic organisation (Kauffman, 2000, 2012; Di Paolo et al., 2017; Sharov & Tønnessen, 2021) and in relation to evolution by natural selection (Walsh, 2015; Diogo, 2017), much less is known about the agential aspects of developmental and comparative biological processes, particularly in relation to the theory of life-history and biological scaling (i.e., allometry). In these areas, the basic question arises of *how can the activity and behaviour of organisms as autonomous agents* (Kauffman, 2000; Walsh, 2015; Chap. 8) *affect their comparative life-history, including the scaling relations between various biological functions and structures*? Further, what developmental and agential mechanisms could underlie such variability?

Although rarely cited in current debates, the anti-entropic approach to physiology founded by Ilya A. Arshavsky (1903–1996) and his colleagues (Arshavsky, 1972, 1982; Arshavsky et al., 1987) may cast some light on these questions. This school, while focusing on the non-equilibrium mechanisms and regularities of developmental physiology, arrived at a broader concept of bioenergetic processes which seems well compatible with modern, agential notions of biological work and its self-sustaining effects in living systems (Kauffman, 2000; Di Paolo et al., 2017). In fact, in the field of bioenergetics, Arshavsky's research programme has been recognised as a major forerunner to modern approaches in the study of biological self-organisation and autocatalysis (Igamberdiev, 2018). At the same time, in terms of biological agency, the relevance of this early fundamental work and its possible biosemiotic implications have not yet been sufficiently analysed.

In modern literature, an agent is defined minimally as a system whose work (energy expenditure) contributes to maintaining the constraints that define its own dynamic organisation, and hence its capacity to act as a self-regenerating and self-reproducing system (Kauffman, 2000, see Chap. 8, 9, 10; Kauffman & Clayton, 2006). Moreover, it is yet to be seen how such minimal concepts of self-regenerative work in natural agents (Di Paolo, 2005; Kauffman & Clayton, 2006; Ruiz-Mirazo & Moreno, 2012) could be expanded to encompass both the semiotic capacities of organisms (Kauffman, 2012; see Chap. 8; Sharov & Tønnessen, 2021) and the dynamics of their physiological and developmental mechanisms. While this chapter will focus mainly on the latter question, it will also touch upon its implications for the former where appropriate.

Interestingly, although the bioenergetic functions and capacities of organisms have been the subject of much research, the organisational and biophysical specifics which may distinguish organisms from lifeless systems have seldom been the focus of detailed research or experimental study in this context (cf. Koch & Britton, 2008, 2018; Careau et al., 2014). In particular, the issue of energy metabolism and its variable scaling relations with other biological traits remains a central, yet particularly challenging problem in both comparative biology and allometric theory (Harrison, 2017; Genoud et al., 2018). Could an approach based on agential dynamics and non-equilibrium mechanisms help us reframe these issues in more systemic terms, and could new empirical or conceptual tools be proposed on such an extended basis? As shown in this chapter, current research trends and questions seem to point in this direction.

In fact, a holistic and agential research programme could directly respond to the need for more biologically realistic and multi-mechanistic models in current comparative and allometric frameworks (Hulbert, 2014; Glazier, 2015a). Furthermore, a historical and theoretical biological perspective may importantly contribute to research methodology in the field, particularly when considering the often overlooked diversity of early modelling approaches and concepts in comparative and allometric study (Hulbert, 2014; Glazier, 2014, 2018). In the context of biological agency, such perspectives are important also in terms of raising new research questions. For example, how can a focus on organismal agency help to rethink the basic problem of the neural and informational regulation of processes responsible for developmental and allometric growth in organisms (Glazier, 2015a), where the effects of behavioural choices and/or neuro-hormonal systems become most evident (e.g., in phenotypic plasticity and niche construction, cf. Diogo, 2017). While not all these questions can be addressed in the current chapter, they help to outline the methodological challenges and background against which to explore the anti-entropic, or negentropic, approach.[1]

Viewing the activity of organisms as an entropy-countering, non-equilibrium factor underlying development may have wide implications for philosophical considerations of organismal agency, but also for comparative modelling. Here, the

[1] These two terms are used here synonymously, following Arshavsky (1982).

different types and strategies of energy metabolism in mammals present particular interest. As is well known, the developmental patterns and lifestyles of more altricial (i.e., earlier, more immaturely-born) vs. more precocial eutherians (later and more maturely born placentals) can exhibit a remarkable variation in terms of their energy provision, general physiology and morphology (Werneburg & Spiekman, 2018). Nevertheless, the mechanisms and origins of these differences remain relatively elusive, including the potential role of phenotypic plasticity in their ontogenetic and evolutionary divergence (Kurismaa, 2021a, b). Here, it seems that the anti-entropic approach may be of specific interest, as it analyses how the ontogenetically plastic expression of the phenotype may fundamentally depend upon the motor activities of the developing organism. Such ontogenetic plasticity was observed in diverse taxa, including predominantly altricial and precocial eutherian species, specifically. On an extensive empirical and experimental basis, it became apparent here that *key life-history patterns, such as longevity and precociality, can be altered considerably during mammalian ontogenesis, and this can involve coordinated shifts in the relative scaling of major functional and structural traits even without natural selection. Instead, such ontogenetic plasticity is based on changes in the biological work and aerobic capacity induced in the developing animal* (i.e., its rate and form of entropy production) (Arshavsky, 1967, 1972, 1982). Here, it seems that this approach could still shed new light on the roles of organismal agency and bioenergetic activity in the emergence of phenotypic variations and plasticity in extant species; in addition, it could also help to specify new models for testing developmental and evolutionary hypotheses in this context (cf. section on evolutionary hypotheses). This would be particularly relevant considering that the basic question of how organisms' developmental plasticity and activity may impact the evolution of biological scaling relationships remains largely unexplored (Casasa & Moczek, 2019), particularly in mammals (Glazier, 2014, 2015a).

To illustrate these issues and their practical implications, we shall focus on a fundamental but relatively paradoxical issue in comparative research (section on experimental modelling). As is well known, virtually all larger eutherian mammals belong to the precocial group (Clauss et al., 2014), where it is common to observe more differentiated brain and body plans, a higher degree of encephalisation (larger brain-to-body mass ratio), and other features associated with a more evolved and derived life mode, such as a longer lifespan and a lower number of offspring (Shea, 2007). At the same time, it has remained largely unclear how, in comparison to the more ancestral altricial taxa, these more recent mammalian groups (which include humans) could achieve their higher bioenergetic capacities, including aerobic scope (for definition cf. footnote 8) (Raichlen and Gordon, 2011; Harrison, 2017). This seems paradoxical because with increasing body size, the resting (or basal) metabolism of an organism tends to slow down, which should lead to reduced energy availability per unit mass in the animal and a corresponding reduction of its working capacities due to biophysical supply constraints.[2] However, this model may require considerable qualifications to be ecologically valid and applicable. Below, we outline how Arshavsky's non-equilibrium framework enables a reformulation of this fundamental issue and seems to propose a novel, agential solution to the above-mentioned energetic paradox.

In fact, this early approach may shed new light on the basic open question of how larger, typically more precocial animals can attain reduced rates of baseline (resting state) metabolism, and why this reduction does not necessarily reduce their working capacities. On the contrary, and in good agreement with recent models (cf. Pontzer, 2018), Arshavsky presumed that a reduction of baseline metabolism is an inevitable precondition, both physiologically and biophysically (thermodynamically), for attaining the higher working capacity typical of larger and/or more precocial organisms. If this explanation proves valid it could significantly clarify current basic debates on bioenergetics. Furthermore, their scope could be expanded by including the concept of agential work in organisms.

Finally, in the last section, we discuss the possible evolutionary consequences of the anti-entropic approach. In particular, we look at the possibility that *increased aerobic expenditure (in larger organisms) may not only necessitate general energetic trade-offs, but also a downregulation of energy allocation to specific functional systems, such as the reproductive one. An internal compensatory limitation on the reproductive energetics of animals could, in turn, importantly affect their developmental evolutionary patterns* (Arshavsky, 1982; Arshavsky et al., 1985), as certain recent findings also suggest (Pontzer, 2018). In fact, if backed by further evidence, such internal energetic trade-offs evoked by biological work could offer a novel explanation of the prolonged gestation characteristic of precocial animals and/or of their reduced number of offspring. Such notions, drawing on the anti-entropic theory, could also be relevant for modern epigenetic and biosemiotic approaches to evolution, by highlighting a possible new case of 'plasticity-led' developmental evolution in mammals (Arshavsky, 1985; Levis & Pfennig, 2020; Pfennig, 2021). As seen below, this epigenetic scenario may involve intra-species recognition as one of its (semiotic) mechanisms (Lambert & Spencer, 1995; Kull, 2016).

Against this background, which touches upon several fields, efforts to overcome the legacy of object-theories in biology (Walsh, 2015; Sharov & Tønnessen, 2021) may well benefit from the anti-entropic approach reviewed and elaborated here. Ultimately, expanding this framework could be relevant to rethinking the established genocentric and physicalistic research traditions in their respective fields and to grounding new, agential and semiotic models in biology more generally.

---

[2] The energetic consequences of this allometric principle are discussed in above. It refers to the principle that due to biophysical and geometrical constraints, as an organism grows in size, its volume increases as a function of the third order; while its surfaces, which govern energy absorption, and elimination, can only increase as a function of the second order.

On Life History and Biological Scaling Variations in Mammals

The developmental patterns and lifestyles of mammals exhibit a remarkable diversity of morphological, physiological, as well as behavioural traits. This is particularly evident among placental mammals (eutherians) classified as predominantly altricial or precocial (Shea, 2007; Werneburg & Spiekman, 2018). Regardless of extensive research on the subject, however, this diversity continues to pose many questions and challenges, both regarding its evolutionary emergence and in terms of the developmental regulation underlying predominantly altricial or precocial traits in particular species and environments (Kurismaa, 2021a). Although these comparative criteria are equally relevant for discussing the human form, it is not discussed in this chapter, partly because it combines both precocial and altricial-like features in a singular pattern or type, and partly because the topic was not pursued by Arshavsky in detail (cf. Kurismaa, 2021a).

The noted comparative issues were first comprehensively studied by Adolf Portmann, a pioneer of semantic biology (Portmann, 1967, 1990; Kurismaa, 2021a; see also Chap. 14). In subsequent work, they have been mainly analysed from the neo-Darwinian perspective, with its emphasis on natural selection and its effects on population dynamics over time (Reznick et al., 2002). In both traditions, the differences between altricial and precocial animals is typically characterised in terms of offspring's state at birth, variations in lifespan and reproductive patterns (e.g., gestation length and offspring number) (Shea, 2007), but also other morpho-functional features, including energy metabolism and encephalization. Interestingly, although belonging to very different developmental systems, these features and trait complexes often vary in more or less coordinated ways in animals (Table 12.1; Fig. 12.1).

For example, influential neo-Darwinian models analysing the fast vs. slow life history (or the related r/K strategies) (Reznick et al., 2002; Shea, 2007) have shown that many of the relevant traits (Table 12.1) can be (co-)responsive to ecological pressures associated with natural selection. Nevertheless, from a population genetic perspective, it remains unclear why cross-generational processes of fitness maximisation of particular heritable traits tends to result in more systemic morpho-functional variations in animals (Table 12.1). Answering this question may require distinct developmental and epigenetic lines of evolutionary analysis (West-Eberhard, 2003; Pfennig, 2021). In fact, the relevant traits and/or their integration may be to a high degree developmentally plastic, and hence, responsive to modification not only by natural selection but also by developmental mechanisms governing organismic responses during individual ontogeny, as seen below.[3]

[3]The concept of phenotypic plasticity is defined broadly here, as an organism's ability to (adaptively) alter its phenotype in response to changes in environmental conditions, regardless of whether the expressed changes are more or less reversible or concern physiological, life history, or morphological characteristics (Kelly et al., 2012). When this plasticity is expressed, primarily or solely, in particular age periods, we speak of developmental plasticity.

Table 12.1 Ontogenetic relationships in non-human mammals. Characteristic features of altricial and precocial mammals representing the earlier and later evolved forms of ontogenetic organisation

	Early form of organisation	Later form of organisation
State of the offspring at birth	Altricial: Closed sense organs, limited mobility and thermoregulation	Precocial: Open sense organs, advanced mobility and thermoregulation
Gestation length	Short (e.g., 20–30 days)	Long (over 50 days)
Number of young per litter	Large (e.g., 5–22)	Usually 1–2 (rarely 4)
Relative brain size	Small, considerable post-natal growth	Large, moderate post-natal growth
Examples	Many rodents, marten-like carnivores	Ungulates, seals and whales, apes and prosimians

Adapted from Portmann (1990: 22), Martin (1976), and Shea (2007)

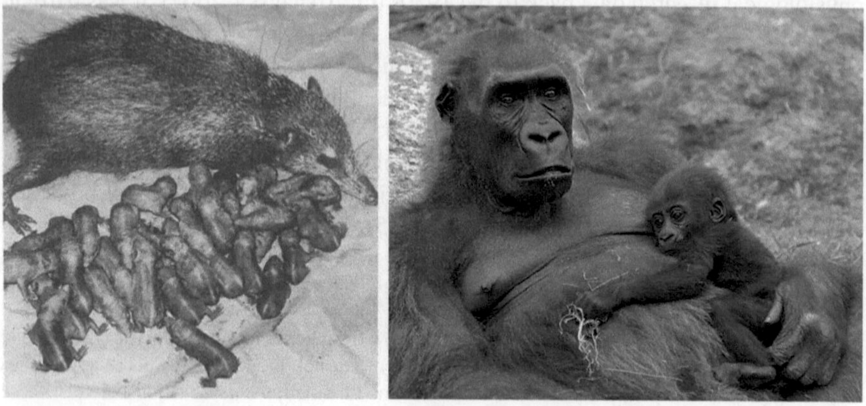

Fig. 12.1 An illustration of contrasting developmental modes in two mammalian taxa. Left: The highly altricial tailless tenrec, with numerous immature, highly dependent offspring (Louwman 1973; reproduced with the permission of Blackwell publishing). Right: The precocial gorilla with a single, well-developed offspring (reproduced with the kind permission of its author Boris Diakovsky ©)

At this point, it should be noted that in recent theorising, it is increasingly appreciated that natural selection can never operate directly on individual alleles or the genotype *per se*. On the contrary, any type of selection must act on traits that make up the organism's phenotype, which is almost always developmentally and ecologically variable in its expression due to epigenetic mechanisms and diversity (Uller et al., 2020; Pfennig, 2021; see Chap. 9). Furthermore, it is now understood that epigenetic processes and plasticity can actively bias, constrain, and modify the pathways available for evolution on the genetic level (West-Eberhard, 2003; Lema, 2020), and this calls for a new dimension of evolutionary analysis and theoretical integration (Perry et al., 2018; Pfennig, 2021). Respective theoretical issues are mentioned here given their expectable relevance to interpreting the altricial–precocial divergence, and the role which phenotypic plasticity may play in its developmental and evolutionary origins, in the light of current views and Arshavsky's early analyses (Arshavsky, 1972, 1985).[4]

In the last few decades, the problems raised by the anti-entropic approach have thus found renewed attention, at least on a theoretical and general methodological level. At the same time, relatively little is known of the implications of an epigenetic and developmental evolutionary approach to the problems of life history, as well as more broadly, with respect to the relative scaling relations between biological traits, as analysed in allometric frameworks (cf. section From plasticity to agency). We shall explore the anti-entropic framework with particular reference to this context.

Within the anti-entropic approach, mammalian developmental plasticity and divergence were investigated both mechanistically and theoretically to understand the broad principles, as well as specific developmental mechanisms of eutherian life history and allometric variation. While the beginnings of this line of comparative physiological research date back to the 1930s (cf. Arshavsky, 2002), it was particularly intensively pursued between 1960s and 1990s, when Arshavsky and his co-workers formulated their specific non-equilibrium approach to physiology and explored the general and theoretical biological implications of this framework (Arshavsky, 1960, 1966, 1972, 1982; Arshavsky & Mezhevikhina, 1992; Arshavsky et al., 1992).[5] The relatively scarce reception of this work in modern theorising may partly owe to methodological and sociohistorical reasons (which cannot be elaborated here in more detail, but cf. Nadin, 2015), and which would deserve further analysis in a separate study.[6]

---

[4] Generally, instead of a primarily one-way causal arrow from genes to the phenotype (i.e., genetic determinism characteristic of neo-Darwinian models) (Reznick et al., 2002), recent approaches increasingly model evolutionary mechanisms in terms of a reciprocal causation between these levels (Noble, 2021). Here, similarly to Arshavsky's epigenetic interpretation (1985), the role of developmental and phenotypic processes is considered potentially decisive in terms of exposing – or not – certain gene variants to ecological selection (West-Eberhard, 2003; Uller et al., 2020). Furthermore, according to some recent models (Shapiro, 2011), epigenetic processes may be actively involved in constraining the likelihood of particular genetic mutations in DNA base pair sequences, and thus in regulating genomic evolution. Such findings and epigenetic views lend new actuality to the problems analysed in this chapter.

[5] The results published by this group include nearly 250 scientific works by Arshavsky (between 1928 and 1995, most cited in full in Arshavsky, 2002) and probably a comparable number of publications by his co-workers, including numerous Ph.D. dissertations (cf. Arshavsky, 1982). Some relevant papers available in English and not cited in the relatively comprehensive bibliography of Arshavsky (2002), include Arshavsky and Demenshtein (1991), Arshavsky et al. (1975), and Arshavsky et al. (1976).

[6] For instance, Arshavsky's approach was influenced significantly by several frameworks outside the Western tradition. They include the theory of biologically stable non-equilibrium states by E. Bauer (1982 [1935]; cf. Igamberdiev, 2018); the evolutionary biological school of A.N. Severtzov (the teacher of I. Schmalhausen) and, most specifically, the physiological school of Arshavsky's teacher A.A. Ukhtomsky (1875–1942), who with his concept of physiological dominants began to develop non-equilibrium and agential concepts in neurophysiology (including the

Of note, the attention of the anti-entropic approach is focused not only on variations in metabolic scaling and its typological aspects, as pursued in some other early works (e.g., von Bertalanffy, 1951; cf. Glazier, 2005). Rather, its main focus is on developmental plasticity and the specific relation between allometric variability and organisms' bioenergetic and motor activity. In this context, experiments spanning diverse developmental processes in various species, ecological settings, and age periods all seemed to indicate that the formation of the bioenergetic phenotype is both highly plastic and responsive to the animal's motor activity. In fact, the same conclusion could be reached independently of the specific sources of bioenergetic stress – e.g., motor, hypoxic, thermal, or other – and of the particular species and studied taxa. However, most of the relevant experiments were carried out on mammals, including mice and rats, rabbits and hares, squirrels, chipmunks, dogs and cats (and in some cases, also large-horned cattle and other domesticated mammals, in addition to non-mammalian species) (Arshavsky, 1967, 1972, 1982).

In all these conditions, optimally increased exposure to bioenergetic stressors in early life was found to induce significant physiological and morphological changes in the observed taxa. In fact, the close interdependence between general developmental processes and motor energy expenditure in organisms seemed to reflect a general principle or rule of bioenergetics, which Arshavsky therefore came to call the *energy rule of skeletal muscles*, and to which we return below (Arshavsky, 1966, 1967). Indeed, subsequent research pointed to the broader basis and significance of this rule, given that also in plant and single-cell organisms, cytoplasmic contractile and motor functions may play a pivotal role in the formation of their bioenergetic phenotype (Arshavsky et al., 1992; Arshavsky & Mezhevikhina, 1992).

Hence, it may be appropriate to speak of a more comprehensive *energy rule of motor systems,* as the authors later proposed (cf. Arshavsky et al., 1992; Arshavsky & Mezhevikhina, 1992). As all these cases testify, *the processes of growth and development cannot be dissociated from the molecular, cellular, or system-level functions of motor and contractile work in organisms. Accordingly, the energetic exploitation and work of motor systems may induce major ontogenetic differences not only in the intensity of organismal energy-processing and metabolism, but also general differences in size, development, and form across the observed species* (Arshavsky, 1972, 1982; Arshavsky & Mezhevikina, 1992; Arshavsky et al., 1992). Such mechanisms of biological scaling and developmental regulation can be analysed more closely based on the concepts of organismal agency and biological

notions of biological work, and the non-linear, autocatalytic, and synergetic aspects of physiological processes, to use modern terms) (cf. Arshavsky & Demenshtein, 1991; Kazansky, 2015; Kurismaa, 2015; cf. also Kurismaa, 2023). The same processual orientation, focusing on developmental regularities and physiological variability rather than a search for abstract, timeless constants in biology, is evident in the anti-entropic approach. Indeed, a synthesis and development of these traditions enabled Arshavsky to approach classical problems of comparative physiology from new developmental and agential perspectives. Nevertheless, these less-known sources of inspiration and their methodological specifics may also help explain the limited reception and adoption of this work by contemporary authors.

non-equilibrium, as discussed below (cf. sections on experimental modelling and evolutionary hypotheses below).

In this context, appreciating the roles of phenotypic plasticity and agency in evolution may depend on a relatively far-reaching reassessment of classical comparative theories. This concerns not only a critical review of genetic reductionism, typical of neo-Darwinian approaches, but also of biophysical reductionism, forming the foundation of classical allometric frameworks (Hulbert, 2014; Glazier, 2005, 2015a, 2018). In fact, against the backdrop of these reductive approaches, it is virtually unknown how the (often notable) plasticity of phenotypic traits may affect the evolution of biological allometries, and more research is needed in this neglected area (Casasa & Moczek, 2019). Moreover, in the specific case of metabolic allometry, it is noted that the high variability of its scaling exponents is well documented, but that the relevant findings have been considerably underappreciated in many classical investigations, especially by theorists (reviewed in Glazier, 2005, 2014, 2018). To a large extent, this may be due to the reductive and physicalist foundations of classical allometric thinking (Glazier, 2005, 2014). Considering such broad challenges, the plasticity-oriented approach by Arshavsky and his school could lend considerable support for new lines of more integrative research.[7]

Interestingly, the question of why the rate of aerobic metabolism varies (scales) in regular but different ways in organisms depending on their body size and life history is far from new; however, despite nearly two centuries of attempts to solve this problem, it is still only partly explained by current allometric frameworks (Harrison, 2017; Glazier, 2018). In fact, the area of metabolic allometries is seen as one of the most controversial and complex fields in allometric research in general (Genoud et al., 2018). According to a recent review, the frequently observed pattern of metabolic rate reduction related to increasing body size – and the various behavioural, ecological, and evolutionary correlates of this negative allometric scaling – 'might be the best-documented and least-understood pattern in biology' (Harrison, 2017: 662). Hence, it may be worthwhile to examine how the anti-entropic framework formulates this issue and attempts to answer it in agential terms (section on experimental modelling).

From this viewpoint, it also seems natural to ask here whether the challenges observed above could be partly due to the almost exclusively physicalist foundations

[7] Interestingly, while developmental and systemic approaches to comparative and allometric modelling have been proposed also by other early authors, paradoxically, their concepts have largely disappeared from modern frameworks, as noted by several scholars in the field (cf. Hulbert, 2014; Glazier, 2018). Such neglect of early works may have various, often interlinked and mutually reinforcing reasons that are relevant also here. For example, this may be due to the original work's appearance in less known or foreign-language journals, the general preference for more recent citations in current papers, and a broader conceptual discrepancy with current models (Glazier, 2018). Although not included in Glazier's reviews, the same factors seem to apply to Arshavsky's school. Indeed, little attention has been paid to the phenomena of independent (re)discoveries and 'sleeping beauties' in science, which are by no means rare; on the contrary, they seem to be characteristic of progress in life sciences, with its frequent detours and discontinuities (Hulbert, 2014; Glazier, 2018; cf. also Agutter, 2019).

of classical allometric theory and the resulting relative exclusion of an organismic line of analysis from its methodological scope. For example, when examining the changing ratios of surface-to-volume in differently sized animals, Arshavsky had early on concluded that the corresponding 'energy law of the surface' (cf. below) is in fact 'neither a law, nor even a rule', but rather an empirically useful heuristic and constraint (Arshavsky, 1982: 3). More recent findings seem to support this assessment also with respect to other physically based explanations of allometric phenomena, such as the current models of resource transport networks (which, as an alternative to the 2/3 of the classical surface rule, feature a 3/4 scaling exponent; cf. Hulbert, 2014; Glazier, 2014, 2018). In this context, an analysis of phenotypically plastic scaling mechanisms may be of particular importance for questions of energy metabolism which, although constrained by universal biophysical factors, may display considerable variation and sensitivity to local contingencies and behavioural factors (cf. Glazier, 2014, 2015a; Casasa & Moczek, 2019). In current understanding, such factors may include informational control of allometric scaling by organisms' central nervous and neuroendocrine regulation systems (Glazier, 2015a, b), which probably also rely on changes in hormonal and humoral homeostasis (Rozanova, 1968; Arshavsky, 1972, 1982). Against this background, *an agential view could re-examine the role of organismal activity in constructing not only organisms' own external ecological niche, through which ecological selection operates, but also in shaping their own internal physiological niche (the internal environment), whose parameters may directly regulate the patterns of allometric growth and energy metabolism during ontogeny* (cf. also Casasa & Moczek, 2019; Lema, 2020).

The central findings of the anti-entropic approach make these ideas highly plausible. As we review in more detail below (section on experimental modelling), optimally enhanced rates of motor and skeletomuscular activity may not only induce increased rates of energy dissipation in the organism, but moreover, also specific anabolic effects that can 'hyper-restore', or overcompensate, the spent energy at a surplus level, thus allowing for the organism's ontogenetic growth and development. This emphasis seems particularly important for modelling agential processes over time, because they not only reproduce the work capacity of living systems (Kauffman, 2000) but can also enhance it during development and growth: in this case, the indexes of biological non-equilibrium typically increase both in cellular bioenergetic, systems-physiological, and morphological terms, as outlined below (cf. two following sections).

Based on these considerations, it is natural to presume that the self-sustaining and autocatalytic effects of biological work would necessarily reflect the degree of surplus anabolic reactions evoked in the animal, which is contingent on its activity in particular developmental and ecological conditions (Arshavsky, 1972, 1982). Such contingencies would by definition contrast with the more universal, biophysical size- and mass-based aspects of metabolic scaling which have so far attracted most theoretical attention. These issues will be further analysed in the section on experimental modelling, where the hyper-restorative mechanisms noted above are discussed in more detail.

Before that, however, we should outline certain key questions and positions in current allometric research on biological scaling and energy metabolism. This provides the context for re-examining the anti-entropic framework and for assessing the modern implications of Arshavsky's proposals both on a mechanistic and theoretical level.

From Plasticity to Agency: Allometric Theory in the Anti-entropic Perspective

Current and classical approaches to the allometric theory have generally focused on four main factors in biological scaling. These factors are usually investigated relatively independently of each other, although in principle they are not necessarily mutually exclusive (Hoppeler & Weibel, 2005; Glazier, 2005, 2014, 2018; Hulbert, 2014).

Firstly, this pertains to the tradition examining the intrinsic, biophysical, and geometrical constraints on energy metabolism. A part of this tradition goes back to the classical work of Max Rubner, who formulated the '*energy rule of the surface*' (Hoppeler & Weibel, 2005) with reference to the relationship between an organism's size and the surface areas governing its energetic exchange with the environment. This approach has helped to explain why larger organisms tend to have slower metabolic rates vis-a-vis their decreasing relative surface area per unit mass (i.e., while increases in the mass and volume take place as a function of the third order, the surface areas subserving heat and metabolic exchange increase at close to the second order, other things being equal). In more recent work, similar biophysical constraints have been specifically linked to the structure of organisms' internal *resource transport networks* and their geometrical features, particularly in the cardiovascular system in animals and the vascular system in plants (Savage et al., 2004). This focus has produced a different scaling exponent for analysing mass-specific metabolic rates (which should be closer to ¾ over body mass), but its biophysical focus and search for a singular, more or less universal scaling relationship has been consistent with the earlier focus on organisms' size and surface area (Glazier, 2014).

Thirdly, there is a more recent line of *system composition theories*, which explore the scaling effects linked to the relative proportion of tissues with higher or lower metabolic rates in specific organisms. As a result of such variations (e.g., in the relative mass and volume of the liver, brain, and muscles), metabolic rates can vary considerably even between organisms of identical or similar size. This line of inquiry is thus more biological in its orientation. Fourthly and finally, there are also various *demand reduction* models, which have started to explore the possibility that in addition to *supply limitations* (e.g., due to the structure of resource transport networks and/or the surface rule), metabolic rates and associated life-history traits may also vary due to biologically specific, adaptive regulation of *energy demand* in

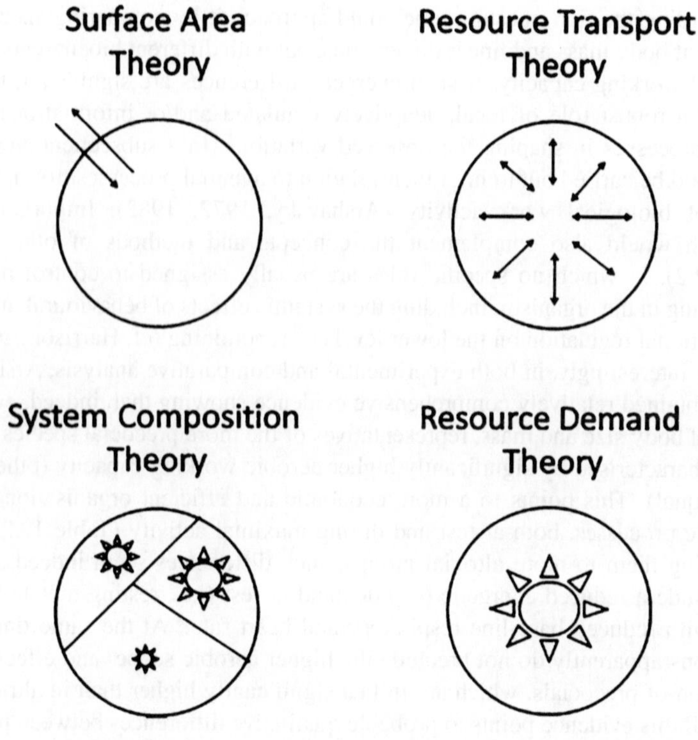

Fig. 12.2 A schematic representation of the four major factors and/or theoretical approaches to metabolic scaling as distinguished by Glazier (2015a, b, 2018) and others (Hoppeler and Weibel 2005; Harrison 2017) and explained in the text. (Figure redrawn from Glazier (2015a, b))

organisms, for instance, depending on their age, behaviour, or ecological conditions (Bergen & Phillips, 2005; Harrison, 2017). This shows that systemic constraints at the organismal and ecological level may have a downward regulatory and causative effect on the intensity of metabolic rate in tissues and cells (Glazier, 2015a, b). These four types of factors and models are schematically illustrated in Fig. 12.2.

In current work, it is an important priority to overcome the contradictions and limitations of each approach, while integrating their respective insights and contributions (Glazier, 2014, 2018). In reality, while the depicted models are in many cases not mutually exclusive, few approaches have so far managed to incorporate two or more of them (Glazier, 2014, 2018). Nonetheless, further work in this direction would be important to attain more complex and multi-mechanistic allometric approaches that would be biologically specific and context-bound. What role could the anti-entropic approach play in this respect?

We should perhaps state this question more precisely and ask what specific processes may constrain or regulate the noted energetic scaling phenomena irrespectively of the purely biophysical differences in organisms' shape and body size. Further, could such regulatory processes also involve organismal agency?

Regarding the first question, one could approach it by comparing species with equivalent body mass and linear dimensions, but with different bioenergetic phenotype and working capacity. If such energetic differences are significant, this may indicate a robust role of local, adaptively regulated and/or informationally controlled processes in shaping the observed variations. In a subsequent step, direct tests could be carried out to infer their relation to agential processes, for instance in terms of biological work activity (Arshavsky, 1972, 1982). Importantly, this approach would also complement the concepts and methods of other models (Fig. 12.2), in which no specific roles are usually assigned to control processes originating in the organism, including the systemic effects of behavioural, neural, or informational regulation on the lower levels of functioning (cf. Harrison, 2017).

Most interestingly, in both experimental and comparative analysis, Arshavsky's group obtained relatively comprehensive evidence showing that, indeed, even with identical body size and mass, representatives of the more precocial species are normally characterised by significantly higher aerobic working capacity (other things being equal). This points to a more economic and efficient organisation of their energetic processes, both at rest and during maximal activity (Table 12.2). When comparing them to more altricial groups, such differences seem indeed systemic and include a reduced energetic (O_2) demand at rest (i.e., resting metabolic rate), along with reduced base-line respiratory and heart rate.[8] At the same time, these reductions apparently do not preclude the higher aerobic scope[9] and effective heat insulation of precocials, which are in fact significantly higher than in altricial animals. All this evidence points to probable qualitative differences between predominantly altricial and precocial species in terms of the energetic provision of their main organs and functional systems, including the nervous, respiratory, cardiovascular, and musculoskeletal systems (Table 12.2).

[8] In line with classical definitions, *basal metabolic rate* denotes the lowest metabolic rate of an adult endothermic organism that is postabsorptive ('fasting'), non-reproducing, and at rest in a thermoneutral environment during the quiescent phase of its daily cycle. Often viewed as the minimum energy required for self-maintenance.

The more general concept of *resting metabolic rate* denotes the lowest metabolic rate of an endotherm resting in a thermoneutral zone but here one or more criteria defining the basal metabolic rate need be not observed. Unlike the basal metabolic rate, resting metabolic rate is applicable in developmental research (involving non-adult organisms).

The term *maximal metabolic rate* designates maximal sustained metabolic rate estimated as either exercise-induced maximal metabolic rate (by oxygen uptake rate, VO_2 max), cold-induced summit metabolic rate (during thermogenesis), or by daily total energy expenditure (rates of CO_2 excretion).

Aerobic scope denotes the animal's ability to increase its aerobic metabolic rate and functional capacity above maintenance level, as expressed by the possible range between resting metabolic rate (or basal metabolic rate) and maximal metabolic rate, either in absolute (maximal minus resting metabolic rate) or factoral terms (maximal metabolic rate divided by resting metabolic rate). In normal environments, the metabolic rate of an organism usually lies in the mid-range between resting and maximal metabolic rate (Careau et al., 2014; Kurismaa, 2021b). As seen in Table 12.2, Arshavsky assessed the aerobic scope in absolute terms.

Table 12.2 Characteristic features of predominantly precocial and altricial mammals as illustrated in four species (Arshavsky, 1967, 1982). The differences in cardiac and humoral regulation refer to their predominant features in each group

Precocial Altricial	Rabbits (~2500 gr)	Hares (~2500 gr)	Rats (~250 gr)	Squirrels (~250 gr)
Oxygen demand in resting state (O_2 ml/kg/min)	13–15	6–9	30	14–15
Respiration rate (per min)	60–90	12–16	N/A	N/A
Heart rate (bpm)	220–250	60–70	420–460	120–160
Aerobic scope	40–50	240–260	N/A	N/A
Thermal insulation coefficient (°C/kcal/m²/h)	0.353	0.766	0.25	0.719
Cardiac regulation	Sympathetic	Vagal	Sympathetic	Vagal
Humoral homeostasis	Adrenergic	Cholinergic	Adrenergic	Cholinergic
Gestation length (days)	30	54	21	38
Relative brain mass (% body mass)	0.3–0.4	1	0.5–0.7	1.8–2
Life span (years)	4–6	10–12	2.5–3	12–15

From a theoretical perspective, these findings seem well compatible with recent models explored by Pontzer and colleagues (2018), who likewise propose that the total energy expenditure available to an organism may be quite narrowly constrained (e.g., per day; cf. footnote 8), and *any increase in aerobic activity or capacity may thus need to be compensated by downregulation of other functional systems, such as the stress response and reproductive systems, or immune functions*. At the same time, such selective reductions in energetic investment may give organisms also a significant biological advantage in case they suppress nonessential physiological activities and enhance focal, adaptive ones (Pontzer, 2018). While these views and results are presented as relatively novel interpretations of organismic physiology and adaptation, they in fact significantly overlap with those of Arshavsky's group (Arshavsky, 1976).

For example, such energetically costly and dispensable physiological activities can include high sympathetic and HPA reactivity to environmental stressors, as well as a high rate of resting-state metabolic functions (Arshavsky, 1976, cf. also 1968, 1972), which are both typical of more altricial species (Table 12.2). This may help to explain key aspects of their biological scaling relationships, including shorter life-span and increased rates of reproductive energetic investment, as seen below.

In contrast, the observed bioenergetic optimisation seen in precocial animals appears closely related to a modified balance between inhibitory/parasympathetic vs. excitatory/sympathetic nervous regulation, and to a corresponding shift in their internal environment reflecting the humoral and hormonal homeostasis (Table 12.2), to which we return below (section on experimental modelling). Thus, the observed reduction of the baseline resting metabolic rate, and the downregulation of cardiac and respiratory rate, seem to be directly associated with general changes in the internal milieu (Table 12.2) that develop alongside with the organism's working capacity. We shall return to this topic further below to show how the various biological scaling models considered above (Fig. 12.1) can be integrated using the anti-entropic approach, with interesting implications for current work.

The findings and concepts considered above have been presented in some detail because they concern certain theoretical assumptions and questions which guide not only comparative and allometric research, but the biophysical and reductive traditions in modern life sciences more generally. Indeed, given their significance for the concepts of biological agency and work, such conceptual aspects should be addressed in some more depth before returning to the issues of comparative bioenergetics in the anti-entropic approach.

On Agency and the Problem of Biological (Hyper-Restorative) Work

Indeed, the discussions and findings noted above (Table 12.2) raise several controversial issues regarding the nature of bioenergetic work in organisms which may, at first sight, seem somewhat paradoxical. For example, how should we explain findings which indicate that activity, and especially intense activity (indexed by aerobic scope), may be mechanistically linked to increased longevity and organismic resilience?

As pointed out early on (Arshavsky, 1970, 1972), such observations could be interpreted as contradicting not only the received 'wear and tear' theories of stress (Selye, 1956; cf. Arshavsky, 1976; Hostinar & Gunnar, 2013) but even the basic physical requirements stemming from the Second Law of Thermodynamics. These requirements place limits on the energy budget and potential entropy production of any system, living or not (Arshavsky & Rabinovich, 1979; Arshavsky, 1983).

According to the Second Law, the work capacity and potential energy of any closed system tend to decrease over time, along with the system's internal order or organisation. Conversely, the system's original state of higher energy becomes dissipated in the form of heat or, in the case of a machine, also its working effects (e.g., transfer of force over a distance) (Kauffman, 2000; Kauffman & Clayton, 2006; Deacon, 2011). However, as energy is transferred or transformed, increasing amounts of it are lost – and this applies also to work-performing mechanisms. Not being exempt from the second law, work-performing mechanisms are, other things being equal, subject to entropic decay in proportion to their rate of activity and exploitation (Arshavsky & Rabinovich, 1979). In this context, work corresponds to positive entropy flow, which increases the system's disorder and decreases its energetic gradient or work potential (Arshavsky & Rabinovich, 1979; Arshavsky, 1983, 1970). *When we transfer this entropic interpretation to living systems, their specific biological far-from-equilibrium state in no way precludes these effects; on the contrary, work could be expected to dissipate such systems more rapidly, given that the far-from-equilibrium state corresponds to a state of increased order and self-organisation, but also to the condition of maximal energy dissipation rates* (Prigogine & Stengers, 2018; Arshavsky & Rabinovich, 1979; Koch & Britton, 2008, 2018).

Therefore, while recognising the importance of the nonlinear and self-organising features associated with non-equilibrium thermodynamic systems (e.g., dissipative structures) in biological models (Michaelian, 2022; cf. Chap. 8), from the anti-entropic perspective they do not address the core of bioenergetic workings in organisms, including the specifics of their non-equilibrium organisation and hyper-restorative capacity (Arshavsky, 1982, 1983; Arshavsky & Rabinovich, 1979). In this respect, Arshavsky's approach is also qualitatively different from that developed by Koch and colleagues (Koch & Britton, 2008, 2018; Koch et al., 2011, 2012), whose analysis of non-equilibrium dynamics and its comparative differences in mammals take as their starting point an analogy with Prigogine's dissipative structures (Koch & Britton, 2008, 2018). It is important to bear this issue in mind, even as both approaches associate higher aerobic capacity in organisms with improved nonspecific disease resistance, overall resilience to stressors, and longevity. Moreover, both Arshavsky's and Koch's group still represent a small minority in searching for new integrative models of health and disease with reference to the non-equilibrium state of living systems, even if it is conceptualised in a different way (Arshavsky, 1972, 1976, 1982; Koch & Britton, 2008, 2018; Koch et al., 2012).

More recent developments in agential theory could doubtless contribute to further development of the anti-entropic approach. In one of the most influential accounts, Kauffman (2000) has defined an agent, in a minimal sense, as a natural or artificial system that can perform at least one cycle of self-regenerative work. This means that the energy it releases during the catabolic phase of its activity is neither randomly dissipated (as heat), nor used solely for external purposes, but instead, it is also deployed in reconstruction of the internal constraints that permit the work cycle to occur in a functionally useful way in the system in the first place (Kauffman, 2000, 2012). In other words, 'work begets constraints beget work' (Kauffman, 2000). From this perspective, a system which satisfies this circular requirement of self-reconstruction can be said to 'act on its own behalf,' in a minimal sense: for example, it can grow and reproduce based on the specified organisation of its internal work cycles. Various extensions to this concept have been proposed in subsequent models and approaches (Letelier et al., 2011; Deacon, 2011; Ruiz-Mirazo & Moreno, 2012; Di Paolo et al., 2017), including semiotic ones (cf. Chap. 8; Kauffman, 2012; Sharov & Tønnessen, 2021).

Such extensions are important for considering the natural emergence of increasingly complex forms of agency, which is also relevant to the anti-entropic framework. In the semiotic scenario, for example, with further increases in an agent's complexity, the conditions under which it can attain its autocatalytic closure become more diverse and dependent on its regulation by biochemically arbitrary messengers: for instance when a transmembrane molecule is triggered to unleash a cascade of intracellular reactions in response to a chemically arbitrary set of extracellular molecules which simply share the characteristic of binding to the same particular receptor site. At this point, besides a physico-chemical type of analysis, a semiotic or relational description of the agent's organisation becomes inevitable if one wants to interpret the system's functions and global behaviour, because these become increasingly indeterminate in biophysical terms (Kauffman, 2012).

In this chapter, we will consider other areas where the concepts of recognition and semiotic mediation may need to be included in the analysis of agential processes (as seen below, cf. section on evolutionary hypotheses). For the moment being, let us just note that the issue of hyper-restorative work and optimal (anabolic) stress, which are central to the current framework, apparently have not attracted further specific attention in modern accounts of agency (semiotic or others). This left open the key question of how to account for the observation that, in living systems, the work they perform may lead to decreasing entropy production at rest and, conversely, to increasing degrees of internal organisation and anti-entropic resilience. The basic normative and anticipatory aspects of semiotic functioning may be also closely tied to the maintenance of such non-equilibrium energetic dynamics (Deacon, 2011). In fact, many developmental and evolutionary processes seem to reflect this hyper-restorative pattern and may depend on it (as seen in two following sections).

Besides these theoretical questions on anti-entropic processes, it seems that also on a mechanistic level the above-mentioned types of anabolic stressors have so far found relatively little consideration in current comparative and evolutionary work. For example, neither earlier works on 'allostatic load', which analyse the *costs* of stress on later development (McEwen & Gianaros, 2011), nor more recent studies, which include the potential *benefits* of stressors in calibrating endocrine, metabolic, and immune systems to current or future environments (i.e., 'adaptive calibration models') (Ellis & Del Giudice, 2019) make any reference to such optimal, hyper-restorative type of stress. The latter would refer to the potentially stimulating effect of internally or externally triggered stress reactions on organismal growth and development, where physical and physiological stressors can play an indispensable anabolic role for normal functioning (Arshavsky, 1968, 1976; cf. also Igamberdiev, 2018).

In fact, these questions have far-reaching implications not only for physiology but also for developmental and evolutionary theory, as is increasingly recognized (Mattson, 2010; Schirrmacher, 2021). For this reason, it is appropriate to examine more closely the evidence underlying these anti-entropic concepts and generalisations. Here, alongside correlative evidence from ontogenetically divergent groups (Table 12.2), we shall consider some comparative experiments on the mechanisms of hyper-restorative stress, particularly where they touch upon the relevant problems of biological agency and metabolic scaling (for a more extensive treatment, cf. Arshavsky, 1972; Kurismaa, 2021b).

Experimental Modelling of Work Capacities and Phenotypic Plasticity

At this point, some of the main experimental models investigated by Arshavsky's group shall be outlined in more detail. Of specific interest here is a series of key experiments, where normally highly altricial animals, such as rabbits, were subjected to various ecological stressors directly affecting their motor workload. As seen below (Table 12.3), the organisms developing under such experimental conditions displayed, after 4–6 months of training, phenotypic characteristics significantly different from those of control animals in terms of physiological, biochemical, and morphological indexes, all of which showed a clear shift towards precociality (Arshavsky, 1982; cf. Kurismaa, 2021b).[9]

One of the most important mechanisms here has to do with the mildly hypoxemic[10] internal environment of these animals that forms under increased

Table 12.3 Physiological, morphological, and biochemical characteristics of experimental male animals developing under increased bioenergetic (motor) workload

Index	Rabbits Controls	Rabbits Experimental	Hares
RMR O_2 demand (ml/kg/min)	13–15	7–10	6–9
Respiratory rate (per min)	60–90	30–40	12–16
Cardiac rate (bpm)	220–250	130–140	60–70
Cardiac regulation	Sympathetic	Vagal	Vagal
Humoral homeostasis	Adrenergic	Cholinergic	Cholinergic
Lifespan (years)	4–6	4.5–7	10–12
Liver weight (g)	96.9 ± 1.3	79.1 ± 2.8	50.0 ± 0.9
Liver weight (% of body weight)	3414 ± 0.109	2383 ± 1.10	1950 ± 0.09
Liver glycogen (mg %)	5100 ± 213	6800 ± 477	6700 ± 274
Muscle weight (g)	1254 ± 41	1600 ± 36	1394 ± 54
Muscle weight (% of body weight)	43.7 ± 0.7	47.6 ± 0.7	54.1 ± 1.7
Muscle glycogen (mg %)	510 ± 15	930 ± 53	848 ± 59
ATP, mmol/1 g (muscles)	7.23 ± 0.98	8.88 ± 0.1	9.96 ± 0.58
Lactic acid in muscles (mg %)	73.07 ± 4.58	50.0 ± 4.0	26.85 ± 2.26
Protein in muscles (g %)	22.0 ± 0.1	23.0 ± 0.3	23.5 ± 0.12
Blood acetylcholinesterase (mcmol/ml/min)	1.36	0.91	0.65

Data summarised from Arshavsky (1982: 193–195, 1967: 425; biochemical data by L.A. Siryk)

[9] In these experiments, rabbits, who otherwise dwell in burrows, were subjected to an intense daily schedule of locomotor activity (swimming in water, 24–26 °C), whose duration and timing aimed at optimal levels of physiological stress and activity. The workloads were age-specific, started at 4 weeks of age and lasted for 4–6 months (or, in a special series, until the end of the animals' life).

[10] In contrast to hypoxia, which refers to low oxygen content in tissues, hypoxemia refers to low oxygen concentration in the blood.

musculoskeletal work. Understanding the broad mechanisms of its emergence, while somewhat technical, is also theoretically relevant in order to see how the anti-entropic approach integrates the basic questions of energetic supply constraints and demand reduction, as outlined above in terms of the distinct allometric modelling approaches and factors (Fig. 12.2).

On the one hand, it is natural that when a developing organism is exposed to conditions of intense work, it results in a temporary decrease of O_2 saturation in the blood relative to its increased demand in the working organs. At the same time, Arshavsky demonstrated here that when this mismatch (or temporary supply restriction) takes place systematically and within certain optimal boundaries, it can lead to a qualitative shift in the organism's internal environment. This developmental shift involves switching its regulation from a predominantly adrenergic background, typical of mainly altricial animals, to a more energy-efficient cholinergic one, characteristic of the more precocial species (cf. below; Arshavsky, 1972, 1982). In more precise terms, this shift was associated with reduced acetylcholinesterase activity in the blood (corresponding to a reduced O_2 demand in the resting state) and a related shift to alkalinity within the acid–base balance, according to the authors (cf. Rozanova, 1968; Arshavsky, 1972, 1982).[11]

It thus seems, at first sight somewhat paradoxically, that physiologically optimal O_2 supply limitations induced by increased activity during early development led to an increase in the experimental organisms' aerobic capacities and far-from-equilibrium indexes. This increase can be observed in their maximal metabolic rate and

[11] It is assumed here that the effects of acetylcholine are determined by distinct factors that depend on whether it acts as a tissue hormone or a synaptic mediator (e.g., in the neuromuscular synapse). In the latter case, there is a close correspondence between acetylcholine release and its enzymatic degradation rate, whereas in the former case, instead of such parallelism, high acetylcholinesterase activity can *reduce* the effects of acetylcholine in the blood and tissues, where it plays non-synaptic roles. In this case, higher levels of blood acetylcholinesterase are associated with higher catecholamine concentrations (adrenaline, noradrenaline) and with increased sympathetic-adrenal activity, which condition the adrenergic features of homeostasis. This is characteristic of newborn and more altricial mammals. In contrast, in all mammals the inhibitory central and peripheral mechanisms of parasympathetic innervations and vagal tone are slower to develop, and the same applies to the corresponding cholinergic features of homeostasis which are typical of more precocial species. The latter typically show a reduced level of catecholamines (adrenaline and noradrenaline) and acetylcholinesterase in the blood (Rozanova, 1968).

Although the mechanisms have not yet been entirely explained, Arshavsky suggests that *for the excess acetylcholine concentration in tissues to be anabolically effective, as in precocial animals, it must occur in conditions of physiological hypoxemia whose extent is activity dependent* (Arshavsky, 1972). For example, while in the postnatal ontogeny of all mammals the decrease in natural respiratory rate significantly precedes the corresponding developmental decrease in the heart rate, the level of this mismatch – and resulting hypoxemia – is activity dependent. According to Arshavsky, such supply mismatches lead to an economisation and downregulation of resting-state O_2 consumption only under sufficiently high workload (Arshavsky, 1972, 1982). It would be interesting to compare this hypothesis with current studies on so-called intermittent metabolic switching, which investigate activity-induced temporary stress responses at a molecular level (but generally with less attention to developmental processes and plasticity) (Mattson et al., 2018; cf. also Raichlen & Gordon, 2011).

aerobic scope measures, but also in various other biochemical, physiological, and structural changes (Table 12.3). *As a result of bioenergetic work, predominantly altricial organisms thus acquire features which in many ways approximate those of their more precocial relatives, both in terms of the resting state and aerobic working indexes and in relation to their life-history traits and morphology* (Table 12.3). From a schematic perspective (introduced above, see Fig. 12.2), *the basic mechanism of these processes seems to follow the logic of adaptive 'demand reduction' resulting from self-induced but intermittent 'supply limitations' in the metabolic delivery networks, in this case the supply limitation of O_2 and nutrients to the working organs and systems under optimal stress.*

While the exact cellular and molecular mechanisms involved in these processes were relatively poorly understood at the time of Arshavsky's work, within current research the model of 'intermittent metabolic switching' developed by Mattson et al. (2018) may be particularly relevant here. This work has generalised how periodically alternating periods of energetic balance, with negative (e.g., training, fasting) and positive (rest, inactivity) phases, can bolster the resistance of organisms to stress, injury, and disease. In fact, on the system level, one of the most robust and readily discernible expressions of such mechanisms of intermittent metabolic switching may depend on inhibitory parasympathetic innervations (Mattson et al., 2018), as also observed above (Tables 12.2 and 12.3). For instance, as a result of endurance training, a lowering of the resting heart rate and blood pressure can be induced via brainstem (cholinergic) neurons which innervate the heart and increase its vagal tone (Billman et al., 2015). The cellular and molecular details of such processes are becoming well understood in many cases (Mattson et al., 2018), and this may shed new light on the findings reported above (Tables 12.2 and 12.3), including the precise mechanisms of optimal anabolic stress by which precocial traits may emerge in altricial organisms on a molecular, cellular, and systems physiological level (cf. also Raichlen & Gordon, 2011; Raichlen & Alexander, 2017).

On the other hand, even as such mechanistic details are being worked out in many cases, the exact regimens and species-specific variations of intermittent metabolic switching optimal for enhanced neural health and bioenergetic capacities throughout life have so far not been closely studied and would, according to the authors (Mattson et al., 2018), require a broader set of experimental models and approaches. In particular, such approaches should ideally include not only various strains of (laboratory-bred) mice and/or rats (Mattson et al., 2018), but also a wider range of predominantly altricial and precocial organisms with their qualitatively different physiological and morphological adaptations and different (allometric) relations to human ontogeny (Workman et al., 2013; Rivers & Ashton, 2013; reviewed in Kurismaa, 2021a, b).

It is important to note here that from the anti-entropic perspective, the physiological processes which lead to adaptive demand reduction and/or energetic reorganisation through intermittent supply limitation are not restricted to organisms' postnatal development. On the contrary, it is predicted that they constitute a general biological mechanism at the centre of various developmental processes, including mammalian embryonic, foetal, and prenatal development (Arshavsky, 1968, 1972, 1982). For example, it has been long known that above-normal amounts of nutrition and O_2 provided to the developing embryo and foetus do not support its optimal growth and development, and on the contrary, can significantly retard them and lead to the birth of physiologically immature and/or morphologically reduced offspring with suboptimal chances of survival (as found in, e.g., experiments on pregnant rabbits; cf. Arshavsky, 1968, 1972). On the other hand, optimal hypoxemic and nutritive stressors can significantly increase offspring mass and size, for instance from 30–50% up to 80–100% in some cases, as shown in the same line of work (Arshavsky, 1968, 1972), where rabbits growing under partial fasting or exposure to hypoxia during the last trimester developed significantly larger heart, lungs, and brains, both in absolute and relative terms, and a significantly larger relative mass of skeletal muscles (to total body mass) (Arshavsky, 1968, 1972, 1982). Such findings are mechanistically entirely consistent and interpretable, for example, given that the relevant optimal stressors are known to induce increasingly intense intrauterine respiratory and generalised motor reactions in the foetus, which leads to an amplification of anabolic processes (i.e., the circulation and transcapillary transport of nutrient- and oxygen-rich blood over the placental barrier (Arshavsky, 1960, 1968, 1972) in response to the experienced supply limitation.[12]

Nonetheless, there is a principled distinction between the anabolic processes operating in the ante- and postnatal periods according to the anti-entropic perspective. Specifically, Arshavsky proposed that during the antenatal period, optimal episodic stressors primarily lead to increases in the protoplasmic mass and size of the organism (i.e., the *1st type of anabolism*), while in postnatal development, the main type of surplus anabolism leads to an accumulation of the non-equilibrium bioenergetic potential of the cells, tissues, and organ systems (the *2nd type of anabolism*). In contrast to the first type, which starts at conception and is energetically highly expensive due to its anaerobic nature, the second type is aerobic and established together with the standing posture (which requires tetanic anti-gravitational motor reactions). Importantly, this transition from the first to the second type of surplus anabolism is, according to Arshavsky, characterised by a gradual minimisation and optimisation of energy use, including a reduced relative amount of O_2 and nutrient demand at rest (Arshavsky, 1972, 1982). And most notably, *this gradual economisation and energetic downregulation of maintenance costs, characteristic of the second type of anabolism, is particularly pronounced in precocial(ised) eutherians – even as at the same time, the energetic potential of their skeletal muscles and organ systems becomes increased by all main cellular and systems criteria, including the cellular membrane potential, glycogen concentration (in the muscles and the liver), ATP and total protein content (in muscles and other tissues), as well as the*

[12] Conversely, excess supply of nutrients and hyperoxia suppress such generalised motor reactions along with their critical circulatory and anabolic functions during this stage of development (Arshavsky, 1960, 1968).

systems-physiological indexes of the cardiac and respiratory system (Tables 12.2 and 12.3) (cf. Arshavsky, 1972, 1982).[13]

Interestingly, it was also reported that alongside changes in its biochemical content (cf. above, Table 12.3), the amount of circulating blood is likewise significantly higher in the precocial(ised) group of mammals (for instance reaching 10.5% of adult body mass in hares, in comparison to 6.5% in rabbits, independently of the relative volume of their vascular systems) (Arshavsky, 1982). Such observations seem particularly important considering that in contrast to physical and dissipative systems directly open to environmental physico-chemical energy gradients – all energy-assimilating and transforming processes in the organism are primarily of a physiological nature and occur in its internal milieu. In fact, all such internal energetic conversions are of a metabolically closed biomolecular type (Letelier et al., 2011), in the sense of being catalysed by enzymatic molecules that are synthesised by the organism itself (Cornish-Bowden et al., 2007) and in conditions set by its own internal milieu, blood composition, and volume (Arshavsky, 1972). *Therefore, while from a general thermodynamic viewpoint, the primary sources of raw material and energy are in the organism's ecological surroundings, from a physiological viewpoint they are located within the organism itself, in its internal milieu and blood supply.* Both in terms of its composition and volume, such supply appears to be highly activity-dependent and to develop together with the biochemical and physiological features of the skeletal muscles (cf. Arshavsky, 1972, 1982). In this context, the extent to which organismal activity and entropy production can affect its metabolic as well as physiological and morphological scaling relations may be more evident.

These observations are relevant, especially given that in current research the basic question of how increased aerobic and functional capacity may be attained in more precocial and/or larger species seems to be without a clear answer. In fact, given the theoretical difficulties this issue entails, it is perhaps natural that certain influential recent models seem to predict the very opposite energetic patterns of metabolic scaling. For example, in one of the most detailed accounts examining the

[13] It should be noted (Kurismaa, 2021b) that such comparative facts do not in any way contradict general physical principles, including the second law of thermodynamics – according to which an increased energetic potential accumulating in any living system must be compensated by increased entropy production in the environment (Prigogine & Stengers, 2018; Arshavsky, 1983; Michaelian, 2022). In fact, *in precocial animals the production of entropy in resting state is minimised* (as noted above), *among other things by direct thermal insulation* (cf. Table 12.2). Related effects can be seen experimentally, as *in the artificially precocialised group, increased work capacity emerges on the background of reduced stable body temperature compared to controls* (by 1–1.8 °C) (Arshavsky, 1972). Similarly, the surplus anabolic processes of energy restoration and macroerg accumulation in the muscles after work occur in conditions of reduced muscle temperature (by 1–2 °C) as well as other systemic changes (e.g., membrane hyperpolarisation, muscle cell hyper-relaxation to a length exceeding the initial one, etc.) (cf. Arshavsky, 1972). In this regard, the novel contribution of the anti-entropic framework is to point out that the optimisation of energetic functions in cells, tissues, and organs may importantly depend on baseline energy dissipation minimisation via adaptive changes in the internal milieu of the organism as a whole (Arshavsky, 1972; cf. Igamberdiev, 2018; Kurismaa, 2021b).

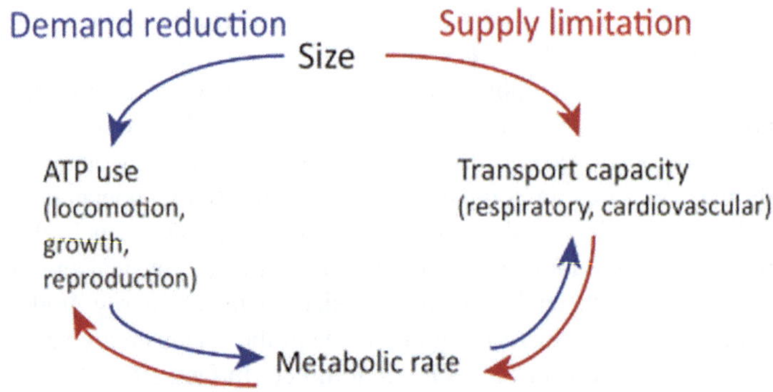

Fig. 12.3 Alternative causal explanations for the declining mass-specific metabolic rate in larger animals according to Harrison (2017). With increases in size, reduction of metabolic rate may arise mainly via supply constraints on the transport capacity of nutrients and O_2 or, alternatively, via adaptive and biologically specific reduction in their demand due to natural selection acting on life history traits. (Image redrawn from Harrison (2017))

adaptive and biologically specific demand reduction aspect of metabolic scaling in animals, it is assumed that smaller, more frequently altricial homeotherms must be selected for enhanced 'locomotor capacity,' along with other energetically costly functions of fast life history. Conversely, it is posited that larger and usually more precocial species have evolved strategies to minimise overall energy expenditure and working capacity as part of their safety-maximising and slower life history (Harrison, 2017). This account summarises valuable evidence in favour of biologically evolved and demand-driven metabolic allometries. Paradoxically, though, it also makes predictions which appear difficult to maintain, in proposing that larger and more precocial organisms have a generally reduced aerobic scope and energetic capacities – including ATP reserves and availability – and that their ATP-demand is reduced generally, not only in the resting state (as Arshavsky presumed) (Fig. 12.3). Nevertheless, these proposals are not backed by obvious evidence and directly contradict the extensive findings by Arshavsky's group (1967, 1972, 1982; Tables 12.2 and 12.3), which point to increased aerobic capacity and energetic efficiency of larger, predominantly precocial organisms – in good agreement with various subsequent findings and models (e.g., cf. Raichlen & Gordon, 2011; Alexander, 2005). How to interpret such fundamental discrepancies and contradictory viewpoints?

In part, these inconsistencies and theoretical issues may arise from focusing on organismal mass and size as the prime determinants of metabolic scaling (as in Harrison, 2017) or, conversely, from neglecting its potentially vast variations in similarly organised animals. As shown specifically in the anti-entropic framework, such variations may be fundamental even among identically sized and genetically closely related organisms (Arshavsky, 1972, 1982).

We can also consider here the size-specific relations of altricial and precocial animals. While among animals of small body size, both altricial and precocial species are widely represented, among mid-sized animals there are less altricial species, and among larger eutherians we find only precocial species (Clauss et al., 2014). There seems to be nothing in Harrison's model to explain this uneven distribution of life history traits. Rather than assuming – as he suggests (Harrison, 2017) – that this is the result of natural selection favouring overall minimisation of energy-intense and locomotor functions in larger, longer-living species (Fig. 12.3), Arshavsky's approach draws attention to the specific aspects of their physiology, life history, and ecology. The latter may be consistent with the demand-reduction strategy only as far as this pertains to resting state metabolic processes, while maximal metabolic rates under intense work should not undergo such reduction. In fact, as seen above, they may usually shift in the opposite direction, namely towards demand-based extremization (maximal metabolic rate) and increased aerobic capacity (aerobic scope). Failure to consider this possible energetic trade-off and contrastive functional significance of resting and maximal metabolic rate seems to be the second aspect of this model that may lead to the above-mentioned paradoxical results (Harrison, 2017) and disagreement with the findings of Arshavsky (1972) and related systemic approaches (cf. below; Raichlen & Gordon, 2011; Pontzer, 2018). Thus, if the 'metabolic rate' on Fig. 12.3 were to be broken down into resting metabolic rate and maximal metabolic rate as two qualitatively distinct categories, the latter would probably behave very differently both in relation to organisms' size and life history variations in the altricial–precocial spectrum.

Importantly, in developing his demand-focused model, Harrison admits that 'O_2 delivery limitations that cause the hypometric scaling of metabolic rate might occur at particular developmental stages or under conditions that have not yet been evaluated, and compensatory adaptations to limited O_2 supply in larger animals might yet be found' (Harrison, 2017: 659). What Harrison does not consider is the possibility that the adaptive, ecologically specific demand reductions observed in larger organisms (e.g., reduced resting metabolic rate and slower life history traits) may have, at least in part, originated in the phenomena analysed above, namely episodically present optimal stressors and supply limitations, such as the episodic O_2 restrictions, in precocial animals. Such supply limitations could bolster growth and development by intermittent metabolic switching or related anabolic processes, the mechanisms of which are currently closely studied (e.g., in the hormetic models of anabolic stress) (Mattson, 2010; Mattson et al., 2018; Pontzer, 2018; Schirrmacher, 2021). In this context, it is clear that *what is called above a 'demand reduction' model should be perhaps more fittingly called a 'demand modulation' or 'demand regulation' model, where not only the adaptive demand-related downregulation but also possible episodic increases in mass-specific metabolic rates could be addressed in larger organisms and seen as a potential driver of ontogenetic growth and development.*

From an agential and developmental perspective, it is therefore apparent that not only differences in the size and morphological composition of organisms (cf. Fig. 12.2, Table 12.2) but also phenotypic plasticity, and its associated physiological features, such as blood composition, ought to be considered when framing predictions and hypotheses about metabolic scaling.[14] In this context, the anti-entropic approach presents a developmental and mechanistically multimodal framework that naturally integrates the contrastive allometric principles and regularities discussed above (cf. Fig. 12.2) on the one hand, and shows their intimate connection to the topics of organismal agency and ontogenetic plasticity on the other. Such a perspective could contribute to more integrative models and heuristics in the field, but also to advancing testable predictions and hypotheses in comparative and evolutionary research more broadly, as seen below.

Evolutionary Hypotheses and Outlooks

The observed high degree of developmental plasticity analysed above (Table 12.3) naturally does not, on its own, offer any indication of its evolutionary role or significance. In this context, it is interesting to note that according to the observations of Arshavsky and his team, the experimental altricial animals ceased to interbreed with the control animals after the six-month study period. In part, this may have owed to the notable morphological differences that had developed between the two groups over this period (Arshavsky, 1982; Arshavsky et al., 1985). At the same time, this observation is also interesting from a semantic or semiotic perspective, as advanced by Portmann (Jaroš & Klouda, 2021), and which emphasises the semantic and perceptual aspects of animal morphology. In modern terms, such phenotypic aspects could be also related to sexual selection and other perceptually based processes (Chap. 14), which may act as a recognition-based mediator in evolutionary change (Brejcha et al., 2019; Kleisner, 2022). In particular, such evolutionary scenarios have been analysed in relation to the recognition concept of speciation (Lambert & Spencer, 1995; Kull, 2016), which posits that prior to the formation of genetic reproductive barriers, the changes which lead to reproductive isolation in a population may be primarily phenotypic and semiotic in nature – e.g., reflecting the way an organism recognizes its conspecifics and potential mates. This could be one direction in which the semiotic aspects of this issue could be explored further, while also bearing in mind the empirical testability of the anti-entropic hypothesis of ontogenetic divergence (Arshavsky, 1966, 1968, 1972, 1982; Arshavsky et al., 1985; cf. Kurismaa, 2021b).

[14] From this perspective, similar qualifications may be relevant also to the 'supply limitation' models, which tend to assume primarily (if not exclusively) biophysical constraints on transport rates and energy exchange capacity, for instance due to differences in relative surface area and/or internal resource transport network structures (Fig. 12.2). Nevertheless, nothing should preclude other, organismic types of supply limitations from affecting metabolic allometries, such as the episodically increased local work rates considered above (section on experimental modelling). Such local and activity dependent limitations may be related to natural ecological stressors, which through biological work and motor responses may stimulate growth and development in the organism via both the first and the second type of anabolic processes (Arshavsky, 1972, 1982).

In a developmental evolutionary context, the bioenergetic reorganisation seen above in the experimental altricial animals, especially the significantly reduced resting metabolic rate (Table 12.3), would expectably also affect their reproductive and parental energy investment. Indeed, it is documented that reproductive and parental energy investments scale nonuniformly to the body size of mammals and may be shaped by diverse ecological factors affecting the resting metabolic rate (Jackson et al., 2014). Although for methodical reasons, the experiments reported above focused mainly on male mammals (cf. Tables 12.2 and 12.3), compatible results on phenotypic plasticity were obtained also in more limited studies involving females, pointing to the generality of the observed developmental mechanisms and bioenergetic processes (Arshavsky et al., 1981).

This raises interesting questions, especially in the light of recent studies showing that not only anatomical and morphological limitations (e.g., pelvic dimensions), but also energetic physiological constraints may directly regulate gestation length in mammals, including nonhuman primates and humans (Dunsworth et al., 2012; cf. Kurismaa, 2021a). To the best of our knowledge, however, the effects which an ontogenetically markedly reduced resting metabolic rate may have on eutherian gestation length have not been experimentally explored yet. At the same time, in the light of physiological and mechanistic considerations (Dunsworth et al., 2012; Jackson et al., 2014; Pontzer, 2018) such effects do appear likely and might have important evolutionary implications.

For example, one could ask whether activity-driven ontogenetic reduction in maintenance metabolic expenditure could lead to a correspondingly prolonged gestation and/or to a reduction in offspring number in precocial(ised) organisms. A third possibility, namely reduction in offspring size, is theoretically also possible; however, it seems to contradict available evidence (cf. the anabolic, growth-promoting effects of optimal stressors on mammalian intrauterine development; section on experimental modelling) (Arshavsky, 1972). However, the two aforementioned scenarios, involving a longer gestation period and/or fewer offspring due to energetic restrictions on resting-state metabolism, would deserve further investigation. This seems warranted also considering the evidence on the lower foetal growth constants observed in precocial animals, that is, their lower maternal energetic investment in comparison to gestation in altricial species (Martin & MacLarnon, 1985). In this case, longer intrauterine growth and development could be a natural corollary of activity-induced down-regulation of the baseline rates of maternal energy expenditure in precocial eutherians.

It is important to note here that the fundamental role of skeletal muscles in regulating the normal patterns of growth and development was confirmed in various studies. These involved not only overt locomotor activity and physical work, but a wide variety of other physiological stressors affecting the skeleto-muscular system (e.g., including early cold exposures to induce muscular thermogenesis, restriction on caloric intake, hypoxemic exposures, etc.). All of these studies point to the same general conclusion, namely that the skeletomuscular system is a key regulator in the formation of eutherian bioenergetic phenotypes (Arshavsky, 1982). It is important to keep this in mind when assessing the anti-entropic approach as a general developmental and evolutionary physiological framework. In this context, the possibility of transgenerational epigenetic inheritance of altered gene expression patterns accompanying increased skeletal muscle training (Hargreaves, 2015) also presents interest and, if reconfirmed, might provide further support for the anti-entropic perspective on eutherian developmental divergence (Arshavsky, 1985).

As noted above, the mechanisms which shape the heterochronic reproductive and developmental patterns in altricial and precocial animals remain relatively poorly understood and, instead of developmental and physiological models, the relevant research has mostly focused on ecological and population genetic aspects of these traits (Reznick et al., 2002; Flatt & Heyland, 2011; Montiglio et al., 2018). However, with no alternative (or complementary) mechanistic and proximal models for experimental elaboration, comparative evolutionary analyses across such diverse taxonomic groups and systems present a considerable challenge (Table 12.1; Fig. 12.2). To overcome these difficulties, one possible approach could draw on the protocols and experimental paradigms of the anti-entropic framework as specified in numerous publications by Arshavsky and colleagues (1967, 1972, 1982; Arshavsky et al., 1985, 1987).

As noted above, the mediation and limitation of energy allocation to reproductive functions by somatic energy expenditure in organisms is predicted also by the constrained total energy expenditure framework recently developed by Pontzer et al. (2016; Pontzer, 2018). This framework enables a more detailed analysis of the possible causes of activity-induced downregulation in reproductive functions – including the phenomena of reduced fecundity and/or fertility, and of lowered levels of reproductive hormones (oestrogen, progesterone), all of which can accompany increased energetic workload and exercise in females (Pontzer, 2015, 2018). Such coupled systemic effects suggest that instead of simply reflecting the results of limited food intake or energy imbalance, downregulation of the energetic provision of certain functional systems may reflect specific compensatory processes and adaptations accompanying increased work (Pontzer, 2015, 2018). In this context, to arrive at a broader approach to energy metabolism and its regulation, it might be important to analyse the possibility of an activity-driven prolongation of gestation periods and/or the reduction of the number of offspring in precocial(ised) organisms. This strategy may highlight the often-overlooked phenotypic plasticity of such traits and their possibly high reliance on informational types of control in organisms, featuring the nervous and endocrine systems (Arshavsky, 1972, 1982; Glazier, 2015a; Casasa & Moczek, 2019). However, this line of studies may require not only revision of the traditional metabolic models (Pontzer et al., 2016) but also a broader reassessment on allometric and life-history frameworks such as mentioned in this chapter.

Conclusion

The anti-entropic approach to ontogeny developed by I.A. Arshavsky and his school offers an integrative framework for understanding the role of organismal agency and phenotypic plasticity in shaping the developmental and evolutionary trajectories of mammals. By investigating the impact of early life bioenergetic stress and motor work (entropy production) on the differentiation of bioenergetic phenotypes, Arshavsky's work sheds important light on the non-equilibrium mechanisms of developmental processes and, in particular, on the remarkable energetic, physiological, and morphological variety which distinguishes the predominantly altricial eutherian species from more precocial ones. By highlighting the interconnection between phenotypic plasticity and biological agency, these results have significant consequences for understanding not only the variations of metabolic scaling but life history and allometric principles in general. Indeed, in all these areas, the issues of organismal agency and phenotypic plasticity are currently only beginning to be explored and could benefit from earlier findings and insights in the anti-entropic framework.

In addition to reviewing these topics, we have also considered the implications of Arshavsky's approach for certain open questions of metabolic scaling, including the apparent biophysical paradox of increased work capacity in larger/more precocial organisms. This issue, which remains particularly challenging for classical allometric and life history models, can be reformulated and perhaps partly resolved based on the anti-entropic concepts discussed and elaborated here. Finally, the chapter outlined a new hypothesis aimed at explaining the emergence of prolonged gestation and reduced offspring numbers in more precocial species by drawing on the anti-entropic theory and later findings. In general, the integrative and non-equilibrium concepts analysed in this chapter offer a systemic perspective, which could contribute to the development of more biologically realistic and multi-mechanistic models within comparative and allometric frameworks, as well as to new, agential and semiotic models in biology more generally.

Acknowledgements This study received support from the Estonian Science Foundation (ETAg) grant MOBJD1046 and the Czech Science Foundation (GACR) grant 20-16633S.

References

Agutter, P. S. (2019). Why is the history of medicine and biology important? In *A guide to the scientific career: Virtues, communication, research and academic writing* (pp. 331–334). Wiley.

Agutter, P. S., & Wheatley, D. N. (2008). *Thinking about life: The history and philosophy of biology and other sciences*. Springer.

Alexander, R. M. (2005). Models and the scaling of energy costs for locomotion. *Journal of Experimental Biology, 208*(9), 1645–1652.

Arshavsky, I. A. (1960). *The physiology of blood circulation during the intrauterine period*. Medgiz. (In Russian).

Arshavsky, I. A. (1966). The "energy rule of skeletal muscles" and the physiological mechanisms of divergence and evolution in mammals. *The Journal of Evolutionary Biochemistry and Physiology, 2*(6), 511–518. (In Russian).

Arshavsky, I. A. (1967). *Essays on growth physiology*. Medizina. (In Russian).

Arshavsky, I. A. (1968). Adaptive and homeostatic mechanisms in the development of physiologically mature and immature organisms. In G. Newton & S. Levine (Eds.), *Early experience and behavior* (pp. 299–337). Charles C. Thomas.

Arshavsky, I. A. (1970). Some basic regularities of ontogenesis in relation to the problems of aging and longevity. In V. I. Mahinko (Ed.), *Molecular and functional foundations of ontogenesis* (pp. 203–223). Medizina. (In Russian).

Arshavsky, I. A. (1972). Musculoskeletal activity and rate of entropy in mammals. In G. Newton & A. H. Riesen (Eds.), *Advances in psychobiology* (Vol. 1, pp. 1–52). Wiley.

Arshavsky, I. A. (1976). The biological and medical aspects of the problems of adaptation and stress in the light of ontogenetic physiological evidence. In *Current problems of modern physiology* (pp. 144–191). Nauka. (In Russian).

Arshavsky, I. A. (1982). *Physiological mechanisms and regularities of individual development* (Foundations of the negentropic theory of ontogeny). Nauka. (In Russian).

Arshavsky, I. A. (1983). On specific peculiarities of transition processes in ontogenesis of mammals. In I. Lamprecht & A. I. Zotin (Eds.), *Thermodynamics and kinetics of biological processes* (pp. 461–472). Walter de Gruyter.

Arshavsky, I. A. (1985). On the basic principles and tasks of evolutionary physiology in light of comparative ontogenetic studies. *The Journal of Evolutionary Biochemistry and Physiology, 21*(2), 163–170. (In Russian).

Arshavsky, I. A. (2002). *We all have equal rights before one-another* (Recollections of A.A. Ukhtomsky. The meaning and fate of the ethical dominant). Pedagogical center "Experiment". (In Russian).

Arshavsky, I. A., & Demenshtein, A. A. (1991). A.A. Ukhtomsky's principle of dominanta (on the problem of creation of neurocomputers). In A. V. Holden & V. I. Kriukov (Eds.), *Neurocomputers and attention: Neurobiology, synchronisation, and chaos* (Vol. 1, pp. 11–19). Manchester University Press.

Arshavsky, I. A., & Mezhevikhina, L. M. (1992). Analysing the causes and mechanisms that determine the emergence of individual development in animals beginning from the zygote (on the role of the cytoskeleton). *Biophysics, 37*(5), 969. (In Russian).

Arshavsky, I. A., & Rabinovich, E. Z. (1979). Thermodynamics of open systems and the problem of individual development. In G. R. Ivanitskij (Ed.), *Methodological and theoretical problems of biophysics* (pp. 108–120). Nauka. (In Russian).

Arshavsky, I. A., Darinsky, N. V., Li, L. A., & Siryk, L. A. (1975). Comparative ontogenetic features of skeletal muscles in physiologically mature and immature mammals. *Developmental Psychobiology, 8*(2), 117–128.

Arshavsky, I. A., Arshavskaya, E. I., & Praznikov, V. P. (1976). Motor reactions during the antenatal period correlated with the periodic change in the activity of the cardiovascular system. *Developmental Psychobiology, 9*(4), 343–352.

Arshavsky, I. A., Rozanova, V. D., & Savkiv, T. G. (1981). The characteristics and regularities of individual development in males and females in the light of the energy rule of skeletal muscles. *Journal of General Biology, 42*(5), 698–707.

Arshavsky, I. A., Nemetz, M. G., & Rozanova, V. D. (1985). Experimental modeling of the potential mechanisms of morpho-physiological transformation in wild animals during domestication. In L. V. Davletova (Ed.), *Morphology and genetics of the wild boar* (pp. 73–87). Nauka. (In Russian).

Arshavsky, I. A., Nemetz, M. G., Rabinovich, E. Z., Rozanova, V. D., & Siryk, L. A. (1987). The principle of the dominant in individual development and in determining the transformation of homeostasis and resilience in various age periods. In D. M. Grozdinsky (Ed.), *Resilience and homeostasis in biological systems* (pp. 112–129). Naukova Dumka. (In Russian).

Arshavsky, I. A., Kalevich, A. E., & Kefeli, V. I. (1992). Analysing the role of the cytoskeleton in the growth and development of the plant embryo. *Biophysics, 35*(5), 983. (In Russian).

Bauer, E. S. (1982 [1935]). *Theoretical Biology*. VIEM. (In Russian). Republished in English by Akadémiai Kiadó, Budapest, 1982.

Bergen, Y. V., & Phillips, K. (2005). Size matters. *The Journal of Experimental Biology, 208*(9), i–iii.

Billman, G. E., Cagnoli, K. L., Csepe, T., Li, N., Wright, P., Mohler, P. J., & Fedorov, V. V. (2015). Exercise training-induced bradycardia: Evidence for enhanced parasympathetic regulation without changes in intrinsic sinoatrial node function. *Journal of Applied Physiology, 118*(11), 1344–1355.

Brejcha, J., Pecháček, P., & Kleisner, K. (2019). Complementarity of seeing and appearing. In *Cognitive architectures* (pp. 13–30). Springer.

Careau, V., Killen, S. S., & Metcalfe, N. B. (2014). Adding fuel to the "fire of life": Energy budgets across levels of variation in ectotherms and endotherms. In *Integrative organismal biology* (pp. 219–233). Wiley.

Casasa, S., & Moczek, A. P. (2019). Evolution of, and via, developmental plasticity: Insights through the study of scaling relationships. *Integrative and Comparative Biology, 59*(5), 1346–1355.

Clauss, M., Dittmann, M. T., Müller, D. W., Zerbe, P., & Codron, D. (2014). Low scaling of a life history variable: Analysing eutherian gestation periods with and without phylogeny-informed statistics. *Mammalian Biology, 79*(1), 9–16.

Cornish-Bowden, A., Cárdenas, M. L., Letelier, J. C., & Soto-Andrade, J. (2007). Beyond reductionism: Metabolic circularity as a guiding vision for a real biology of systems. *Proteomics, 7*(6), 839–845.

Deacon, T. W. (2011). *Incomplete nature: How mind emerged from matter*. WW Norton & Company.

Di Paolo, E. A. (2005). Autopoiesis, adaptivity, teleology, agency. *Phenomenology and the Cognitive Sciences, 4*(4), 429–452.

Di Paolo, E., Buhrmann, T., & Barandiaran, X. (2017). *Sensorimotor life: An enactive proposal*. Oxford University Press.

Diogo, R. (2017). *Evolution driven by organismal behavior. A unifying view of life, function, form, mismatches and trends*. Springer.

Dunsworth, H. M., Warrener, A. G., Deacon, T., Ellison, P. T., & Pontzer, H. (2012). Metabolic hypothesis for human altriciality. *Proceedings of the National Academy of Sciences, 109*(38), 15212–15216.

Ellis, B. J., & Del Giudice, M. (2019). Developmental adaptation to stress: An evolutionary perspective. *Annual Review of Psychology, 70*, 111–139.

Flatt, T., & Heyland, A. (Eds.). (2011). *Mechanisms of life history evolution: The genetics and physiology of life history traits and trade-offs*. Oxford University Press.

Genoud, M., Isler, K., & Martin, R. D. (2018). Comparative analyses of basal rate of metabolism in mammals: Data selection does matter. *Biological Reviews, 93*(1), 404–438.

Glazier, D. S. (2005). Beyond the '3/4-power law': Variation in the intra-and interspecific scaling of metabolic rate in animals. *Biological Reviews, 80*(4), 611–662.

Glazier, D. S. (2014). Metabolic scaling in complex living systems. *Systems, 2*(4), 451–540.

Glazier, D. S. (2015a). Is metabolic rate a universal 'pacemaker' for biological processes? *Biological Reviews, 90*(2), 377–407.

Glazier, D. S. (2015b). Body-mass scaling of metabolic rate: What are the relative roles of cellular versus systemic effects? *Biology, 4*(1), 187–199.

Glazier, D. S. (2018). Rediscovering and reviving old observations and explanations of metabolic scaling in living systems. *Systems, 6*(1), 4.

Hargreaves, M. (2015). Exercise and gene expression. *Progress in Molecular Biology and Translational Science, 135*, 457–469.

Harrison, J. F. (2017). Do performance–safety tradeoffs cause hypometric metabolic scaling in animals? *Trends in Ecology & Evolution, 32*(9), 653–664.

Hoppeler, H., & Weibel, E. R. (2005). Scaling functions to body size: Theories and facts. *The Journal of Experimental Biology, 208*(9), 1573–1574.
Hostinar, C. E., & Gunnar, M. R. (2013). The developmental effects of early life stress: An overview of current theoretical frameworks. *Current Directions in Psychological Science, 22*(5), 400–406.
Hulbert, A. J. (2014). A sceptics view: "Kleiber's Law" or the "3/4 Rule" is neither a law nor a rule but rather an empirical approximation. *Systems, 2*(2), 186–202.
Igamberdiev, A. U. (2018). Hyper-restorative non-equilibrium state as a driving force of biological morphogenesis. *Biosystems, 173*, 104–113.
Jackson, G., Mooers, A. Ø., Dubman, E., Hutchen, J., & Collard, M. (2014). Basal metabolic rate and maternal energetic investment durations in mammals. *BMC Evolutionary Biology, 14*(1), 1–7.
Jaroš, F., & Klouda, J. (2021). *Adolf Portmann: A thinker of self-expressive life*. Springer.
Kauffman, S. A. (2000). *Investigations*. Oxford University Press.
Kauffman, S. (2012). From physics to semiotics. In S. Rattasepp & T. Bennett (Eds.), *Gatherings in biosemiotics 11* (pp. 30–46). Tartu Semiotics Library.
Kauffman, S., & Clayton, P. (2006). On emergence, agency, and organization. *Biology and Philosophy, 21*(4), 501–521.
Kazansky, A. B. (2015). Agential anticipation in the central nervous system. In *Anticipation: Learning from the past* (pp. 101–112). Springer.
Kelly, S. A., Panhuis, T. M., & Stoehr, A. M. (2012). Phenotypic plasticity: Molecular mechanisms and adaptive significance. *Comprehensive Physiology, 2*(2), 1417–1439.
Kleisner, K. (2022). Semiotic fitting, co-option, and the art of life. *Biosemiotics, 15*(1), 31–35.
Koch, L. G., & Britton, S. L. (2008). Aerobic metabolism underlies complexity and capacity. *The Journal of Physiology, 586*(1), 83–95.
Koch, L. G., Kemi, O. J., Qi, N., Leng, S. X., Bijma, P., Gilligan, L. J., ... & Wisløff, U. (2011). Intrinsic aerobic capacity sets a divide for aging and longevity. *Circulation research, 109(10), 1162–1172.*
Koch, L. G., & Britton, S. L. (2018). Theoretical and biological evaluation of the link between low exercise capacity and disease risk. *Cold Spring Harbor Perspectives in Medicine, 8*(1), a029868.
Koch, L. G., Britton, S. L., & Wisløff, U. (2012). A rat model system to study complex disease risks, fitness, aging, and longevity. *Trends in Cardiovascular Medicine, 22*(2), 29–34.
Kull, K. (2016). The biosemiotic concept of the species. *Biosemiotics, 9*(1), 61–71.
Kurismaa, A. (2015). Perspectives on time and anticipation in the theory of dominance. In *Anticipation: Learning from the past* (pp. 37–57). Springer.
Kurismaa, A. (2021a). Revisiting basal anthropology: A developmental approach to human evolution and sociality. In F. Jaroš & J. Klouda (Eds.), *Adolf Portmann: A thinker of self-expressive life* (pp. 89–118). Springer.
Kurismaa, A. (2021b). The negentropic theory of ontogeny: A new model of eutherian life history transitions? *Biosemiotics, 14*(2), 391–417.
Kurismaa, A. (2023). From integrative biology to the nerve impulse: Rethinking neural information and semiotics in functional systems perspective. *Adaptive Behavior, 31*(1), 87–101.
Lambert, D. M., & Spencer, H. G. (Eds.). (1995). *Speciation and the recognition concept: Theory and application*. Johns Hopkins University Press.
Lema, S. C. (2020). Hormones, developmental plasticity, and adaptive evolution: Endocrine flexibility as a catalyst for 'plasticity-first' phenotypic divergence. *Molecular and Cellular Endocrinology, 502*, 110678.
Letelier, J. C., Cárdenas, M. L., & Cornish-Bowden, A. (2011). From L'Homme machine to metabolic closure: Steps towards understanding life. *Journal of Theoretical Biology, 286*, 100–113.
Levis, N. A., & Pfennig, D. W. (2020). Plasticity-led evolution: A survey of developmental mechanisms and empirical tests. *Evolution & Development, 22*(1–2), 71–87.

Louwman, J. W. W. (1973). Breeding the tailless tenrec Tenrec ecaudatus at Wassenaar Zoo. *International Zoo Yearbook, 13*(1), 125–126.

Markoš, A., & Švorcová, J. (2019). *Epigenetic processes and the evolution of life*. CRC Press.

Martin, R. D. (1976). The bearing of reproductive behavior and ontogeny on strepsirhine phylogeny. In W. P. Luckett & F. Szalay (Eds.), *Phylogeny of the primates* (pp. 265–297). Springer.

Martin, R. D., & MacLarnon, A. M. (1985). Gestation period, neonatal size and maternal investment in placental mammals. *Nature, 313*(5999), 220–223.

Mattson, M. P. (2010). The fundamental role of hormesis in evolution. In M. P. Mattson & E. J. Calabrese (Eds.), *Hormesis: A revolution in biology, toxicology and medicine* (pp. 57–68). Humana Press.

Mattson, M. P., Moehl, K., Ghena, N., Schmaedick, M., & Cheng, A. (2018). Intermittent metabolic switching, neuroplasticity and brain health. *Nature Reviews Neuroscience, 19*(2), 81–94.

McEwen, B. S., & Gianaros, P. J. (2011). Stress-and allostasis-induced brain plasticity. *Annual Review of Medicine, 62*, 431–445.

Michaelian, K. (2022). Non-equilibrium thermodynamic foundations of the origin of life. *Foundations, 2*(1), 308–337.

Montiglio, P. O., Dammhahn, M., Dubuc Messier, G., & Réale, D. (2018). The pace-of-life syndrome revisited: The role of ecological conditions and natural history on the slow-fast continuum. *Behavioral Ecology and Sociobiology, 72*(7), 1–9.

Nadin, M. (Ed.). (2015). *Anticipation: Learning from the past*. Springer.

Noble, D. (2021). The illusions of the modern synthesis. *Biosemiotics, 14*(1), 5–24.

Pattee, H. H., & Rączaszek-Leonardi, J. (2012). *Laws, language and life: Howard Pattee's classic papers on the physics of symbols with contemporary commentary*. Springer Science & Business Media.

Perry, B. W., Schield, D. R., & Castoe, T. A. (2018). Evolution: Plasticity versus selection, or plasticity and selection? *Current Biology, 28*(18), R1104–R1106.

Pfennig, D. W. (Ed.). (2021). *Phenotypic plasticity & evolution: Causes, consequences, controversies*. Taylor & Francis.

Pontzer, H. (2015). Constrained total energy expenditure and the evolutionary biology of energy balance. *Exercise and Sport Sciences Reviews, 43*(3), 110–116.

Pontzer, H. (2018). Energy constraint as a novel mechanism linking exercise and health. *Physiology, 33*(6), 384–393.

Pontzer, H., Durazo-Arvizu, R., Dugas, L. R., Plange-Rhule, J., Bovet, P., Forrester, T. E., et al. (2016). Constrained total energy expenditure and metabolic adaptation to physical activity in adult humans. *Current Biology, 26*(3), 410–417.

Portmann, A. (1967). *Zoologie aus Vier Jahrzehnten: Gesammelte Abhandlungen*. Piper.

Portmann, A. (1990). *A Zoologist looks at Humankind*. Cambridge University Press.

Prigogine, I., & Stengers, I. (2018). *Order out of chaos: Man's new dialogue with nature*. Verso Books.

Raichlen, D. A., & Alexander, G. E. (2017). Adaptive capacity: An evolutionary neuroscience model linking exercise, cognition, and brain health. *Trends in Neurosciences, 40*(7), 408–421.

Raichlen, D. A., & Gordon, A. D. (2011). Relationship between exercise capacity and brain size in mammals. *PLoS One, 6*(6), e20601.

Reznick, D., Bryant, M. J., & Bashey, F. (2002). r-and K-selection revisited: The role of population regulation in life-history evolution. *Ecology, 83*(6), 1509–1520.

Rivers, J. R., & Ashton, J. C. (2013). Age matching animal models to humans-theoretical considerations. *Current Drug Discovery Technologies, 10*(3), 177–181.

Rozanova, V. D. (1968). *Essays on experimental developmental pharmacology*. Medizina. (In Russian).

Ruiz-Mirazo, K., & Moreno, A. (2012). Autonomy in evolution: From minimal to complex life. *Synthese, 185*(1), 21–52.

Savage, V. M., Gillooly, J. F., Woodruff, W. H., West, G. B., Allen, A. P., Enquist, B. J., & Brown, J. H. (2004). The predominance of quarter-power scaling in biology. *Functional Ecology, 18*(2), 257–282.

Schirrmacher, V. (2021). Less can be more: The hormesis theory of stress adaptation in the global biosphere and its implications. *Biomedicine, 9*(3), 293.

Selye, H. (1956). *The stress of life*. McGraw-Hill.

Shapiro, J. A. (2011). *Evolution: A view from the 21st century*. Pearson education.

Sharov, A. A., & Tønnessen, M. (2021). *Semiotic agency*. Springer.

Shea, B. T. (2007). Start small and live slow: Encephalization, body size, and life history strategies in primate origins and evolution. In M. J. Ravosa & M. Dagosto (Eds.), *Primate origins: Adaptations and evolution* (pp. 583–623). Springer.

Uller, T., Feiner, N., Radersma, R., Jackson, I. S., & Rago, A. (2020). Developmental plasticity and evolutionary explanations. *Evolution & Development, 22*(1–2), 47–55.

von Bertalanffy, L. (1951). Metabolic types and growth types. *The American Naturalist, 85*(821), 111–117.

Walsh, D. M. (2015). *Organisms, agency, and evolution*. Cambridge University Press.

Werneburg, I., & Spiekman, S. N. F. (2018). Mammalian embryology and organogenesis. In F. E. Zachos & S. N. Asher (Eds.), *Handbook of zoology: Mammalia. Mammalian evolution, diversity, and systematics* (pp. 59–116). De Gruyter.

West-Eberhard, M. J. (2003). *Developmental plasticity and evolution*. Oxford University Press.

Workman, A. D., Charvet, C. J., Clancy, B., Darlington, R. B., & Finlay, B. L. (2013). Modeling transformations of neurodevelopmental sequences across mammalian species. *Journal of Neuroscience, 33*(17), 7368–7383.

Chapter 13
Organisms as Agents in Zoosemiotic Perspective: The Case of Umwelt Reversion

Nelly Mäekivi

Abstract In a zoosemiotic inquiry, alloanimal agency is accepted as a natural property of all animals. The aim of zoosemiotics is to analyse the behaviour of alloanimals and ecological relations through the emic perspective. This chapter shows certain changes that take place in an animal's perceptual world and are mirrored in their operational world. The zoosemiotic analysis will be based on the Umwelt theory and work chiefly with a case study of reintroduction of the European mink (*Mustela lutreola*) and interviews conducted with local people in 2019 on the island of Hiiumaa in Estonia. The agency of animals becomes especially important when substantial changes take place in their environment, their food resources, or social relations; these are changes to which animals must adapt according to their Umwelt. In the context of reintroduction of the European mink, animal agency is central because *ex situ* environment was replaced with an *in situ* environment. What will be proposed and formulated is a special case of Umwelt transition, an 'Umwelt reversion', where the reintroduced minks return to their 'natural' social relations and food resources. This indicates that meaning carriers in the environment can change due to the agency of the mink.

Keywords Animal agency · Choices · Umwelt reversion

Introduction

The issue of agency, with its empirical and philosophical implications, has long been part of the scholarly debate (see, e.g., MacFarland & Hediger, 2009; Rees, 2017). Some areas of research, such as posthumanism, human-animal studies, or critical animal studies, include in their investigation also the query into alloanimal, i.e., other-than-human, animal agency. One of such research fields is zoosemiotics,

N. Mäekivi (✉)
Department of Semiotics, University of Tartu, Tartu, Estonia
e-mail: nelly.maekivi@ut.ee

J. Švorcová (ed.), *Organismal Agency*, Biosemiotics 28,
https://doi.org/10.1007/978-3-031-53626-7_13

which investigates how animals live in their environments, what choices they make, and what is meaningful for them. Zoosemiotics defines animal agency as a meaningful interaction between the environment and alloanimals and argues that meaning making is an inherent characteristic of all animals. All animals are thus agents[1]: they actively participate in semiosis, make choices, solve specific problems, have motivational interests, and create and change relations with other animals and the environment according to their Umwelten (see Chap. 8).

When we set out to scrutinise the emic perspective, that is, the agency of alloanimals as experienced by themselves, against the etic position, that is, against our interpretations of the animals' agency, we always face the question of how much we are actually imposing the semiotic activity on the animal. Nevertheless, it seems that we have various cues and hints to understanding the emic perspectives of other animals, for instance based on the way they live in the world or based on the (environmental) objects and other subjects that carry specific meanings for the animals.

The aim of chapter is to analyse, through the operational world of alloanimals, the changes that take place in their perceptual world. The zoosemiotic analysis presented here is based on a case study of reintroduction of the European mink (*Mustela lutreola*). Understanding of alloanimal agency becomes crucial when substantial changes within the animal's environment, diet, or social relations force the animals to adapt according to their Umwelt. In the context of reintroduction of the European mink, the animal's agency was thus central, because the *ex situ* environment was replaced with those animals' *in situ* environment, i.e., life in the wild. We shall build on the concept of 'Umwelt transition' (Tønnessen, 2009), analyse the lasting and fundamental changes within the life of the alloanimal, and consider all this from an ontogenetic (individual) perspective. What we will propose is a special case of Umwelt, an 'Umwelt reversion', where the reintroduced minks return to their 'natural' social relations and diet. This indicates that the meaning attached to objects in the environment is malleable due to the minks' agency. A part of this case study, we also make use of interviews conducted with local people in 2019 on the island of Hiiumaa to support findings from scientific literature on the European mink.

Alloanimal Agency in a Zoosemiotic Inquiry

The definition and conceptual interpretation of the necessary requirements for agency are issues which have been discussed in a glossary project undertaken by Tønnessen (2015), where it was discovered that the minimum requirements for agency agreed upon by semioticians are 'goal-directedness, self-governed activity, processing of semiosis (i.e., sign use) and choice of action' (Tønnessen, 2015: 138; see Chap. 8). These attributes are seen as applying to all animals. Three of the four

[1] In this article, we refer to 'agent' and 'agency' in a synonymous fashion (see Tønnessen, 2015 for further discussion), where agents – i.e., subjects possessing agency – are the alloanimals.

abovementioned criteria are echoed in other research fields, where they are likewise considered constitutive of animal agency. Goal-directedness is seen as an important factor in agency and interpreted as defining intentional activity, where each organism forms a 'centre of subjectivity' with its own motivation (Steward, 2009: 229). Similarly, self-governed activity and choice of action shift the focus on autonomy, will, and the 'self' of the organism, with some authors going even further and arguing for self-awareness in alloanimals (see, e.g., de Waal, 2019). What stands out in the glossary project – and what is unique to semiotics – is the inclusion of processing of signs, or semiosis, as the requirement for agency. In a semiotic inquiry, the prerequisite of discussing the agency of any organism is closely tied to the notion that the thresholds of life and semiosis coincide (Sebeok, 1986: 15; Sebeok, 2001: 3). Semiosis here amounts to the claim that all organisms use and interpret signs (where a sign is something that stands for something else: object, feeling, etc.) which they are biologically predisposed to create and interpret. Organisms use signs to gather information from their environment and to communicate both within their own and with other species (see also Sebeok, 2001). Thus, signification and communication are always sign processes. Kull and Favareau (2022) describe semiosis pertaining to animal organisms as interpretation which creates optionality (based on subjective space and creation of cognitive maps) and habits (based on learning). This takes us closer to alloanimal agency as understood within the semiotic paradigm. Semiosis, or a sign process, is not just one of the parts of alloanimal agency: it is its very foundation. It supports all the other manifestations of agency.

The area of semiotics that deals with alloanimal semiotic activity is zoosemiotics, a discipline on the interface of ethology and semiotics (Sebeok, 1968). Because it sees all animals as capable of signalling, communication, and interpretation, that is, as beings embedded in relations, zoosemiotics takes the agency of alloanimals for granted. It views agency as a necessary precondition of any subject whose specific sign systems and semiosis we wish to investigate. Zoosemiotics is interested in alloanimals' active use of environment, their meaning-making, and their communication. In other words, zoosemiotics studies how alloanimals make sense of their surroundings, their companions, how they perceive, and how they act based on these perceptions. Alloanimals *are* agents: 'Animals are treated as active participants in semiosis, that is, as interpreters of signs, and as being related to other animals and the environment through perception-action cycles' (Maran et al., 2016: 10). This also implies that when it comes to alloanimals in zoosemiotics, the concept of agent is synonymous with the concept of a subject.

Communicative capabilities and subjectivity of alloanimals were recognised even before the word 'agent' started to be used. For example, the cryptosemiotician Heini Hediger wrote at length about animals having subjective experiences, aversions, and attractions to various elements in their environment and other animals (including humans) in it. He stated clearly that 'No animal should be treated as a mere physical reagent' (Hediger, 1950: 82). Thomas Sebeok, the founder of zoosemiotics as a discipline, proposed a platform for studying animal communication to help investigate the basic assumption of zoosemiotics, namely that all animals have specific communication issues which they are actively trying to solve. He was also interested in the place of zoosemantics (the study of meaning of messages) in animal communication (Sebeok, 1972, 1990). Both of the aforementioned authors made, at one point or another, references to Jakob von Uexküll, a biologist who, like Hediger and Sebeok, rejected mechanistic approaches to alloanimals. He advocated the idea that the biological constitution and functioning of alloanimals should be viewed as the means and ways of interacting with the world (von Uexküll, 1992[1934]: 324). For von Uexküll (1992)[1934]: 319–320), the notion of 'animal' was synonymous to 'subject', that is, an entity that perceives, acts, and cannot be interpreted merely mechanistically. Uexküll focused on alloanimals' subjective realities (see, e.g., von Uexküll, 1982; 1992[1934]). The most prominent current zoosemiotician, Dario Martinelli (2010), also uses in discussions of signification, representation, and communication in zoosemiotics the term 'subject' (and to a lesser degree 'agent'). We could thus argue that for zoosemiotics, the question is not *whether* alloanimals have agency but *how* we can access it.

Accessing Alloanimal Agency in a Zoosemiotic Inquiry

Every empirical study that deals with alloanimals has to face the challenge of accessing agency. This holds also of case studies in zoosemiotics, where the task takes the form of efforts to answer questions such as 'What, why, and how something is meaningful to this animal?' or 'How should we describe the animal's sign production and communication processes from the animal's own perspective?' These are just a few examples of issues that are relevant to descriptive or applied (zoo)semiotics.

It is one thing to accept alloanimal agency, but to access the perspective or the subjective stance of another animal is no easy task, especially given that explanations of real-life phenomena should be well-supported by actual findings. Fortunately, zoosemiotics, although it adopts a semiotic methodology to analyse various phenomena, is an interdisciplinary science partly embedded in the natural sciences, such as ethology, zoology, comparative psychology, and others. As a result, it can work with findings from those fields. Several researchers working in zoosemiotics (e.g., von Uexküll, 1992[1934], 1982; Martinelli, 2010; Maran et al., 2016; Mäekivi, 2018) have emphasised the importance of incorporating accounts of animals' physiology, behaviour, signalling, ecological relations, etc., into zoosemiotic studies. This is because inclusion of these further accounts can aid our efforts to access the agency of an alloanimal by approximating its emic perspective. We shall discuss some of the core concepts of zoosemiotics in the following.

Umwelt

One of the central concepts of zoosemiotic study of alloanimals is 'Umwelt'. This notion was introduced by Jakob von Uexküll to describe the subjective reality or the world as perceived by the subject, that is, the animal. Crucially, Uexküll says that 'each subject lives in a world composed of subjective realities alone' (von Uexküll, 1992[1934]: 383). Although we are unable to access an animal's internal subjective reality through his or her Umwelt directly, it is a tool that affords us with an accurate estimate based on the physical constitution of the animal (von Uexküll, 1992[1934]: 338–339). A similar idea is voiced by Martinelli in his discussion of ways to approximate the emic perspective of an animal (instead of settling for just the etic perspective of the researcher). He takes a strong stance against the mind/body dualism: 'we *do have* clues as to how we might emically study nonhuman species. To start with, we can scientifically study the sensorial organs of animals, the Merkwelt,[2] not only the Wirkwelt.[3] In other words, the way an organism interacts with the environment is largely due to the way s/he perceives it' (Martinelli, 2010: 84). This idea of glimpsing into subjective world of a subject is further elaborated by Umwelt mapping, 'where knowledge about the animal's perceptual organs, body plan, ways of living, and ecology are taken into account for building hypotheses about the structure and contents of its umwelt' (Maran et al., 2016: 31). In this approach, findings from natural sciences enable us to map or model the subjective realities of alloanimals under observation. Various factors influence the shape and form of alloanimal Umwelt: the alloanimal's species (e.g., its biological constitution, communication channels, sign repertoire), cultural traditions (for instance, when the recording of alarm calls of crows are played, crows in France congregate against the enemy while crows in America flee; see Sebeok, 1990: 24), age, gender, personal dispositions (e.g., telling apart inedible animals by taste; see Payne et al., 2004: 848), and the animal's physiological state (e.g., health). All these aspects can provide relevant information when we delve deeper into an animal's Umwelt.

The Functional Circle

Alloanimal Umwelt is the sum of all the functional circles[4] of the subject: the functional circle thus constitutes the core of zoosemiotic analysis of agency. The concept of functional circle, introduced by von Uexküll (1992)[1934]: 320–326; 1982: 31–33), refers to interactions between the subject and the object, and can be consid-

---

[2] Perceptual world (see, e.g., Uexküll 1992: 320).

[3] Effector world (see, e.g., Uexküll 1992: 320).

[4] An overview of the concept of a functional circle should also make it clear that since plants do not have receptor and effector organs, they 'are not able to construct and be in command of an Umwelt' (von Uexküll, 1982: 33).

ered a model of semiosis (Krampen, 1997). The alloanimal interacts with objects via perceiving them by his or her organs of perception and the objects are acted upon by the subject's effector organs. Although the subject and the object 'constitute a systematic whole' (von Uexküll, 1992[1934]: 324), the connection between them is not mechanistic. The concept of a functional circle allows us to describe the dynamic relation between perception and action because it clearly illustrates that an alloanimal is not a passive recipient observing his or her surroundings but an actor (or 'operant') within its Umwelt.[5] Uexküll considered the fact that some alloanimals have more complex Umwelten than others but arrived at four functional circles that necessarily belong to every subject's Umwelt: those related to food, enemy, the physical medium, and partner (von Uexküll, 1992[1934]: 145). We will further discuss some of these functional circles in our case study.

Meaning Carriers

The object perceived and acted upon within a functional circle is a meaning carrier. This means that all objects in an alloanimal's Umwelt are selected and interpreted for a specific purpose, such as 'Only blossoming flowers have a meaning for them [bees]; buds do not' (von Uexküll, 1992[1934]: 351). This in turn indicates that alloanimals do not enter into relationships with neutral objects: they immediately transform them into meaning carriers due to the meaning which they attribute to them (von Uexküll, 1982: 26–27). One object can thus take on different meanings (e.g., food, shelter) and the same meaning can be attributed to different objects (e.g., object of play), and this takes place in correlation with the complexity of the alloanimal's Umwelt. Uexküll stated that the number of meaning carriers for the alloanimal may increase: 'It grows within the individual life span of every animal that is able to gather experiences. For each new experience entails a readjustment to new impressions. Thus, new perceptual images with new functional tones are created' (von Uexküll, 1992[1934]: 359). That such changes can take place in alloanimal Umwelten via incorporation of new meaning carriers, reconfiguration of existing meaning relations, and formation of new functional circles is crucial to our case study of the European mink.

Methodology of the Case Study

We shall perform an Umwelt analysis to highlight the changes which take place in the functional circles of European mink (*Mustela lutreola*) when the animals are moved from *ex situ* to *in situ* environment. We will build on the concept of Umwelt

[5] See von Uexküll, 1982: 32 or 1992[1934]: 324 for a figure depicting a model of a functional circle.

transition (Tønnessen, 2009) to explain how meaningful relations of the mink change once the animals are reintroduced into their once native habitat. We shall focus mainly on the functional circles of food and enemy to demonstrate the potential benefits of this approach for other similar studies on agency in situations where profound changes take place in alloanimal Umwelten. One crucial aspect to consider here is that our discussion of mink agency is not based on a particular animal's life cycle but on a generalisation of the reintroduced animals. Nevertheless, we believe this first step offers a valuable insight into dealing with alloanimal agency, as suggested by zoosemiotics, in empirical settings.

Our study is based mainly on scientific literature pertaining to the reintroduction of European mink, but we shall also use interviews carried out with stakeholders on the Estonian island of Hiiumaa. Although the purpose of these semi-structured interviews made with 95 people from Hiiumaa was to investigate their attitudes to the minks and their reintroduction, the interviews also offered some insights that can be used to support our claims (details of the methodology for sample selection, interview questions, stakeholder groups, etc. are covered in Mäekivi et al., 2021: 15–20, 45–47). Some of the interviews are particularly relevant because they were made with people who took care of the minks during their reintroduction (lent their land for building mink enclosure and feeding the minks in it) and could thus provide important information about the behaviours they noticed in the animals. The use of local (ecological) knowledge in species conservation in conjunction with or as a basis of comparison with scientific data (see e.g., Gilchrist et al., 2005; Ulicsni et al., 2020) or even as a primary source information (see e.g., Anadón et al., 2008) is an approach that attracted much attention in recent decades. Although this experience-based knowledge has been dismissed by some scientists who claimed that it is unsubstantiated or anecdotal, it has proven its worth in inaccessible and remote areas (Brook & McLachlan, 2008). We used local knowledge together with scientific literature and in our case study, the two information sources turned out to support and complement each other.

The European Mink *in situ* and *ex situ*

The European mink (*Mustela lutreola*) is listed as critically endangered in the IUCN red book.[6] The mink went extinct in the natural habitat in Estonia in 1996, but first successful captive breeding was accomplished already in 1986 (Maran et al., 2017). Current efforts to preserve this species in Estonia combine captive breeding and reintroduction. Species conservation laboratory at the Tallinn Zoological Gardens carries out both of these activities by breeding the animals in the zoo and reintroduc-

[6] See: https://www.iucnredlist.org/search?query=mustela%20lutreola&searchType=species

ing them to the island of Hiiumaa (Maran et al., 2009: 1686). The mink are, except for the mating season and offspring rearing, solitary animals. They are small, slender, and semi-aquatic animals who inhabit the banks of rivers and streams (Maran et al., 2009; Haage et al., 2017). Except for the males during the mating season, they do not venture much over a couple hundred meters from the river (Maran & Põdra, 2009: 6), but along the riverbanks, the females' home range can be up to 3.6 km and the males' range up to 17 km (Garin, Zuberogoitia, et al., 2002b). Their life span is relatively short, about 3–4 years (Maran & Põdra, 2009: 6). Their *in situ* diet is composed mainly of fish, crustaceans, amphibians, but also small mammals such as water voles, birds, and insects (Maran et al., 1998; Maran & Põdra, 2009: 6; Sidorovich et al., 2010). Since they are mostly active during the night (Garin, Aihartza, et al., 2002a; Ortiz-Jiménez et al., 2021), it is relatively safe to assume that they tend to have little contact with people. Their hidden lifestyle is further supported by the fact that the mink is sensitive to anthropogenic disturbances, such as noise, and the animal is a prey species to aerial and terrestrial predators such as raptors and dogs or foxes, respectively (Ortiz-Jiménez et al., 2021). Our interviews had moreover shown that at least two of the reintroduced European minks were killed by dogs (Mäekivi et al., 2021: 35).

The above information about the life of European minks *in situ* becomes relevant when we compare it with minks living *ex situ*. The enclosures of minks who are being prepared for reintroduction in the Tallinn Zoological gardens are sized 25–50 m^2 with natural vegetation (trees and ground cover) and ponds with a surface area of 2–6 m^2 and depth of one meter. The animals are fed once a day: they receive (defrosted) rodents, fish, chicken, or minced meat with supplemental vitamins (Haage et al., 2017) and occasionally also live prey (Maran et al., 2017: 376). Their contact with people is kept to a minimum (only for necessary care), and if necessary, the minks are trained to swim (Maran et al., 2009: 1687). The feeding and training take place during the keepers' working hours (personal communication with Tiit Maran[7]): as a result, the minks are active during the day. The enclosures are moreover fenced from the sides and above (personal observation at the Tallinn Zoo), so that neither terrestrial nor aerial predators can prey on the minks. When a hard-release method (see next section) was used to reintroduce 12 animals to the island of Saaremaa in 2012, most minks fell prey to foxes (personal communication with Martin Silts[8]). A total of 77% of reintroduced minks that died in 2000–2003 were killed by predators (Maran et al., 2017: 309) because they did not have the requisite skills to avoid predation.

[7] Previous head of conservation centre and currently the director of the Tallinn Zoological Gardens.
[8] Specialist on reintroduction of the European mink to the island of Saaremaa.

Reintroduction of the European Mink in Estonia

Reintroduction of the European mink to the island of Hiiumaa took place between 2000 and 2016.[9] Since that time, a self-sufficient population has been established and there was no need to reintroduce any more captive-bred animals (personal communication with Tiit Maran). In fact, in 2019 all minks caught during monitoring had been born in the wild (see Mäekivi et al., 2021: 7–8). In reintroducing the mink, two protocols were followed: the hard-release and the soft-release method. The hard-release method means releasing animals raised in captivity at a natural site (i.e., near rivers or canals with lush vegetation and prey) without providing them with any food or shelter (Maran et al., 2009: 1687; Maran et al., 2017: 378). Soft-release of minks to the island of Hiiumaa entailed building onsite enclosures on the riverbanks (so that part of the enclosure was in a natural watercourse and there was natural vegetation in the enclosure), transportation of pregnant females together with a nesting box to those enclosures, feeding the animals (mother and the offspring born onsite) with (defrosted) rodents and minced meat. During the dispersal period of the litter, the enclosure gate would be left open to allow the animals to come and go as they wished, but feeding continued. After dispersal of the litter, when minks stopped returning to the enclose, care in the form of feeding etc. was discontinued (Maran et al., 2017: 378–379). Since we have access to interview materials pertaining to the soft release of minks, we shall discuss it and analyse the changes in minks' Umwelten, in particular the point of a major alteration in their environment, that is, when they were reintroduced to their once native habitat.

Analysis of the Umwelt Reversion

From the comparison of European minks' lives in captivity and in the wild described above, we can see several differences in the meaning carriers of food and in how the minks related to people. This case study enabled us to see what new meaning carriers were introduced to the minks' basic functional circles, which objects transformed their meaning, and thereby also the implications of these changes for the agency of the reintroduced minks.

Although we have already introduced the concept of Umwelt as a subjective world and noted that different meaning carriers can be added to the alloanimals' Umwelten during their lifetime, Uexküll in his analysis did not consider what happens to alloanimal Umwelten when changes in their environment take place. In fact, he held a static view of the environment and assumed that the environment and the alloanimal are in harmony and the effector organs are contrapuntal or perfectly suited to the environment (von Uexküll, 1982: 53–57; see also Chap. 8). But our case study clearly indicates a possibility of decoupling the environment and the

[9] For a detailed timeline, see Mäekivi et al., 2021: 7.

alloanimal. In the case of the minks, this went hand in hand with major changes in the minks' Umwelten. We thus need a more contemporary, though still Uexküllian, solution to tackle the issue. Tønnessen (2009) directed his attention to this issue and came up with the notion of 'Umwelt transition', which he defined as a 'lasting, systematic change, within the life cycle of a being, considered from an ontogenetic (individual), phylogenetic (population-, species-) or cultural perspective, from one typical appearance of its Umwelt to another' (Tønnessen, 2009: 49). This concept allows us to assign due importance to changes in the environment when analysing its effects on the Umwelten of the reintroduced minks.

Our research turned out to involve a special case of Umwelt transition, which we decided to call 'Umwelt reversion'.[10] We define Umwelt reversion as a process where an animal or a population of animals return to a previous state (configuration) of Umwelt after experiencing a considerable Umwelt alteration (e.g., regarding the four main functional circles of companion, food, enemy, and environment). We could describe Umwelt reversion as a situation where the Umwelt of an animal or animal population was altered by external factors (e.g., captive conditions, habituation, other anthropogenic factors) which caused considerable changes in the Umwelt. Upon reverting the external factors to their previous state, the Umwelt is then transformed so that the previous meaning carriers and meanings are reacquired.

We shall analyse the functional circles of food and, to a certain extent, also enemy (we will deal only with humans, not with all mink predators) to explain Umwelt reversion in the case of the European mink. We would like to stress here that in our case study, mink agency forms the background against which the Umwelt transition from *ex situ* to life *in situ* took place because there were external (human) influences constraining the choices that minks could make. In particular, as long as they were in captivity, for instance, the minks were more passive because they were fed and experienced (and became used to) human company. As we are about to demonstrate, with the soft release the minks' agency was more evident because Umwelt reversion took place through choices which the animals themselves had made.

Umwelt Reversion in the Functional Circle of Food

In the brief description of *ex situ* and *in situ* mink diet, we can see clear differences: in captivity, they had little live prey, fish, or crayfish. After the soft release, they did not receive any minced meat or vitamin supplements, which led to a re-arrangement of the relation between the meaning carrier 'food' and food as an object.. For this

[10]Although the concept of 'reversion' carries some connotations to genetics, where this concept is used to describe cases where alterations or mutations are reversed. We do not see it as a problem for our Umwelt analysis, where the reversion can take place solely through the activity of the subject, i.e., animal.

reason, food, as of the main functional circles for any alloanimal, deserves here some further attention.

As noted above, in the soft release scenario the minks went through a transition period when while still fed with (defrosted) rodents and minced meat, but could also catch occasional insects, amphibians, or even fish that happened to pass through the wire mesh fencing of their enclosures. One of the interviewees (M, 70, landowner), a man who had a mink enclosure on his land, noted that the stream flowing through the enclosure contains pikes, river burbots, and eels, that is, several fish species preferred by the mink (see Maran & Põdra, 2009: 20). The same interviewee also confessed that his son, who regularly goes fishing, was taking a large part of his catch to the minks. Here, we can see a tendency to offer minks more natural food. There was also a counterexample where a mink family after leaving their enclosure on the island settled in the harbour's seawall next to a pizza restaurant. The owner (M, 69, restaurant owner) told us several different stories about the summer after the minks' release, saying that 'the mink wanted to drag away the flour bag but because her teeth are sharp, she tore the bag and the whole kitchen was covered in flour'. The mink also dragged some pizza crusts under the hood of a car and stole food from customers' tables. This kind of desire for anthropogenic food sources was echoed also in other accounts. For instance, one (M, 70, landowner) recalls the mink climbing on the table in their garden during his son's birthday, while another (F, 70, retired pensioner) remembers feeding the mink grilled chicken. Perhaps the oddest story was about the mink tasting vodka. This was shared by an interviewee who said that 'Because the mink liked to play with the vodka bottle, the builders played a trick on her by pouring a bit on the ground. She tasted it and ran to the steam room to drink some water' (M, 50, landowner, livestock farmer). There were also several recorded cases of the minks raiding henhouses (F, 45, environmental officer, personal communication with Tiit Maran), which is also an unusual food source for the mink.

These examples show that there was a wide range of objects which carry the meaning of 'food' for the reintroduced captive-born European minks. It seems that they freely incorporated new objects to the functional circle. What is interesting, though, is that the described incidents involved mainly minks born at the breeding facility and brought to the island – and not their offspring born in Hiiumaa.[11] In fact, the offspring were too cautious of humans to take advantage of anthropogenic resources (see the following section). The claim that captive-born minks use unusual food sources upon reintroduction but cease to do so after some time is also supported by relevant literature (e.g., Põdra et al., 2013; Maran et al., 2017).

In Fig. 13.1, we have recapitulated the process of changes in the diet of the European mink, that is, we have listed objects which for the minks carry the meaning of 'food' in different phases of Umwelt transition.

[11] Some females were equipped with radio-tracking devices, which made them easily recognisable, and all minks caught during monitoring were equipped with a data chip (F, 55, veterinarian; personal communication with Tiit Maran).

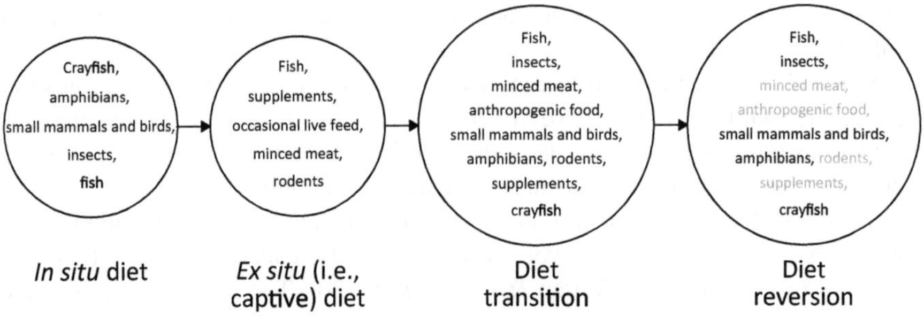

Fig. 13.1 Umwelt reversion in relation to the functional circle of food

We can see typical foods sought by wild mink populations in the circle of '*in situ* diet'. When the animals were caught and transported to a captive environment, which was a major Umwelt transition in the animals' lives, their diet was also largely replaced with a new one. As a result, the meaning of 'food' had changed and referred to different objects. Given that the minks survived and thrived, we can assume that the connection between the meaning carrier and object is not very strong. When the minks were reintroduced – another major Umwelt transition – and kept in enclosures on the island, new meaning carriers were added to their Umwelten. After leaving the enclosures, they could try out further kinds of food and as a result, during this transition period, the diversity of objects in the functional circle of food was the largest. This demonstrates the minks' agency in this Umwelt transition.

What is interesting is that after the transitional period, after the minks became accustomed to life in the wild, we can see that they excluded certain objects as food, for instance the anthropogenic resources and items representative of the captive environment. The relationship between the meaning carrier and the object in this functional circle thus became similar to or even identical with conditions characterising minks living *in situ* environment. In this case, the Umwelt reversion was evident and regarding the meaning carriers and objects in the functional circle of food, we can see that after a transitional period the minks reacquired a suitable configuration of this aspect in their Umwelten to manage in their native habitat. It is important to stress again the active role of the minks in this Umwelt reversion: although anthropogenic food resources were still available to them (by continued feeding, raiding of henhouses, visits to human settlements), they interpreted the environment and opted for other food sources, more compatible with their native habitat.[12]

[12] One might argue that 'natural' food sources are more easily attainable for the reintroduced mink and that is the reason for choosing them, but our emphasis is rather on the agential aspect then reasons for executing their agency.

Umwelt Reversion in the Functional Circle of Enemy: The meaning of Humans

In the Umwelten of the European mink, enemy can take a variety of shapes. Enemies can be the avian or mammalian predators we mentioned when introducing the conditions of minks' life *in situ*, it can be species which compete with the mink for food resources and habitat (especially the American mink *Neovision vision*; see Sidorovich et al., 2010; Maran et al., 2017),[13] or it can be humans. In any case, the 'enemy' is related to negative affective states in the mink which are communicated by avoidant behaviour, i.e., avoidance of whatever is identified as enemy.

Let us now discuss how the meaning of humans changed during the Umwelt transition. First of all, we should take a look at it because humans were closely related to the functional circle of food. Secondly, relationships to humans have a major impact on reintroduction of alloanimals whose Umwelten have overlaps with human Umwelten (see, e.g., Lestel, 2014). This in turn indicates that there are possibilities for meaningful interspecies interactions.

Heini Hediger claimed that humans are the universal enemy for almost all species that live in the wild, and this is expressed by an escape reaction similar to that for any enemy approach (Hediger, 1950: 19). We briefly mentioned that minks living *in situ* were weary of human disturbances and tended to stay away from human habitats, but many examples related to changes in the functional circle of food prove otherwise. In the interviews, we find additional illustrative instances. For example, one interviewee (M, 70, landowner) who took care of minks in their island enclosure described: 'They heard us approaching from afar. The mother was waiting next to the fence until we started throwing the mice over it.' Another interviewee (M, 45, nature photographer) stated that 'humans were definitely not strangers to her, otherwise she wouldn't steal flounders from a bucket that was one meter from me', while yet another (M, 69, restaurant owner) added: 'She was too used to humans. When we had guests during the summers, she would climb on the table […]. She was used to people feeding her.' The interviewee's insight, i.e., his linking the affinity to humans with food, is important because it has a clear connection with captive rearing of these specific minks. In the breeding facility and in the island enclosures, they were fed by human caretakers, which is why it was normal for them to associate humans with food. Humans thus acquired a positive meaning.

There are some indications to the effect that only the females who were brought from the zoo to Hiiumaa had this positive association, whereas their offspring, born on the island, did not. In the interviews, we thus find statements such as: 'only the mother came into the house, the cubs stayed away. When I wanted to approach them, they would hide […]. The mother was different from the cubs' (M, 69, restaurant owner) or 'after they left the enclosure, we never saw them again' (F, 33, livestock farmer). We can thus say that the offspring, who had much less contact with

[13] They were all caught before reintroduction of the European mink (Maran et al., 2017: 373; Mäekivi et al., 2021: 7–8).

humans, were cautious of people and exhibited hiding and escape reactions. This also features in the interviews: 'an animal born in the wild gets a negative experience and will not go to our trap the second time' (M, 52, environmental officer); 'it is much harder to catch wild-born minks than those raised in the zoo' (F, 45, environmental officer); and 'there are very few streams on whose banks I don't see mink footprints, but now they are no longer human companions' (M, 45 nature photographer). The last quote indicates that although there are many minks on the island,[14] they hide from humans. Only a handful of interviewees managed to catch a glimpse of the mink in the past few years and during this entire period, there were no recorded instances of mink making its way to human settlements (personal communication with Tiit Maran).

We show the meaning of humans in the Umwelt of the European mink in different phases of Umwelt transition in Fig. 13.2.

In the case of food, we discussed that items belonging to this functional circle and thus carrying this meaning had changed in the course of reintroduction. In the case of humans, the object itself did not change but the meaning attributed to humans did. In particular, there was a shift in the functional circle in which humans belong to the mink Umwelt. We said that it is natural for humans to have a negative meaning for any animal who lives in the wild and we also showed that for animals living in captivity, humans had a strong connotation with food and therefore acquired a positive meaning. Additionally, we saw that after some time, perhaps already with the first generation of island-born minks, humans lost this positive connotation and regained the negative meaning associated with belonging to the functional circle of enemies. A reversion thus took place because the meaning of humans became once again similar or identical to that in previous natural conditions. This is an appropriate Umwelt transition for a life in a natural habitat.[15] Two major Umwelt transitions took place with respect to humans: when the animals' environment changed from captivity to living in the wild and vice versa. Although we have information about the transition from *ex situ* to *in situ*, where we can follow the change of meaning of

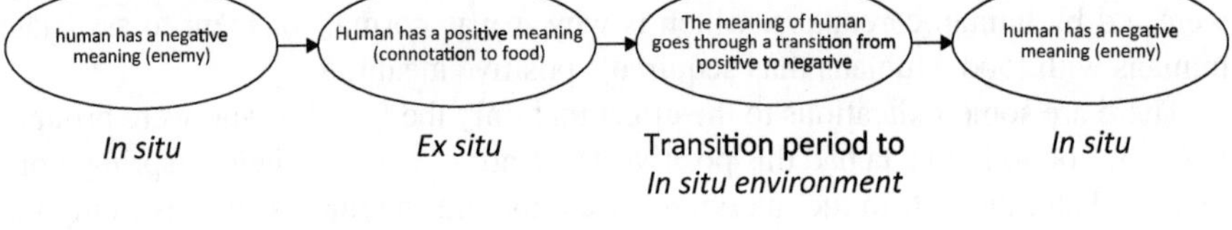

Fig. 13.2 Umwelt reversion in relation to humans

[14] According to 2019 monitoring results app. 190–210 animals in the autumn (personal communication with Tiit Maran).

[15] Finding their way to human settlements ended fatally for the mink at least on two occasions according to our interviews. In both cases, the minks' boldness was misinterpreted as having rabies (F, 45, environmental officer; F, 33, livestock farmer; F, 55, veterinarian).

humans from positive to negative, we can only assume that during transition from *in situ* to captivity, the change went in the opposite direction. Similar to Umwelt reversion with respect to the functional circle of food, in the case of Umwelt reversion regarding the meaning of humans we can see that minks' own agency played a central role. While in the captive environment a relationship between humans and minks was inevitable, *in situ* it was minks' own activity that determined the contacts.

Conclusion

In this chapter, we argued that zoosemiotics does not question alloanimal agency. It focuses on ways to access that agency by analysing meaning relations relevant for the alloanimals. It is more common to discuss animal agency on the level of individuals and one could thus conduct studies similar to the one presented here also on the level of an individual (with tracking devices and personality studies of alloanimals before their reintroduction). Our research, however, concentrated on a specific animal population of the European mink. We showed that through an analysis of their Umwelt, by employing the concepts of functional circle, meaning carrier, and Umwelt transition, one can explain the changes that take place in alloanimals' lives thanks to their agency under conditions of major changes in their environments. We showed that in the case of the European mink, their agency manifested itself mainly through choices they made in their new environment regarding food and relations to humans. In other words, we demonstrated that their functional circles underwent a transformation that can be described as Umwelt reversion, where some basic functional circles that were strongly altered in captivity acquired characteristics common in functional circles of *in situ* populations due to minks' own activities. We based this explanation mainly on observation of reintroduction of animals who reacquired meaning relations to the environment and other living beings similar to their conspecifics in natural habitats. We can see that mink's agency on population level is strongly dependent on the environment they inhabit and we can to some degree access it through the ways in which they interact with their environment. Although our analysis touched upon only two of the four basic functional circles, a similar investigation could be undertaken for the functional circles of physical environment and partner. Future analysis pertaining to these areas would help us achieve a more thorough understanding of the Umwelten of reintroduced European minks.

Zoosemiotics has valuable tools and methods to access alloanimal agency, especially in empirical studies in conjunction with findings about alloanimal behaviour coming from various natural sciences and, in our case, also local knowledge. Therefore, we propose that a zoosemiotic analysis should be part of any research into relations between alloanimals and their environment. Especially in this age of rapidly changing environments, it is important to be better equipped in mitigating possible complications stemming from lack of understanding how alloanimals interpret their environments and what is meaningful for them.

Acknowledgements This research is supported by grants PRG314 'Semiotic fitting as a mechanism of biocultural diversity: instability and sustainability in novel environments (1.01.2019–31.12.2023)' and Australian Research Council Discovery Projects scheme (DP200103404). Special thanks to Riin Magnus and Maie Kiisel with whom we conducted the interviews, and to the interviewees for providing us with valuable knowledge.

References

Anadón, J., Giménez, A., Ballestar, R., & Pérez, I. (2008). Evaluation of local ecological knowledge as a method for collecting extensive data on animal abundance. *Conservation Biology, 23*(3), 617–625.

Brook, R., & McLachlan, S. (2008). Trends and prospects for local knowledge in ecological and conservation research and monitoring. *Biodiversity and Conservation, 17*(14), 3501–3512.

de Waal, F. (2019). Fish, mirrors, and a gradualist perspective on self-awareness. *PLoS Biology, 17*(2), e3000112.

Garin, I., Aihartza, J., Zuberogoitia, I., & Zabala, J. (2002a). Activity pattern of European mink (*Mustela lutreola*) in southwestern Europe. *Zeitschrift für Jagdwissenschaft, 48*, 102–106.

Garin, I., Zuberogoitia, I., Zabala, J., Aihartza, J., Clevenger, A., & Rallo, A. (2002b). Home ranges of European mink *Mustela lutreola* in southwestern Europe. *Acta Theriologica, 47*(1), 55–62.

Gilchrist, G., Mallory, M., & Merkel, F. (2005). Can local ecological knowledge contribute to wildlife management? Case studies of migratory birds. *Ecology and Society, 10*(1), 20. http://www.ecologyandsociety.org/vol10/iss1/art20/. Accessed 20 Apr 2021.

Haage, M., Maran, T., Bergvall, U., Elmhagen, B., & Angerbjörn, A. (2017). The influence of spatiotemporal conditions and personality on survival in reintroductions – Evolutionary implications. *Oecologia, 183*, 45–56.

Hediger, H. (1950). *Wild animals in captivity: An outline of the biology of zoological gardens*. Dover Publisher.

Krampen, M. (1997). Models of semiosis. In R. Posner, K. Robering, & T. Sebeok (Eds.), *Semiotics: A handbook on the sign-theoretic foundations of nature and culture* (pp. 247–287). Walter de Gruyter.

Kull, K., & Favareau, D. (2022). Neurosemiotics and the mechanisms of animal semiosis. In A. García & A. Ibáñez (Eds.), *The Routledge handbook of Neurosemiotics* (pp. 15–32). Routledge.

Lestel, D. (2014). Hybrid communities. *Angelaki – Journal of the Theoretical Humanities, 19*(3), 61–73.

MacFarland, S., & Hediger, R. (Eds.). (2009). *Animals and agency: An interdisciplinary exploration*. Brill.

Mäekivi, N. (2018). *The zoological garden as a hybrid environment – A (zoo)semiotic analysis*. University of Tartu Press.

Mäekivi, N., Kiisel, M., & Magnus, R. (2021). *Euroopa naaritsa taasasustamise õppetunnid* [*Lessons from Reintroducing the European mink*]. https://www.flfi.ut.ee/sites/default/files/raport_-_euroopa_naaritsa_taasasusamise_oppetunnid_loplik.pdf. Accessed 24 Apr 2021.

Maran, T., & Põdra, M. (2009). *Euroopa naaritsa Mustela lutreola tegevuskava (2010–2014)* [*Action Plan for the European mink Mustela lutreola (2010–2014)*] https://www.envir.ee/sites/default/files/elfinder/article_files/euroopanaarits_kava_uus.pdf. Accessed 24 Apr 2021.

Maran, T., Kruuk, H., Macdonald, D., & Põlma, M. (1998). Diet of two species of mink in Estonia: Displacement of *Mustela lutreola* by *M. vison*. *Communications from the Mammal Society, 76*, 218–222.

Maran, T., Põdra, M., Põlma, M., & Macdonald, D. (2009). The survival of captive-born animals in restoration programmes – Case study of the endangered European mink *Mustela lutreola. Biological Conservation, 142*, 1685–1692.

Maran, T., Tønnessen, M., Magnus, R., Mäekivi, N., Rattasepp, S., & Tüür, K. (2016). Introducing zoosemiotics: Philosophy and historical background. In T. Maran, M. Tønnessen, & S. Rattasepp (Eds.), *Animal Umwelten in a changing world. Zoosemiotic perspectives* (pp. 10–28). Tartu University Press.

Maran, T., Põdra, M., Harrington, L., & Macdonald, D. (2017). European mink: Restoration attempts for a species on the brink of extinction. In D. Macdonald, C. Newman, & L. Harrington (Eds.), *Biology and conservation of Musteloids*. Oxford Scholarship.

Martinelli, D. (2010). *A critical companion to Zoosemiotics: People, paths, ideas. (biosemiotics 5)*. Springer.

Ortiz-Jiménez, L., Iglesias-Merchan, C., & Barja, I. (2021). Behavioral responses of the European mink in the face of different threats: Conspecific competitors, predators, and anthropic disturbances. *Scientific Reports, 11*(8266), 1–13.

Payne, C., Tillberg, C., & Suarez, A. (2004). *Recognition systems and biological invasions. Annales Zooogici Fennici, 41, 843–858*. Finnish Zoological and Botanical Publishing Board.

Põdra, M., Maran, T., Sidorovich, V., Johnson, P., & Macdonald, D. (2013). Restoration programmes and the development of a natural diet: A case study of captive-bred European mink. *European Journal of Wildlife Research, 59*(1), 93–104.

Rees, A. (2017). Animal agents? Historiography, theory and the history of science in the Anthropocene. *British Journal for the History of Science, 2*, 1–10.

Sebeok, T. (1968). Zoosemiotics. *American Speech, 43*(2), 142–144.

Sebeok, T. (1972). *Perspectives in Zoosemiotics*. Mouton.

Sebeok, T. (1986). *I think I am a verb: More contributions to the doctrine of signs*. Springer.

Sebeok, T. (1990). Communication in animals and men. In T. Sebeok (Ed.), *Essays in Zoosemiotics* (pp. 15–36). Toronto Semiotic Circle.

Sebeok, T. (2001). *Signs: An introduction to semiotics* (2nd ed.). Toronto University Press.

Sidorovich, V., Polozov, A., & Zalewski, A. (2010). Food niche variation of European and American mink during the American mink invasion in North-Eastern Belarus. *Biological Invasions, 12*(7), 2207–2217.

Steward, H. (2009). Animal agency. *Inquiry, 52*(3), 217–231.

Tønnessen, M. (2009). Umwelt transitions: Uexküll and environmental change. *Biosemiotics, 2*, 47–64.

Tønnessen, M. (2015). The biosemiotic glossary project: Agent, agency. *Biosemiotics, 8*, 125–143.

Ulicsni, V., Babal, D., Juhasz, E., Molnar, Z., & Biró, M. (2020). Local knowledge about a newly reintroduced, rapidly spreading species (Eurasian beaver) and perception of its impact on ecosystem services. *PLoS One, 15*(5), e0233506. doi:https://doi.org/10.1371/journal.pone.0233506. Accessed 20 Apr 2021.

von Uexküll, J. (1982). The theory of meaning. *Semiotica, 42*(1), 25–82.

von Uexküll, J. (1992[1934]). A stroll through the worlds of animals and men: A picture book of invisible worlds. *Semiotica, 89*(4), 319–391.

Chapter 14
Agency and Appearance: Reading the Face of Life

Karel Kleisner

Abstract This chapter focuses on a single but essential building block of biological organisation, namely the way living organisms establish their specific relationship between existence and appearance. Although in the following we restrict our interest to the visual dimension, one could conceive of other modalities and their interactions along similar lines. All lineages which led to complex multicellularity have either one or both of the following advanced light-handling skills: (i) light reflection and absorption to create a display of self and/or (ii) the ability to focus the light reflected from other objects on a pigmented retina so as to create an image of the external world. Inanimate objects also reflect light, but only living entities can actively influence their reflection and therefore also their display. Living bodies themselves create the exposed surfaces. Such appearances are therefore in effect an externalisation of an organised self – and that is something that does not apply to the visual aspects of inanimate things. Further, I argue that the organismal capacity of self-representation is intimately related to certain essential characteristics which differentiate life from nonlife. The more developed the self-representational ability, the higher the complexity of organisation, and the further away the organised beings are from nonlife.

Keywords Evolution of appearances · Self-presentation · Exposed surfaces · Biological organisation · Perception · Vision · Epistemic biases

K. Kleisner (✉)
Department of Philosophy and History of Science, Faculty of Science, Charles University, Prague, Czech Republic
e-mail: karel.kleisner@natur.cuni.cz

J. Švorcová (ed.), *Organismal Agency*, Biosemiotics 28,
https://doi.org/10.1007/978-3-031-53626-7_14

Biological Organisation

Organisms are among the most permanent things in the universe. Coherent, self-relating living agents which stand on their own, with the help of or in defiance of others, yet are mutable and open to change. Over the past century, life sciences have undergone rapid technological and methodological development, but the development of our theories seems to proceed at a much slower pace than the technological progress and mass production of scientific facts. Despite some promising attempts, we still do not have a general theory of biological organisation (Minelli & Pradeu, 2014). What we do seem to have is just a fast-running stream of a poorly organised flood of empirical findings. This is not surprising: big data do not necessarily require big theories.

Although organisms *per se* do not usually change faster than our ideas about them, scholars of the past emphasised quite different aspects as the key formative factors of organised living matter and its mutability. Our unsatisfactory current understanding of living matter can, and should, be viewed as a reason for taking the history of biological thought seriously, and carefully reconsidering earlier ways of understanding living organisms in the light of current ideas and the vast empirical evidence we now have at our disposal.

Current information-based and semiotic approaches (see Chaps. 8 and 13) to living matter stress the key role of interconnectivity and communication between all levels of organisation. This perspective finds further support in the currently accelerating revolution in communication and information sharing in human society. Naturally, our theories have always been heavily influenced by sociomorphic reflections of lifestyles, social hierarchy, behavioural habits, and specific forms of competition found in human society. When carefully reconsidered, these reflections may form a suitable base of, or at least contribute to a better understanding of, the functioning of living processes within and between organisms. Such insights are helpful but not sufficient for building a theory of biological organisation. For the same reason, 'Darwinian explanations' likewise provide only a partial solution to the problem.

The current, almost excessive appreciation of the highly networked communication channels, which weave their ways through all layers of biological hierarchy, is likely to lead to an understanding of living entities as diffuse, ungraspable, non-centred, and naturally distributed (see Chap. 7). That seems to contradict the more traditional view of organisms as highly integrated, coherent, totalitarian wholes with a strong control of their parts and subordinate levels (see Chap. 11). On the other hand, the latter characterisation fits only the 'paradigmatic' organisms, that is, mobile and highly integrated multicellular animals such as vertebrates. We ought to realise that both the distributed (multi-centred) and centred view of organisms matters. Biological organisation seems to encompass a wide range of natural manifestations spanning from the cell-dependent 'molecular existence' of viruses to the quasi-independent 'embodied existence' of complex multicellular organisms.

Any theory of biological organisation should integrate a maximum range of manifestations of life to provide a coherent model of reality explicable by a reasonable number of parameters. One crucial step towards this goal is to parametrise agency via various instances of its appearance. Living things act by appearing. Without agency, there can be no autonomy, no functions, no values, no selves, and ultimately no meaning of life.

Inanimate Appearances

Crystals used to attract a vast amount of attention as inanimate siblings of living organisms. In the beginning of the nineteenth century, the science of zoology was strongly inspired by crystallography – so much so that crystallography was even perceived as superior to biological disciplines. The origin of living organisms from inanimate matter by crystallisation was a popular biogenetic scenario. Cairns-Smith's idea of a 'genetic takeover' of primordial mineral-based organisms can be seen as a modern instantiation of the traditional concept of correspondence between organisms and crystals (Cairns-Smith, 1977, 1982). No wonder that crystals used to attract the attention of naturalists so much. They display many properties characteristic of – but not unique to – living systems, such as growth, regeneration, self-reproduction, self-assembly, as well as variation within the same mineral form, that is, a kind of reaction norm or analogy to phenotypic plasticity (Lima de Faria, 1988: 79). The reproduction of crystals can be thought of as involving inheritance patterns. For example, the growth of smectite crystals proceeds by intercalated layers which have the same charge density as the parental ones and can produce up to twenty repetitions, i.e., a kind of mineral generations (Lima de Faria, 1988: 119–120). Crystals are also capable of evolution by mutation, even though they are constrained within the boundaries of predetermined crystal space groups (Arrhenius, 2003). Although crystals do display some analogies with life processes, they are not alive in the sense of being members of the biota. Still, mineral entities do have some key attributes that are commonly considered typical of living beings. To follow up on Cairns-Smith's idea, we could ask a provocative question: Aren't such analogies in fact homologies? Lifeforms might have adopted/inherited these basic kinds of self-sustaining 'behaviour', such as growth, replication, regeneration, and diversification from crystals, and eventually become masters at it. Significantly, the rich array of appearances produced by crystals is possible without genes. Crystals thus teach us that having genes is *not necessary* for either diversification and variation or for the evolution of ordered natural forms. That does not mean that genes do not play an important role in diversification processes, but they are perhaps mediators and enhancers rather than an indispensable factor of evolutionary divergence.

Looking at the celestial sphere, planets exemplify how diversity of inanimate objects can emerge in accordance with a semiregular pattern. Had the planets and their satellites been expelled from the solar system early after its formation, left to wander aimlessly through inter-solar space, they would never have 'evolved' their typical characteristics, such as volcanic activity on Jupiter's moon Io or cryovolcanic activity on Saturn's moon Enceladus. Not just the brute gravitational force of Jupiter, but a combination of tidal forces with the orbital resonance of Europa, make Io a volcanic hell. A huge part of the diversity of planets bound to our star, the Sun, and that of the planets' satellites, is due to the spatiotemporal cycles determined by their orbits. The appearance of the planets' surfaces is substantially affected by their positions in the solar system and by the configuration of interactions with other cosmic objects. If celestial bodies were not gravitationally bound to their elliptical trajectories, they would not have acquired their unique characteristics. Of course, orbital cycles are not the sole cause of planets' diversity. Rather, they synergically affect many otherwise unordered events which ultimately have causative effects. In short, cycles arrange random events into a semiregular pattern.

While the appearance of planets and their satellites is strongly influenced by their position in space, the appearance of living entities has also inward sources of causation, which stem from their complex interaction across many hierarchical levels of organisation. Still, there are some underlying principles shared by both animate and inanimate nature. Variation in the appearance of living beings, embodied in various morphological structures and colour patterns, involves temporal cycles. The temporal cycles help arrange ontogenetical and environmental processes as well as other causative factors into semiregular patterns, which are ultimately observable as self-similar patterns (cyclic parallelism) against a phylogenetic background.

Unicellular Appearances

Many unicellular lifeforms make us wonder about the intricacy of surface patterns produced by just a single cell. But the surface is probably not the most interesting facet of these microscopic creatures. It is not within the scope of this study to comprehensively introduce the vast structural diversity of protists, unicellular eukaryotic organisms which are neither animals, nor plants, nor fungi. Still, several examples from various branches of the protist tree deserve a mention.

Amongst the most elaborate forms, diversified in a number of groups, are radiolarians, which belong to the supergroup of Rhizaria. These protists often possess intricate siliceous skeletons, and their overall design is upgraded by cytoplasm differentiated into an inner endoplasm and outer ectoplasm, whereby the latter can sometimes form various protruding cytoplasmatic projections, such as axopodia and pseudopodia, and which can even be interconnected. The impressive star-shaped body plan is exemplified, for instance, by marine planktonic radiolarians belonging to the Acantharia group, with inner skeletons made from celestine, a strontium sulphate mineral (Decelle & Not, 2015), while within the Orodaria group, one can find almost giant forms, with a diameter up to 5–7 mm, such as *Oroscena huxleyi* (Oroscenidae) and *Cytocladus tricladus* (Thalassothamnidae) (Nakamura et al., 2021).

Different but no less impressive structural organisation is found in coccolithophores (haptophytes, Haptista), which equip their unicellular bodies with variously shaped but mostly disc-like scales (coccoliths) and sometimes even form massive vase-like protrusions known as lopadoliths (e.g., *Scyphosphaera apsteinii*) (De Vargas et al., 2007; Drescher et al., 2012). Coccolithophores play an essential role in the global biogeochemical cycle as leading agents of ocean calcification due to their ability of genetically controlled intracellular biomineralization of calcium carbonate ($CaCO_3$) crystals. The formation of coccoliths (coccolithogenesis) is an actively organised process involving cytoskeletal dynamics. During this process, the nucleation and growth of calcium carbonate crystals is initiated and controlled by protein and polysaccharide cellular products (Brownlee et al., 2015; De Vargas et al., 2007; Taylor et al., 2017). In the course of evolution, these tiny 'Gaian soldiers' of atmospheric CO_2 uptake diversified into over 4000 morphospecies, most of whom had gone extinct and are known to us only as microfossils.

It should be noted that 'unicellular' does not mean necessarily morphostructurally ancient in terms of evolutionary history. For instance, the fossil sediments of the abovementioned coccolithophores can be up to 220 million years old, which corresponds to the Triassic period, but protists – at least according to a recent investigation of acritarch microfossils (Yin et al., 2020) – began to appear at about 2.15–1.95 billion years ago.

Worthy of at least a brief mention are the complex unicellular body plans of ciliophorans, the multiple symmetries of the fractal-shaped cogged cells of some desmids (*Micrasterias)* (Neustupa, 2017), the simple elegance of many diatoms, the chloroplast-filled bodies of some excavate *Euglenas* equipped with a shiny red stigma, or the exquisite morphologies of centrohelids and other heliozoan-like protists (Gast, 2017). Despite their unicellularity, the protist morphospace is immense and their forms utterly beautiful – ready to charm anyone with a microscope.

Multicellular Appearances

In the course of evolution, the increase of the morphological diversity of appearance of living entities went hand in hand with the development of aggregative and communicative tendencies. Multicellularity and sexual reproduction seem to have accelerated the diversification of clades, with multicellularity having a greater effect than sexual reproduction. Ultimately, though, diversification affected the species richness of various groups of the biota (Chen & Wiens, 2021). Interestingly, multicellularity evolved independently about 25 times (Maynard Smith & Szathmary, 1997; Bonner, 2000; Grosberg & Strathmann, 2007; Herron et al., 2013; see also Chap. 11), but complex multicellularity is found only in embryophyte land plants (Embryophyta), florideophyte red algae (Florideophyceae) (Knoll, 2011; Cock & Collén, 2015), laminarialean brown algae (Laminariales), animals (Metazoa), and at least at eight groups of fungi (Nagy et al., 2018).

Most of these groups gave rise to gigantic forms. Extinct and recent metazoan megafauna and the convergent emergence of tree forms in many plant groups are unlikely to surprise us, but there also exist behemoths of quite a different origin. For example, brown macroalgae belonging to genus *Macrocystis pyrifera* can grow to up to 45 meters (and perhaps even more) in length. They form rich underwater ecosystems known as kelp forests. Fungi are special among these groups because they can switch between complex (fruiting bodies) and simple (vegetative mycelium) multicellularity during their lifecycle. The largest known living organism in the Earth is said to be a basidiomycete fungus *Armillaria ostoyae* found in the Malheur National Forest (Oregon, USA): the mycelial mass of this colossus spatially extends about 9 km^2 and weighs about 35,000 tons. Huge size and age of several thousand years is not exceptional among the *Armillaria*, as evidenced by *A. gallica*, formerly *bulbosa* (Smith et al., 1992; Anderson et al., 2018). The largest fruiting body, reaching 10 meters in length, with width of 0.8 meters and weight about 500 kg was found in the giant polypore *Phellinus* *ellipsoideus* (formerly *Fomitiporia ellipsoidea*) on the Hainan Island in southern China (Dai & Cui, 2011).

One important characteristic which accompanies the transition from simple to complex multicellularity is a change in spatial organisation. While in simple multicellular organisms, each cell retains its contact with the external environment (at least during the active phase of its lifecycle), complex multicellular organisms form 'real bodies' characterised by three-dimensional organisation, where only some cells directly interact with the external environment (Knoll, 2011; Knoll & Hewitt, 2011). Such organisation posed some new challenges to the physiology of complex multicellular bodies. It led to the evolution of new means of transport, on both cell-cell and tissue levels, which circumvented the limitations of diffusion and opened the door to a successful diversification of complex multicellular lifeforms (Knoll & Hewitt, 2011). Only in complex multicellular organisms can we start speaking about an actual distinction between the inner and outer dimension of the body. This type of organisation laid the foundations to a further evolution of organic inwardness, complex self-relatedness, individuality represented by exposed organismal surfaces, and eventually even the appearance of self-awareness and more complex forms of experience of the world (see Chaps. 8 and 9).

The first known macroscopic complex lifeforms belong to the fossils of Ediacara, which inhabited pre-Cambrian oceans (app. 579–542 million years ago). Palaeontologists described several assemblages of different ages. The oldest Avalon assemblage (579 Ma) had already evolved an impressive range of morphological types. A comparable level of morphological disparity was retained in the morphospace of younger Ediacara representatives of the White Sea and Nama assemblages (Shen et al., 2008). This rapid evolution of body plans at the beginning of the evolution of Ediacara, which was followed by a taxonomic diversification within a previously established morphospace, resembles the later Cambrian explosion (Shen et al., 2008). In fact, the presence of ancient cholesteroids (unique to animals) extracted from exceptionally well-preserved remains of *Dickinsonia* indicates that the Ediacaran increase in morphological disparity had ushered in the later Cambrian explosion of metazoan body plans (Bobrovskiy et al., 2018). Importantly, it thus seems that the developmental tools for patterning macroscopic complex bodies were available even before the morphological diversification of both Ediacara biota and animals (Evans et al., 2020).

Ediacarans inhabited shallow and deeper seafloors. They were sessile organisms with soft bodies. Their morphological organisation in some instances resembled cnidarians and sponges: they often had a modular, plant-like structure with a repetition of parallel and radial ribs, whorls, spirals, and various frond-like structures. As far as one can infer from preserved fossil records, it seems that these mysterious inhabitants of ancient seas did not evolve any complex sensory organs like eyes, nor did they have any appendages that could serve either for locomotion or as weapons (Droser & Gehling, 2008; Gehling & Droser, 2009; Bobrovskiy et al., 2020), but there is some taphonomic evidence to the effect that at least some of them could move (Liu et al., 2010).

Ediacaran morphology is already characterised by a complex body structure, usually consisting of repeated body elements organised in metameric arrays. These Ediacaran shapes and structures developed long before the appearance of any lifeform with visual perception. Although such morphologies were thus unobserved by other lifeforms and therefore could not play any role in communication or signalling, they had already formed the prerequisite material and organisational foundations for these functions.

The key transition in the evolution of external appearance of organisms dates to a short interval between 520 and 515 million years ago known as the Cambrian explosion. During this brief period in the Cambrian, an immense variety of morphological lifeforms suddenly emerged across most eumetazoan phyla, but especially bilaterians. A recent trilobite study which used the method of Bayesian clock (Paterson et al., 2019) suggests that morphological disparity evolved rapidly over about 20 million years in the last stage of the Ediacaran and at the beginning of Cambrian, followed by a stasis for the rest of the Cambrian. Therefore, at least at the level of animal appearances, the Cambrian explosion seems to be a real event and not just a by-product of the long and little-known evolutionary history of the metazoans during the Neoproterozoic. That does not mean that all major animal clades 'phylogenetically' evolved during this period. Rather, it seems that many members of formerly mainly soft-bodied animal phyla suddenly upgraded their bodies by adding hard biomineralized parts and various other external morphostructural elements, in many cases thus laying foundations to the morphological types of groups that still live around us. In other words, many animal clades donned their 'modern-style jackets' – and in most cases, they wear them to this day. It is unclear what caused the Cambrian explosion. It may have been triggered by progressing oceanic oxygenation or, according to a recent proposal, by marine redox fluctuations that led to altered Mg/Ca ratios, skeleton formation, increased metabolism demands, durophagy and drilling predation, or it may have been due to other environmental/organisational drivers and their combinations (Peters & Gaines, 2012; Wood & Zhuravlev, 2012; Bicknell & Paterson, 2018; Wei et al., 2018).

The Key Role of Vision

The symptom of the Cambrian explosion most relevant to our subject is the emergence of organisms with complex organs of sight. The key role of evolutionary emergence of effective organs of vision and of reflecting surfaces is highlighted in Andrew Parker's 'light switch theory'. According to it, the nearly simultaneous appearance of the first eyes with a lens at around 521 million years ago and of animal surfaces structured as multilayer optical reflectors at about 508 million years ago was not a matter of happenstance (Parker, 2000, 2003, 2011). During a relatively short period of time, we can observe the emergence of highly mobile predators with grasping appendages and mouthparts. Both predators and prey became armoured, and their exposed surfaces developed diffraction grating, which interacted with sunlight and produced structural colours. Moreover, Parker suggests that vision, and its necessary optical elements, emerged in the course of evolution very rapidly, almost suddenly, compared to other senses such as touch, smell/taste, and hearing.

Visual accessibility is a crucial precondition for our topic. The abovementioned microscopic protists, despite their superb design had no (naturally established) connection to the visual system of any other living being. Their unicellular surface structures thus cannot have had a function primarily intended for display, such as we observe in larger macroscopic lifeforms. In this regard, their microscopic appearance was visually (perceptually) irrelevant. Adolf Portmann calls such appearances 'improper phenomena' (*uneigentliche Erscheinungen*) to emphasise the difference between them and phenomena proper (*eigentliche Erscheinungen*), which can enter the senses of organic observers in a natural, non-invasive way (Portmann, 1960; see also Chap. 12). Nevertheless, this inability to form a visual 'cue' did not hinder the propensity of living entities to produce and expose elaborate, sometimes even excessive outer structures, regardless of their purpose, if any.

It should be noted that even in unicellular forms one can find surprisingly complex optic systems. Gymnodinoid dinoflagellates from Warnowiaceae family evolved a most elaborate optical system, an eye-organelle called ocelloid, which is in a sense analogical to the chamber eye of vertebrates. The ocelloid consists of a transparent hyalosome (a focusing lens), ocellar chamber (vitreous body), and a melanosome containing a retinal body placed in a cup of dark pigment (Colley & Nilsson, 2016). The ocelloid can even turn, that is, change its orientation and therefore also the direction of gaze. Whether this optic system can be used for visual detection and predation of other phytoplankton is not clear. Although the optical abilities of dinoflagellate ocelloids can hardly be compared to the multicellular eyes of animals with a nervous system, it shows that elaborate sight structures can evolve even in unicellular Eukaryote settings. Nevertheless, the likelihood that these extraordinary optic devices unique to dinoflagellates had any real impact on the evolutionary formation of the appearance of microscopic organisms is close to zero.

There seems to be a mutual interdependence between intricate forms of visual expression and visual perception, which is conditioned by the evolution of complex multicellularity and advanced organs of vision. The fact that multicellularity combines growth and reproduction enabled the evolutionary emergence of complex exposed surfaces.

Why are surfaces so important? Any interaction or communication takes place on some kind of interface, and any interface logically presupposes the existence of a surface. This applies to microscopic molecular aggregates just as much as it does to intricate macroscopic structures, such as human faces. In human society, it is the fact that we expose the surface of the body part we call a face to the social environment that enables the face to function as a communication interface. In general, though, one can say that the construction and deconstruction of various surface–interface entities played a role in all evolutionary transitions, from the appearance of the first Archean and prokaryotic cells, diversification of eukaryote clades by multiple serial endosymbiotic events, the emergence of simple and complex multicellularity, and all the way to the evolution of self-awareness and cultural evolution. The biosphere inhabits just a thin layer of the Earth's surface, about 20 kilometres deep, so one could say that Life is a surface phenomenon in its very nature.

Theoretical Limitations and Varieties of Epistemic Bias

To rethink the role of organismal surfaces in evolutionary agency, we first need to critically reconsider several kinds of bias which tend to downplay the ontological significance of animate appearances and thus systematically hinder our appreciation of their explanatory role as a causative factor.

Selection Bias

> *Selection and randomness "explain" every problem for which there is no explanation* (Lima de Faria, 1988)

Does selection based on random variation explain organic design and teleology in nature? Darwin was inspired by artificial selection when explaining the evolution of purposeful adaptations. Artificial selection differs from natural selection simply because in the former, it is a human subject and not the natural setup that selects among the variants at hand. Artificial selection involves a decision on the part of the selecting subject. In this, artificial selection resembles sexual selection (mate choice), where certain traits are preferred by the opposite sex and these traits thus provide a reproduction advantage to carriers of those traits. It is an activity of the animal subject that evaluates and selects between possible variants. But it is also well known that sexual selection can lead to the evolution of traits that can hardly be considered a functional optimisation of existing adaptations. Of course, sexually selected traits have a survival function in the sense that they increase the likelihood of reproduction and thus potential biological fitness of the selected individual. On the other hand, it seems that we have a different notion of function in mind when talking about general survival function, or functions that allow organisms to perform certain particular tasks which increase the effectiveness of its organisation and thus its autonomy. There are many cases where, despite increasing the individuals' chances in a courtship, sexually selected traits (exaggerated structures or sexual ornaments) in fact decrease their functional coherence. This is nothing new: ever since Darwin, we know that sexual selection moves in a direction that often contradicts natural selection. This conflict is usually brushed off by noting that everything in living nature is a trade-off. That may well be true, but from an epistemological perspective it is not a satisfactory explanation.

To navigate our way around this issue, let us consider 'adaptations to environment', that is, adaptations such as burrowing legs, biting mandibulae, grasping clamps, or seeing eyes. We can call them instrumental adaptations because they help perform some practical tasks which contribute to the immediate self-maintenance of their bearers. Naturally, their (proximate) functioning is also important for the organisms' (ultimate) purpose, which is to increase their chances of reproducing and passing their characteristics onto future generations via their offspring. The important thing to consider is that artificial selection, a blueprint for the mechanism of natural selection, usually does not lead to the emergence of adaptive structures (traits that enhance instrumental functions) that would lead to a more effectively organised living entity. In other words, artificial selection does not generate structures which are likely to contribute to a higher functional coherence (and autonomy) of an organism. In fact, one often observes quite the opposite. The domestic forms of animals are often unable to survive outside the niches of human culture because the traits we have selected contribute to the autonomy of human organisms at the expense of the domesticated forms.

Sexual selection usually exploits pre-existing forms to produce excessive ones. That, however, tends to lead to a decreased effectiveness of the original function (prolongation of feathers, colouring of integument patches, more pronounced asymmetry of claws) and may be therefore maladaptive in the context of the basic function. It should be stressed that both artificial and sexual selection are processes which give rise to the most elaborate characteristics of animals and plants. Certain examples of the creative power of artificial selection, such as the various breeds of dogs and other domestic animals, have been achieved rapidly over just dozens of generations and are notorious, but evidence to the effect that the two most prominent kinds of selection lead to the emergence of functional traits is at least equivocal.

Why do we expect the selection principle to be behind the generation of all purposeful adaptive structures? Selection undoubtedly contributes to the maintenance and modification of functional traits but the commonly held view according to which neo-Darwinian classical mechanism based on selection from random variation explains the evolution of organismal appearances, including the adaptive forms, is a matter of belief rather than direct evidence.

Competition Bias

Does competition play a key role in evolutionary processes across the different levels of biological hierarchy? Some time ago, one colleague told me during an informal debate that he is firmly convinced that all problems can be solved by competition or, to be more precise, by the selection pressure to which individuals are exposed by increased competition for resources. It is a frustrating worldview, but not surprising. All notions in evolutionary biology have one common denominator, namely that the actors of evolution are primarily unscrupulous super-competitive liars. Life on our planet is represented by occasionally aggregative but naturally selfish entities. The vast increase in the complexity of living entities in the course of evolution is nothing but a result of never-ending efforts to cheat and outcompete others. Reproduction and proliferation are the sole goals of all living matter.

Some form of competition is often helpful but on its own not sufficient to solve a problem within a natural system. There are always other circumstances that affect it, such as for instance processes occurring at higher hierarchical levels. Two imaginal discs in a pupa may compete for the growth factor but when one of them develops into a wing much larger than the other, the resulting butterfly will not be able to fly properly. Such an individual will be easily outcompeted by conspecifics simply because of lack of downregulated competition on a subsidiary level. One might claim that competition at higher levels will ultimately solve this issue by eliminative selection on low-quality individuals, so the competition principle solves this case quite well.

I do not agree with such reasoning, mainly because it does not explain why butterflies with asymmetric wings are not actually seen in nature. In fact, one could ask more generally: why is the relative production of misfits in nature so low? Probably this is due to selection pressures acting on the developmental precision of processes which produce adult structures such as wings. What is the problem then? We can still treat these feedback loops as a result of the mechanisms of variation, competition, and selection. But all this could not operate without some hierarchy (or, in some cases, heterarchy). When conceived of as a multilevel (or multidomain) issue, one might in fact call this organisation. All of the above converge into a simple observation: Competition without organisation is useless! Living entities do compete, but the processes which led to increased complexity were not solely due to competition against neighbouring entities (cf. Chap. 11). The current diversity of appearances of living entities and its evolutionary history shows rather that life on Earth is ambitious – but not necessarily competitive.

Molecular-Genetic Bias

There is a strong bias, or tendency, to study local adaptations of organisms solely from a genetic perspective. Biological variation is commonly seen as a result of genetic variability generated by mutations which continuously emerge within populations. Nevertheless, changes in the frequency of alleles within populations over time cannot, on their own, explain the observed patterns of morphological and behavioural variation. That does not mean that recent large association studies, together with transcriptomics, metabolomics, and epigenomics, do not provide us with relevant information about evolutionary processes. Still, all this vast amount of information would not be meaningful without reference to a biological form and its behavioural role. Molecular explanations preferentially explain the molecular part of the story, which makes the resulting picture epistemically incomplete. An explanation we ought to aim for should be sensitive to the level of biological hierarchy it tries to explain. An explanation of patterns on a phenotype ideally ought to integrate elements observable on the level of morphology with the most proximate levels, such as in this case behaviour and development. Behaviour reflects the various interactions of morphological characters with their environment. This interaction is ultimately the factor which defines the survival value of, in this case, the abovementioned patterns. Development, on the other hand, provides a mechanistic understanding how the morphological characters came into existence, and thus also for instance how costly their development is in a particular environment.

Science-Fashion Bias

As highlighted above, molecular genetic approaches to life sciences are of key importance because they allow us to get, so to speak, under Life's skin. They enable the study of causal processes behind morphogenesis, behaviour, reproduction, development, and so on. These approaches played a pivotal role in disentangling the phylogenetic relationships between organisms in instances where morphological traits were confusing or missing. Still, instrumental utility and current popularity of any approach should not be mistaken for actual epistemic explanatory value.

The recent disproportionate focus on the most progressive metagenomic approaches, barcoding, or the study of complex interactomes does not bring just positive results. It also imposes a huge tax on our understanding of biological reality. Regardless of how much money is pumped into this segment, we are in effect putting all of our eggs in the wrong basket. In other words, the belief that exclusive focus on approaches will lead to a deeper understanding of organisation of living matter is an illusion. We are merely uncovering yet other layers of reality with small or no increase in epistemic value. We are accumulating gargantuan amounts of information but that does not take us much closer to the unique characteristics that distinguish life from nonlife. In sum, we are gathering knowledge about a vast number of processes which serve organic agency, but that tells us little about how agency emerges and how it is maintained.

Where to Go Further with the Interpretation of Organismic Appearances

Exploring Marginal Phenomena

Most researchers ask usual questions, which are easy to test and answer. It may be that this helps to find financing for the project and promote careers. But it happens at the expense of interest in general biological theorising, which is somewhat fossilised despite the abovementioned rapid progress in research techniques and analytical methods. Perhaps biologists are too attached to their standard model of evolution. One way to move on would be to focus on phenomena which resist explanation within the standard model. If we want to test the limits of explanatory frameworks, we should look at the boundary phenomena. A rich source of examples which are difficult to satisfactorily explain purely by the neo-Darwinian model are the various types of mimicry and related phenomena (Wickler, 1968; Komárek, 2003; Kleisner & Markoš, 2005; Maran, 2017;).

Neo-Darwinian adaptive scenarios often give the impression that they are always right just because they make a sense within the theory. I would suggest that if an explanation is plausible, it is not sufficient that it be correct within a theory: it must also be specific and complete. It should explain the class of traits we aimed to explain. On one hand, we should not single out some properties of objects which resist or do not fit our model. On the other hand, we should also avoid an excessively broad scope of our explanation, because no simple model can cover all biological phenomena. Moreover, the explanation should include all, or rather other, possible reasons for the emergence of a trait or its function, which may be contrasting (e.g., cryptic vs. aposematic function of the same colour pattern). Whether an explanation is plausible or not seems to be a matter of taste rather than testing.

One type of explanation of animal display that is often extended beyond its due scope is the case of protective coloration. There is no doubt that the warning function does explain the striking black and yellow patterns of wasps, their Müllerian (protected) co-mimics, and Batesian (harmless) mimics. But there are many conspicuously coloured animals where the warning function as the main explanation of their display much less convincing. For instance, many species of colonial tunicates have strikingly coloured surfaces which resemble the aposematic (warning) coloration that serves to deter potential predators. Ascidians often have an intricately patterned tunic made from polysaccharides and proteins. It covers their colonial bodies and forms a unified visual impression of the entire colony, which is then perceived as a kind of optic whole (Fig. 14.1a, b). Nevertheless, salps, which are a family of tunicates, usually have transparent bodies. This might suggest a cryptic function, but they frequently form floating chains that can be visually detected by a string of serially arranged individual guts, the only opaque part of the body (Fig. 14.1c). Who are the predators of tunicates? Tunicate predators include flatworms, gastropods, crustaceans, starfish, and fish, while their larvae are often eaten by cnidarians (corals) and various species of fish. Most of these predatory lifeforms,

Fig. 14.1 Examples of colonial tunicates. Ascidians – *Botrylloides sp.* (**a**), *Eudistoma sp.* (**b**) – have an intricately patterned tunic made from polysaccharides and proteins, which covers their colonial bodies and forms a unified visual impression of the entire colony. The transparent bodies of salps – *Salpa sp.* (**c**) – often form floating chains that can be visually detected as a string of serially arranged individual guts, the only opaque part of the body

perhaps except for fish and some crustaceans, do not have complex visual perception and colour vision.

Tunicates evolved an array of defensive mechanisms, which are for the most part based on being unpalatable due to various combinations of low energy content, poor digestibility, chemicals, acidity, and mechanical toughness of not just the tunic but also individual body part compartments (Tarjuelo et al., 2002; Holland, 2016). These protective characteristics do not seem to have any clear relation to the ornamented surfaces of many tunicate species. I would suggest that in case of animal appearances, such kinds of functional attributions should be kept only for those cases where we have evidence from direct observation and/or experimental testing. This would help support the frequently made observation that one particular structure, pattern, or display may adopt more than one functional role, most likely due to a series of co-optional events in the course of evolution.

Apart from the tunicates, there are also reasons to doubt whether some cases of striped black and yellow patterns have a solely warning function, that is, that they warn of danger. The black-and-yellow striped opisthosoma of the common wasp spider (*Argiope bruennichi*) immediately warns us that it is dangerous – that is, after all, also coded in the spider's English name. Although this explanation is likely correct, it need not be complete. Similar striped or banded patterns, however conspicuous, may be cryptic against a background of scattered colours, such as grass. A similar explanation based on distance dependence of the cryptic and warning function has already been tested in the case of caterpillars of swallowtail butterflies (Tullberg et al., 2005) and a similar explanation has also been suggested for the 'dazzling' coloration of *Heliconius* butterflies and their co-mimics and models, but it may apply to many other aposematic species (Kleisner & Saribay, 2019). This context-dependent functional switch of the same structure is possible due to organismal agency. It is the agent's action, no matter whether it be a sign bearer, perceiver, or both, that determines which function prevails and when.

Although there are many astonishing examples of mimicry, I would like to briefly mention here the case of the leaf-mimicking climbing plant (*Boquila trifoliolata*), not just because it seems to be one the most bizarre cases of mimicry published in past two decades but because it is a stunning example of plant agency. The leaves of a *Boquila* vine develop a shape that resembles its supporting tree. *Boquila trifoliolata* morph is capable of mimicking not only the leaves of its tree host but also those of surrounding trees, with which it need not have any physical contact. When a single *Boquila* individual passes across several species of supporting trees, it can mimic more than one host. Moreover, a recent study demonstrated its ability to imitate even the leaves of an artificial plastic supporting plant (White & Yamashita, 2022). Several mechanisms have been suggested to explain the similarity between *Boquila* vine and the surrounding foliage, for instance based on chemical volatile semiomolecules emitted by host plants, horizontal gene transfer mediated by airborne microbes (Gianoli & Carrasco-Urra, 2014), or even plant vision based on plant ocelli (Baluška & Mancuso, 2016; White & Yamashita, 2022). The plant vision hypothesis seems to be the most plausible explanation for the result about mimicking an artificial plant, but nevertheless it should be corroborated by various modifications of experimental settings involving the climbing vine and artificial supporting model.

Avoidance of herbivores was also proposed as a reason for why the vine mimics the appearance of the host plant's leaves (Gianoli & Carrasco-Urra, 2014). If protection against herbivores is the main evolutionary driver of this plant's mimicking capacity, one could wonder why it is not more frequent strategy in plants. A possible explanation of its rarity may be that such real-time morphing ability requires a combination of phenotypic plasticity that allows heterophylly with some other mechanism based on sensation of its own surroundings, such as plant vision or some other information channel.

Regardless of the particular mechanism, *Boquila trifoliolata* somehow sense the morphological pattern (shapes and colours) in their closest surroundings and accommodate the morphogenesis of their leaves accordingly. Regardless of whether this helps the vine to avoid herbivory, it is a fascinating capacity. This example may thus inspire a re-examination of other mimicry cases, for instance the close resemblance between some insects and plant parts, while testing the role of the optical channel in establishing the formal similarity, which is in many cases breath-taking, and one can hardly imagine it being established by a selection process acting upon random genetic mutations.

Surface Friendly Ways of Thinking

In the ecological context, surfaces are not just a matter of characteristics assessed by physical measurements. They must be measured relative to the animal, and they are unique to that animal species. In the twentieth century, we find the most influential reflections of this approach in the Umwelt theory of Jakob von Uexküll (see Chaps. 8 and 13) and in Adolf Portmann's semantic morphology. Both of these thinkers stressed the significance of species-specific perception and the uniqueness of ways of experiencing the world, which depends on the organism's body plan and its eco-physiological context. Modern versions of these and similar ideas were later elaborated by various theorising biologists and biosemioticians (Hoffmeyer, 2008; Kull 2010, 2020; Kleisner & Maran, 2014; Magnus, 2021).

A similar conception based on the ecological approach to visual perception appeared also within psychology. Experiences with tests on combat fighter pilots led James Gibson to a belief that perception studied in a laboratory does not correspond to the complex, real-world situations. Gibson's original psychological concept of 'affordance' has been adopted by numerous disciplines spanning from architecture design to artificial intelligence (Chong & Proctor, 2020). According to Gibson, '*the affordance of anything is a specific combination of the properties of its substance and its surfaces taken with reference to an animal*' (Gibson, 1977: 67). Affordances are neither subjective contents nor objective entities. Rather, they arise at the intersections between the properties of things and capabilities of observers. What the environment affords to animals, including human beings, is not only a list of inanimate things such as the terrain, potential shelters, material for tools, or other potentially usable objects, but also animal and human displays, such as human faces (Gibson, 1979). Gibson's theory of affordances outlines a promising approach to the study of organismal surfaces and the potential functional roles they afford. Unfortunately, among contemporary biologists working on the field of sensory and behavioural ecology Gibson's ideas are largely unknown.

Of course, affordances do not cover all problems related to our understanding of visual reality. Nevertheless, this concept nicely fits in with the current biosemiotics and social semiotics (for review of various modelling approaches, see Olteanu, 2021), and it might potentially be also useful in the realm of biology, mainly due to its straightforwardness and simplicity. Moreover, the cues and signals of biological signalling theory could be viewed as examples of affordances, though each at a different level of ritualisation and thus evolutionary significance for the species (Maynard Smith & Harper, 2003). On top of that, there is a functional similarity between Portmann's concept of phenomena proper (*eigentliche Erscheinungen*), i.e., surfaces that may potentially enter the senses of animal/human perceivers, and Gibson's affordances. In particular, while phenomena proper have the potential to afford some usage/function/meaning, phenomena improper (*uneigentliche Erscheinungen*) do not.

Some of Gibson's ideas also resemble Uexküll's concept of counterpoint (*Einpassung*) (von Uexküll, 1928), and possibly also later concepts of ecological and semiotic fitting (Kull, 2020). This is nicely exemplified by his own words:

> What the male affords the female is reciprocal to what the female affords the male; what the infant affords the mother is reciprocal to what the mother affords the infant; what the prey affords the predator goes along with what the predator affords the prey; what the buyer affords the seller cannot be separated from what the seller affords the buyer, and so on. (Gibson, 1979: 135).

When speaking about water, for instance, one can say it affords us washing, bathing, or drinking, but not respiration. Water surface provides support to small and light insects, such as water striders, but the same surface does not afford support to large animals with dense tissues (Gibson, 1979: 131). At this point, Uexküll would rather say that the body plan of a water strider is counterpointing the water surface.

Compared to both Uexküll and Portmann, Gibson puts more emphasis on the 'properties of things taken with reference to an observer' but explicitly avoids the properties of observer's experiences. In the Uexküllian tradition, properties of things and properties of observers are thought to be more reciprocal and interdependent, while Gibson's affordance emphasises the aspect of things which afford regardless of this relationship being impossible without reference to an observer. Both conceptions, however, stress the mutual aspects of perception. For example, Portmann himself called his approach to the study animal displays 'aesthetic' (referring to broad sense perception by all senses as well as the mind), while others viewed it as a kind of engaged 'participatory research' (Von Wahlert, 1999) that is guided by rational means but does not treat the exposed surfaces of organisms as objects in a simple physicalistic sense. In a similar vein, Gibson (1977: 67) wrote that '… the affordance of anything is a specific combination of the properties of its substance and its surfaces taken with reference to an animal'. This allows him to deconstruct the subject–object relation:

> An affordance is not what we call a 'subjective' quality of a thing. But neither is it what we call an 'objective' property of a thing if by that we mean that a physical object has no reference to any animal. An affordance cuts across the dichotomy of subjective-objective and helps us to understand its inadequacy. (Gibson, 1977: 70).

A sensitive re-examination of the interface between the observer and the observed is one of the most critical goals of current life sciences. Subject–objectivity cannot be suddenly erased from our conceptual space without radical consequences, especially for traditional scientific practice. Nevertheless, both the perceived/experienced and objectively measured matrices of reality should be considered if we want to construct plausible models of biological space and time.

I do not wish to make any bold statements about the need for a new conception of living nature. We need something much less grandiose but operational.

Traditionally, science has been describing the reality of the world we share with other organisms as a kind of universal matrix whose parameters can be repeatedly measured and are independent of any particular observer. We, humans, establish the existence of objective reality, and it exists as such only in this sense. Therefore, objective reality is by no means the same as universal reality.

A promising example of modelling that takes on board the abovementioned observations is the attractor field model (AFM) to the evolution of mimicry (Brejcha et al., 2021). According to AFM, it is likely that we can better distinguish between objects in our environment which are numerous than between those which are rare (Tanaka & Corneille, 2007). Experimentally created stimuli which combined, in equal parts, various characteristics of typical and atypical (rare) objects were, statistically conclusively, evaluated by observers as being more similar to the atypical object. This could be extrapolated also to nonhuman observers. Based on the predictions of AFM, Kleisner and Saribay (2019) suggested that functional mimicry evolves more often among rare species in a given locality than among species common in that environment. This is because, due to the dynamics of the AFM, rare species do not necessarily need to be as similar to their model as the more common species would to establish functional mimicry. The observer, here the predator, identifies them with lower efficiency and is more likely to generalise them as representatives of one taxon. This phenomenon, and thus also the possible application of our model, far exceeds the field of biological mimicry. In fact, it seems to generally apply to any recognition process at any level of biological organisation, be it immune recognition of pathogens or the recognition of suitable mating partners, including, for example, known cases of reduced ability to distinguish among individuals of ethnicities one does not usually encounter, that is, the phenomenon described in psychology as the other-race effect (Europeans often have a problem distinguishing among particular Asian faces – and vice versa). In general, this approach allows us to conceive of (and mathematise) a reality where we experience objects in a given environment differently depending on the frequency of interacting with them. In other words, the frequency with which the senses of a given organic subject interact with an object changes the perceived characteristics of the given object. But the frequency of mutual encounters is just one possible factor (an example of AFM implementation) affecting the perceived characteristics of the object. Any set of conditions can be explored, and dimensionality of the model can be arbitrarily extended.

Regardless of which approach to the interpretation of organic appearances we ultimately adopt, it is evident that the functions of organismic exposed surfaces cannot be explained without reference to the perceiver. Exposed surfaces of organisms interact with other organisms' perception to form semiautonomous relational entities called 'semantic organs' (Kleisner, 2015), which participate in biological reality as operational heritable evolutionary units. The overall functioning of a semantic organ is co-determined both by intrinsic sources of variation (genetic, developmental, morphogenetic) and by variation due to the cognitive and perceptual capacities of an external agent. The approach suggested above, which is based on mapping the population of objects from a morphospace to the perceptual space and backwards (as exemplified by AFM) allows us to conceptualise and model the functioning of semantic organs while taking into account the interdependence between the objectively measurable and the perceived (lived) matrices of the same reality.

Conclusions

Repetition is not only the mother of wisdom but also of biological organisation. Variation due to cyclical spatiotemporal patterns underlies various aspects of organismic appearance, including metameric and modular patterns of many body plans. With the rise of multicellularity, modularity (see Chap. 11) and metamerism resulted in various arrangements of macroscopic bodies, which in some cases led to the emergence of gigantic forms. Moreover, increase in size is necessarily linked to increased complexity (Bonner, 2004). The larger the organism, the greater is the necessary specialisation of its body parts – and that goes hand in hand with differentiation and increase in the number of various cell types. The evolutionary emergence of new functionalities is always based on reinterpreting pre-existing information, while the already established developmental pathways, resources, and control algorithms are often repurposed for innovative functions (Sharov, 2016). The rise in complexity was followed by the evolution of various exposed surfaces and outer structures, which further diversified and adapted to various functions due to series of processes of co-option and selection. This opened the door for a reciprocal evolution of meaningfully perceived surfaces and their counter-organs of perception. Once the organismic surface structures were exposed and recognised by others, Life on our planet literally got its face. We may speculate that this largely happened with the Cambrian explosion (Parker, 2000: 2003). Due to the complementarity of expression and perception, the exposed surfaces of organisms ultimately became semi-autonomous entities with their own evolution (Kleisner & Markoš, 2005; Kleisner & Maran, 2014; Kleisner, 2015; Švorcová & Kleisner, 2018; Brejcha et al., 2019).

Natural selection can modify and fine-tune pre-existing structures and their functions by exploiting phenotypic plasticity. Moreover, selection helps to maintain the functional structures when they contribute to the functional coherence of an organism. Of course, functional coherence is itself based on selection and only functionally well-integrated units can pass the tests posed by various events so that their properties are passed to the subsequent generations. But selection does not specifically explain organisation. It only states that 'better-organised' entities are favoured. The fact of survival itself is, however, due to successful organisation.

Organismic appearances reflect both recent functional roles and traces of inner dynamics which accumulated in the course of history of the relevant lineage. Evolutionary processes do not stop, vanish, or turn in the opposite direction immediately after the 'evolutionary force' (be it selection, developmental dynamics, or something else) ceases to act. Life processes maintain their momentum, as can be seen in various atavisms, homologous developmental pathways, but also in persisting behavioural and cognitive habits. Efforts to subsume organismal appearances and their perception solely under the simple principle of utility are thus misleading.

To contemplate the organic form, we therefore ought to return to morphology – which most people believe to be an obsolete science – and try to rethink its legacy. Rethinking morphology and adopting its view of living beings might open new horizons for understanding biological reality. In addition to the phylogenetic view of evolution, it may steer our attention to the anagenetic perspective.

At the beginning, we mentioned that living things act by appearing. Inanimate things do not act in order to appear; they merely happen to appear. Organisms co-opt new qualities based on affordances in their environment, which allow them to modify and afford characteristics (appearances) that make other subsequent co-options possible. Living entities act in order to appear thanks to the intricate organisation of their components; organisation which combines both mechanistic-like and sign-related ways of recursive causation. For this reason, animate appearances are the product of the organic selfhood – and that is something we cannot say about the appearances of inanimate things, which lack any organisation based on self-relating. Organismic appearances thus do not serve primarily and/or exclusively just the purpose of reproduction. The ability to reproduce is rather the consequence of specific organisational settings typical of living entities and self-representation is just another consequence of the same organisation.

References

Anderson, J. B., Bruhn, J. N., Kasimer, D., Wang, H., Rodrigue, N., & Smith, M. L. (2018). Clonal evolution and genome stability in a 2500-year-old fungal individual. *Proceedings of the Royal Society B: Biological Sciences, 285*(1893), 20182233.

Arrhenius, G. O. (2003). Crystals and life. *Helvetica Chimica Acta, 86*(5), 1569–1586.

Baluška, F., & Mancuso, S. (2016). Vision in plants via plant-specific ocelli? *Trends in Plant Science, 21*(9), 727–730.

Bicknell, R. D. C., & Paterson, J. R. (2018). Reappraising the early evidence of durophagy and drilling predation in the fossil record: Implications for escalation and the Cambrian explosion. *Biological Reviews, 93*(2), 754–784.

Bobrovskiy, I., Hope, J. M., Ivantsov, A., Nettersheim, B. J., Hallmann, C., & Brocks, J. J. (2018). Ancient steroids establish the Ediacaran fossil Dickinsonia as one of the earliest animals. *Science*. (New York, N.Y.), *361*(6408), 1246–1249.

Bobrovskiy, I., Hope, J. M., Golubkova, E., & Brocks, J. J. (2020). Food sources for the Ediacara biota communities. *Nature Communications, 11*(1), 1261.

Bonner, J. T. (2000). *First signals: The evolution of multicellular development. First signals*. Princeton University Press.

Bonner, J. T. (2004). Perspective: The size-complexity rule. *Evolution; International Journal of Organic Evolution, 58*(9), 1883–1890.

Brejcha, J., Pecháček, P., & Kleisner, K. (2019). Complementarity of seeing and appearing. In M. I. Aldinhas Ferreira, J. Silva Sequeira, & R. Ventura (Eds.), *Cognitive architectures* (pp. 13–30). Springer.

Brejcha, J., Tureček, P., & Kleisner, K. (2021). Perception-driven dynamics of mimicry based on attractor field model. *Interface Focus, 11*(3).

Brownlee, C., Wheeler, G. L., & Taylor, A. R. (2015). Coccolithophore biomineralization: New questions, new answers. *Seminars in Cell & Developmental Biology, 46*, 11–16.

Cairns-Smith, A. G. (1977). Takeover mechanisms and early biochemical evolution. *Biosystems, 9*(2), 105–109.

Cairns-Smith, A. G. (1982). *Genetic takeover and the mineral origins of life*. Cambridge University Press.

Chen, L., & Wiens, J. J. (2021). Multicellularity and sex helped shape the tree of life. *Proceedings of the Royal Society B: Biological Sciences, 288*(1955), 20211265.

Chong, I., & Proctor, R. W. (2020). On the evolution of a radical concept: Affordances according to Gibson and their subsequent use and development. *Perspectives on Psychological Science, 15*(1), 117–132.

Cock, J. M., & Collén, J. (2015). Independent emergence of complex multicellularity in the Brown and red algae. In I. Ruiz-Trillo & A. M. Nedelcu (Eds.), *Evolutionary transitions to multicellular life: Principles and mechanisms* (pp. 335–361). Springer Netherlands.

Colley, N. J., & Nilsson, D.-E. (2016). Photoreception in phytoplankton. *Integrative and Comparative Biology, 56*(5), 764–775.

Dai, Y.-C., & Cui, B.-K. (2011). Fomitiporia ellipsoidea has the largest fruiting body among the fungi. *Fungal Biology, 115*(9), 813–814.

De Vargas, C., Aubry, M.-P., Probert, I., & Young, J. (2007). Origin and evolution of Coccolithophores: From coastal hunters to oceanic farmers. In P. G. Falkowski & A. H. Knoll (Eds.), *Evolution of primary producers in the sea* (pp. 251–285). Academic.

Decelle, J., & Not, F. (2015). *Acantharia. In eLS* (pp. 1–10). Wiley.

Drescher, B., Dillaman, R. M., & Taylor, A. R. (2012). Coccolithogenesis in Scyphosphaera apsteinii (Prymnesiophyceae). *Journal of Phycology, 48*(6), 1343–1361.

Droser, M. L., & Gehling, J. G. (2008). Synchronous aggregate growth in an abundant new Ediacaran tubular organism. *Science, 319*(5870), 1660–1662.

Evans, S. D., Hughes, I. V., Gehling, J. G., & Droser, M. L. (2020). Discovery of the oldest bilaterian from the Ediacaran of South Australia. *Proceedings of the National Academy of Sciences, 117*(14), 7845–7850.

Gast, R. J. (2017). Centrohelida and other heliozoan-like Protists. In J. M. Archibald, A. G. B. Simpson, & C. H. Slamovits (Eds.), *Handbook of the Protists* (pp. 955–971). Springer.

Gehling, J. G., & Droser, M. L. (2009). Textured organic surfaces associated with the Ediacara biota in South Australia. *Earth-Science Reviews, 96*(3), 196–206.

Gianoli, E., & Carrasco-Urra, F. (2014). Leaf mimicry in a climbing plant protects against herbivory. *Current Biology, 24*(9), 984–987.

Gibson, J. J. (1977). The theory of affordances. In *Perceiving, acting, and knowing: Toward an ecological psychology* (pp. 67–82). Erlbaum.

Gibson, J. J. (1979). *The ecological approach to visual perception*. Houghton Mifflin.

Grosberg, R. K., & Strathmann, R. R. (2007). The evolution of multicellularity: A minor major transition? *Annual Review of Ecology, Evolution, and Systematics, 38*(1), 621–654.

Herron, M. D., Rashidi, A., Shelton, D. E., & Driscoll, W. W. (2013). Cellular differentiation and individuality in the "minor" multicellular taxa. *Biological Reviews of the Cambridge Philosophical Society, 88*(4), 844–861.

Hoffmeyer, J. (2008). The semiotic niche. *Journal of Mediterranean Ecology, 9*, 5–30.

Holland, L. Z. (2016). *Tunicates. Current biology: CB, 26*(4), R146–R152.

Kleisner, K. (2015). Semantic organs: The concept and its theoretical ramifications. *Biosemiotics, 8*(3), 367–379.

Kleisner, K., & Maran, T. (2014). Visual communication in animals: Applying a Portmannian and Uexküllian biosemiotic approach.

Kleisner, K., & Markoš, A. (2005). Semetic rings: Towards the new concept of mimetic resemblances. *Theory in Biosciences, 123*(3), 209–222.

Kleisner, K., & Saribay, S. A. (2019). The dual nature of mimicry: Organismal form and Beholder's eye. *Biosemiotics, 12*(1), 79–98.
Knoll, A. H. (2011). The multiple origins of complex multicellularity. *Annual Review of Earth and Planetary Sciences, 39*(1), 217–239.
Knoll, A., & Hewitt, D. (2011). *Phylogenetic, functional, and geological perspectives on complex multicellularity* (pp. 251–270).
Komárek, S. (2003). *Mimicry, aposematism and related phenomena. Mimetism in nature and the history of its study*. Lincom Europa.
Kull, K. (2010). Ecosystems are made of Semiosic Bonds: Consortia, Umwelten, Biophony and Ecological Codes. *Biosemiotics, 3*(3), 347–357.
Kull, K. (2020). Semiotic fitting and the Nativeness of community. *Biosemiotics, 13*, 9–19.
Lima de Faria, A. (1988). *Evolution without selection: Form and function by autoevolution*. Elsevier.
Liu, A. G., McIlroy, D., & Brasier, M. D. (2010). First evidence for locomotion in the Ediacara biota from the 565 ma mistaken point formation, Newfoundland. *Geology, 38*(2), 123–126.
Magnus, R. (2021). Novel ecological mismatches in the light of Jakob von Uexküll's and Adolf Portmann's works. In F. Jaroš & J. Klouda (Eds.), *Adolf Portmann: A thinker of self-expressive life* (pp. 71–87). Springer.
Maran, T. (2017). *Mimicry and meaning: Structure and semiotics of biological mimicry (Vol. 16)*. Springer.
Maynard Smith, J., & Harper, D. (2003). *Animal signals*. Oxford University Press.
Maynard Smith, J., & Szathmary, E. (1997). *The major transitions in evolution*. Oxford University Press.
Minelli, A., & Pradeu, T. (Eds.). (2014). *Towards a theory of development*. Oxford University Press.
Nagy, L. G., Kovács, G. M., & Krizsán, K. (2018). Complex multicellularity in fungi: Evolutionary convergence, single origin, or both? *Biological Reviews of the Cambridge Philosophical Society, 93*(4), 1778–1794.
Nakamura, Y., Tuji, A., Kimoto, K., Yamaguchi, A., Hori, R. S., & Suzuki, N. (2021). Ecology, morphology, phylogeny and taxonomic revision of Giant radiolarians, Orodaria ord. nov. (Radiolaria; Rhizaria; SAR). *Protist, 172*(3), 125808.
Neustupa, J. (2017). Asymmetry and integration of cellular morphology in Micrasterias compereana. *BMC Evolutionary Biology, 17*(1), 1.
Olteanu, A. (2021). Multimodal modeling: Bridging biosemiotics and social semiotics. *Biosemiotics, 14*(3), 783–805.
Parker, A. (2000). 515 million years of structural colour. *Journal of Optics A: Pure and Applied Optics, 2*(6), 15–28.
Parker, A. (2003). *In the blink of an eye: How vision sparked the big bang of evolution*. Basic Books.
Parker, A. (2011). On the origin of optics. *Optics and Laser Technology, 43*(2), 323–329.
Paterson, J. R., Edgecombe, G. D., & Lee, M. S. Y. (2019). Trilobite evolutionary rates constrain the duration of the Cambrian explosion. *Proceedings of the National Academy of Sciences, 116*(10), 4394–4399.
Peters, S. E., & Gaines, R. R. (2012). Formation of the "great unconformity" as a trigger for the Cambrian explosion. *Nature, 484*(7394), 363–366.
Portmann, A. (1960). *Neue Wege der Biologie*. R. Piper.
Sharov, A. A. (2016). Evolution of natural agents: Preservation, advance, and emergence of functional information. *Biosemiotics, 9*, 103–120.
Shen, B., Dong, L., Xiao, S., & Kowalewski, M. (2008). The Avalon explosion: Evolution of Ediacara Morphospace. *Science, 319*(5859), 81–84.
Smith, M. L., Bruhn, J. N., & Anderson, J. B. (1992). The fungus Armillaria bulbosa is among the largest and oldest living organisms. *Nature, 356*(6368), 428–431.
Švorcová, J., & Kleisner, K. (2018). Evolution by meaning attribution: Notes on biosemiotic interpretations of extended evolutionary synthesis. *Biosemiotics, 11*(2), 231–244.

Tanaka, J. W., & Corneille, O. (2007). Typicality effects in face and object perception: Further evidence for the attractor field model. *Perception & Psychophysics, 69*, 619–627.
Tarjuelo, I., López-Legentil, S., Codina, M., & Turon, X. (2002). Defence mechanisms of adults and larvae of colonial ascidians: Patterns of palatability and toxicity. *Marine Ecology Progress Series, 235*, 103–115.
Taylor, A. R., Brownlee, C., & Wheeler, G. (2017). Coccolithophore cell biology: Chalking up Progress. *Annual Review of Marine Science, 9*(1), 283–310.
Tullberg, B. S., Merilaita, S., & Wiklund, C. (2005). Aposematism and crypsis combined as a result of distance dependence: Functional versatility of the colour pattern in the swallowtail butterfly larva. *Proceedings of the Royal Society of London B: Biological Sciences, 272*(1570), 1315–1321.
von Uexküll, J. (1928). *Theoretische Biologie*. Springer.
Von Wahlert, G. (1999). Portmann's work in a newer evolutionary perspective. *Revue européenne des sciences sociales, 37*(115), 75–87.
Wei, G.-Y., Planavsky, N. J., Tarhan, L. G., Chen, X., Wei, W., Li, D., & Ling, H.-F. (2018). Marine redox fluctuation as a potential trigger for the Cambrian explosion. *Geology, 46*(7), 587–590.
White, J., & Yamashita, F. (2022). Boquila trifoliolata mimics leaves of an artificial plastic host plant. *Plant Signaling & Behavior, 17*(1), 1977530.
Wickler, W. (1968). *Mimikry: Nachahmung und Täuschung in der Natur*. Kindler.
Wood, R., & Zhuravlev, A. Y. (2012). Escalation and ecological selectively of mineralogy in the Cambrian radiation of skeletons. *Earth-Science Reviews, 115*(4), 249–261.
Yin, L., Meng, F., Kong, F., & Niu, C. (2020). Microfossils from the Paleoproterozoic Hutuo group, Shanxi, North China: Early evidence for eukaryotic metabolism. *Precambrian Research, 342*, 105650.

Index

J. Švorcová (ed.), *Organismal Agency*, Biosemiotics 28,
https://doi.org/10.1007/978-3-031-53626-7

MIX
Papier aus verantwortungsvollen Quellen
Paper from responsible sources
FSC® C105338

If you have any concerns about our products,
you can contact us on
ProductSafety@springernature.com

In case Publisher is established outside the EU,
the EU authorized representative is:
Springer Nature Customer Service Center GmbH
Europaplatz 3, 69115 Heidelberg, Germany

Printed by Libri Plureos GmbH
in Hamburg, Germany